Benchmark Papers on Energy

Series Editor: R. Bruce Lindsay, Brown University
Mones E. Hawley, Jack Faucett Associates

Volume

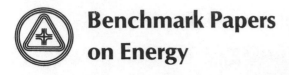

Benchmark Papers on Energy / 10

A BENCHMARK® Books Series

ENERGY IN ATOMIC PHYSICS 1925–1960

Edited by

R. BRUCE LINDSAY

Brown University

Hutchinson Ross Publishing Company

Stroudsburg, Pennsylvania

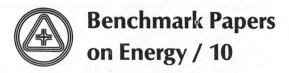

Benchmark Papers on Energy, Volume 10
Library of Congress Catalog Card Number: 82-21248
ISBN: 0-87933-075-9

85 84 83 1 2 3 4 5
Manufactured in the United States of America.

LIBRARY OF CONGRESS CATALOGING IN PUBLICATION DATA
Main entry under title:
Energy in atomic physics, 1925-1960.
(Benchmark papers on energy ; v. 10)
Includes indexes.
1. Nuclear physics—Addresses, essays, lectures.
2. Force and energy—Addresses, essays, lectures.
3. Energy levels (Quantum mechanics)—Addresses, essays, lectures. 4. Quantum theory—Addresses, essays, lectures. 5. Particle accelerators—Addresses, essays, lectures. I. Lindsay, Robert Bruce, 1900-
II. Series.
QC780.E53 1983 539.7 82-21248
ISBN 0-87933-075-9

Distributed worldwide by Van Nostrand Reinhold Company Inc.,
135 W. 50th Street, New York, NY 10020.

CONTENTS

PART III: BEGINNINGS OF QUANTUM STATISTICS

PART IV: ENERGY OF ELEMENTARY PARTICLES

PART V: ENERGY IN THE ATOMIC NUCLEUS

PREFACE

The Benchmark Papers on Energy constitute a series of volumes that makes available to the reader in carefully organized form important and seminal articles on the concept of energy, including its historical development, its applications in all fields of science and technology, and its role in civilization in general. This concept is generally admitted to be the most far-reaching idea that the human mind has developed to date, and its fundamental significance for human life and society is evident everywhere.

One group of volumes in the series contains papers bearing primarily on the evolution of the energy concept and its current applications in the various branches of science. Another group of volumes concentrates on the technological and industrial applications of the concept and its socioeconomic implications.

Each volume has been organized and edited by an authority in the area to which it pertains and offers the editor's careful selection of the appropriate seminal papers, that is, those articles that have significantly influenced further development of that phase of the whole subject. In this way, every aspect of the concept of energy is placed in proper perspective, and each volume represents an introduction and guide to further work.

Each volume includes an introduction by the volume editor, summarizing the significance of the field being covered. Every article or group of articles is accompanied by editorial commentary, with explanatory notes where necessary. Both an author index and a subject index are provided for ready reference. Articles in languages other than English are either translated or summarized in English. It is the hope of the publisher and editor that these volumes will serve as a working library of the most important scientific, technological, and social literature connected with the idea of energy.

The present volume, *Energy in Atomic Physics, 1925–1960,* has been prepared by one of the Series Editors. Its aim is to extend the field of the previous Benchmark volume, *Early Concepts of Energy in Atomic Physics* (Robert Lindsay, editor, 1979), to the developments that have occurred largely in the midsection of the twentieth century. This period has seen the firm establishment of quantum mechanics as the basic theory for the handling of problems in atomic physics. It has also witnessed the enormous development of nuclear physics, as well as great advances in solid-state and liquid physics. It has been marked by the search for and invention of devices for the discovery of elementary particles involved in the constitution of matter. In all of this development, the concept of energy has proved to be the most significant factor. It is hoped that the forty-six papers reproduced in this volume form a representative sample of the seminal work in atomic physics during the period from 1925 through 1960 with special reference to the role of energy. The relatively enormous amount of material published during the period has made the selection of material for inclusion difficult and arbitrary. Many important relevant papers have had to be omitted. Several of these are referred to in the bibliographies at the end of the commentaries on the various groups of articles. It is hoped that these editorial commentaries serve to place the selected papers in proper perspective.

I am deeply indebted to Patricia Galkowski and Caroline Helie of the Sciences Library of Brown University for their gracious and indefatigable help in the location of the source material. I am also very grateful to Rose Ouloosian and Rose Roach of the Department of Physics in Brown University and their corps of efficient typists for the typing of the translated material and the commentaries, as well as for the photocopying of articles being considered for inclusion. It is a pleasure to express my appreciation of the fine collaboration rendered me by the editorial staff of the publishers.

R. BRUCE LINDSAY

CONTENTS BY AUTHOR

ENERGY IN ATOMIC PHYSICS
1925-1960

INTRODUCTION

Previous volumes in the Benchmark Papers on Energy have stressed the historical development of the concept, as represented by constancy in the midst of change, viewing all natural phenomena as examples of the transfer of energy from one place to another or the transformation of energy from one form to another, the total quantity of energy in the universe of our experience remaining constant, no matter what happens. Certain volumes have been devoted to practical applications of the energy idea, such as thermodynamics, coal as an energy source, and various forms of the storage of energy. An earlier volume also dealt with the important role played by energy in the development of atomic physics from early times to the mid-1920s.

The enormous development of all aspects of atomic physics in the contemporary period from 1925 to 1960 strongly suggests the importance of looking at the relation of energy to this development, the subject forming the content of this volume. This period has seen the transition from concern with the energy interaction of atoms and molecules to that connected with the much more powerful energy interactions of the nuclei of atoms and their constituent particles. The term "high energy physics" now describes a very large segment of modern physics, with tremendous technological significance. The introduction of quantum theory into statistical mechanics has played an important role in the understanding of macroscopic phenomena such as liquid and solid-state physics. Throughout all this, the concept of energy enters in vital fashion, and this is also exemplified by the enormous current growth of astrophysics.

The phenomenal proliferation of published material during the last half-century renders the task of selecting seminal literature extremely difficult. The choice of papers to be reprinted is obviously more arbitrary than in the case of earlier volumes, since even the seminal papers have become more abundant. In general, the effort has been made to select the first important paper, even if short, that inaugurated a chain of subsequent publications, although occasionally a review paper has seemed to serve a useful purpose. In every case, the effort is made to show how the concept of energy enters vitally into the content and meaning of the paper reproduced. This means essentially emphasis on the existence or possibility of energy transformation in the phenomena discussed in the papers reproduced. Hence, certain fields, such as spectroscopy, are given less prominence. On the other hand, the discovery of new particles like the neutron and neutrino appear to have an obvious place, since ultimately they participate in reactions involving energy transformation.

For convenience, the contents are divided into seven groups, ranging from the forerunners of quantum mechanics through the basic papers in this field to the exciting areas of nuclear physics and solid-state physics. Emphasis is on the basic scientific aspects. The technological applications are presented in other volumes in the series.

Since space restrictions make it impossible to include all relevant papers, a lengthy bibliography of material connected with the papers being reproduced is added. Lists of such publications are placed at the end of each group of papers.

The commentaries at the beginning of each group of papers endeavor to place the papers of the group in proper perspective and to identify the authors.

ANNOTATED BIBLIOGRAPHY

Born, M., 1951, *The Restless Universe,* 2nd rev. ed., Dover Publications, New York.

Darrow, K. K., 1948, *Atomic Energy,* Wiley, New York.

Physics Today, November, 1981, commemorating the 50th anniversary of the founding of the American Institute of Physics. Section 2 (pages 69–231) contains an excellent review of the development of atomic physics during the period covered by the present volume.

Sproull, R. L., and W. A. Phillips, 1960, *Modern Physics: The Quantum Physics of Atoms, Solids and Nuclei,* 3rd ed., Wiley, New York.

Trigg, G. L., 1975, *Landmark Experiments in Twentieth Century Physics,* Crane, Russak and Co., New York.

Part I

FORERUNNERS OF QUANTUM MECHANICS

Editor's Comments on Papers 1 Through 5

1 **BRILLOUIN**
Hereditary Discontinuous Mechanical Actions Produced by Propagation. Attempt at a Dynamical Quantum Theory of the Atom.

2 **BROGLIE**
Waves and Quanta

3 **ELSASSER**
Comments on the Quantum Mechanics of Free Electrons

4 **DAVISSON and GERMER**
The Scattering of Electrons by a Single Crystal of Nickel

5 **THOMSON and REID**
Diffraction of Cathode Rays by a Thin Film

The great success of the Bohr theory of atomic structure in predicting the precise frequencies of the lines in the Balmer series of the spectrum of hydrogen and the energy exchanges in the collision of certain atoms was somewhat clouded by the rather arbitrary use that Bohr had to make of a mixture of classical, mechanical, and quantum ideas in the application of his fundamental postulates. This undoubtedly stimulated efforts to derive the quantum idea from classical physics or at any rate to provide a more straightforward and plausible presentation of the application of the quantum idea to the motions of particles. One of the most interesting and suggestive efforts in this direction was made by the French theoretical physicist, Marcel Brillouin (1854–1948), for over thirty years professor of mathematical physics at the College de France in Paris. In a hypothesis that essentially foreshadowed the introduction of wave mechanics into atomic physics, he assumed that in its motion an atomic particle emits waves, the crests of which chase the particle around in its orbit, catching up with it at certain definite points. In his analysis, he succeeded in introducing a certain discontinuity, which he interpreted as providing the existence of quanta of energy. The details are presented briefly in his *Comptes Rendus* paper of 1919, which is reproduced in English translation here as Paper 1.

Brillouin's suggestion was taken up by Louis de Broglie (1892-) and led to the famous early paper by the latter in the *Comptes Rendus* of 1923, in which he associated with a moving particle a frequency of vibration such that Planck's constant h multiplied by this frequency is equal to the relativistic energy mc^2 of the particle, where m is the mass of the particle and c the velocity of light in free space. He then associated with the particle a harmonic wave of this frequency and having a velocity c^2/v with respect to a fixed observer, where v is the velocity of the particle in the same reference system. He uses essentially Brillouin's idea to provide stability for the trajectory of the particle and then derives a condition equivalent to the Bohr quantum condition, thus combining the results of special relativity with a stability requirement. An important aspect of Broglie's paper was the fundamental role played by the relativistic energy of the particle. Broglie's work also indicated that the wavelength of the wave associated with the particle is Planck's constant h divided by the momentum of the particle, a result that proved later to be of great experimental significance. His paper was the beginning of quantum wave mechanics. We reproduce it as Paper 2. Broglie followed it up with several other important articles and books, the titles of some of which are given in the bibliography at the end of the papers in Part. 1.

The wave mechanical calculations of Broglie and in particular Einstein's comments thereon stimulated Walter M. Elsasser (1904-), at the University of Göttingen, to suggest that the wave properties of a particle like the electron could be made evident experimentally. Thus, under certain readily available experimental conditions, diffraction effects should be observable in the behavior of slow electrons. He made some rough calculations indicating that the behavior of the slow electrons in the Ramsauer effect[1] can be interpreted in accordance with the wavelength-momentum relationship derived by Broglie. This undoubtedly stimulated further experiments such as those by Davisson and Germer and G. P. Thomson. We reproduce Elsasser's brief note in *Naturwissenschaften* in 1925 as Paper 3.

Elsasser came from Germany to the United States in 1936 and has devoted his attention mainly to geophysics in his professional career. Since retirement from the University of Maryland in 1974, he has been serving recently as adjunct professor at the Johns Hopkins University. Before he left Germany, Elsasser had tried unsuccessfully to carry out experiments to verify the suggestions in the paper noted above. The first successful experimental verification of the Broglie wave-mechanical hypothesis was made by Clinton J. Davisson (1881-1958) and Lester H. Germer (1896-) at the Bell Laboratories in 1927. For several years, Davisson and collaborators had experimented on secondary electron emission from metal surfaces. This led, through an interesting accident, to work on the scattering of electrons from single crystal metal surfaces with the appearance of reflected beams at angles corresponding to Bragg reflection from a crystal lattice of radiation with wavelengths in agreement with Broglie's value. The first definite claim by the pair of investigators appeared as a letter to *Nature* in April 1927. This was followed by a lengthy detailed article in the *Physical Review* for December 1927. We reproduce the letter as Paper 4. The detailed article is referenced in the bibliography at the end of the commentary.

In the meantime, George P. Thomson (1892-1975), then professor at Aberdeen, had been experimenting with the effects observable in the passage of electrons through thin metallic films. With his collaborator, A. Reid, he found a distribution of the transmitted electrons in the form of rings whose radii corresponded to wavelengths agreeing with Broglie's formula. This work was first reported in a letter to *Nature* in June 1927. Together with the work of Davisson and Germer, these investigations led to the enormous development of experimental wave mechanics with valuable applications throughout solid-state physics. We reproduce the joint paper of Thomson and Reid as Paper 5.

REFERENCE

1. W. M. Elsasser, "Comments on the Quantum Mechanics of Free Elections," *Ann. Phys.* **72** (1923):345.

BIBLIOGRAPHY

Bohr, N., H. A. Kramers and J. C. Slater, 1924 On the Quantum Theory of Radiation, *Philos. Mag.* **47**:785-802.

Broglie, L. de, 1923*a,* Quanta, the Kinetic Theory of Gases and Fermat's Principle, *Acad. Sci. C. R.* **177**:631-632.

Broglie, L. de, 1923*b,* Light Quanta, Diffraction and Interference, *Acad. Sci. C. R.* **177**:548-550.

Davisson, C. J., and L. A. Germer, 1927, Diffraction of Electrons by a Crystal of Nickel, *Phys. Rev.* **30**:705-740.

Kramers, H. A., 1924, The Quantum Theory of Dispersion, *Nature* **114**:310-311.

Thomson, G. P., 1930, *Wave Mechanics of the Free Electron,* McGraw-Hill, New York.

1

HEREDITARY DISCONTINUOUS MECHANICAL ACTIONS PRODUCED BY PROPAGATION. ATTEMPT AT A DYNAMICAL QUANTUM THEORY OF THE ATOM

Marcel Brillouin

This article was translated expressly for this Benchmark volume by R. Bruce Lindsay, Brown University, from Acad. Sci. C. R. ***168:****1318–1320 (1919), by permission of the publisher, Gauthier-Villars, Paris*

1. Let us consider a particle that moves in an elastic medium with a velocity much greater than the velocity ω of elastic waves in the medium. Let us suppose that either through natural vibrations or as a result of its displacement through the medium, the particle at each instant emits waves emanating from its instantaneous position as center. If the particle's trajectory is periodic or quasi-periodic and always confined within a sphere of diameter much less than ωT, which is the product of the velocity ω of the wave and its period T, the particle will be rejoined (or caught up with) at each instant a finite number of times by the waves emitted during its anterior motion. It is to this particular circumstance that I wish to call attention.

2. Let x, y, z be the coordinates of the particle at time t; let ds be an element of its trajectory and u be the particle velocity along this element. Let ξ_k, η_k, ζ_k be the coordinates of the particle at the earlier time $t - \tau_k$. The wave emitted by the particle as it passes point M_k will rejoin the particle at time t at its actual position if we have

$$r_k = \omega \int_{M_k}^{M} \frac{ds}{u} = \omega \tau_k \tag{1}$$

where

$$r_k = \sqrt{(x - \xi_k)^2 + (y - n_k)^2 + (z - \zeta_k)^2} \tag{2}$$

If the trajectory is closed, the integral should be taken after a finite number of circuits.

To each actual position of the point there are associated (if u is much greater than ω) a *finite* number of anterior positions M_k of the same point for which the waves reach the actual point at time t.

This can be taken as constituting a hereditary field (following the terminology of Volterra), but it is discontinuous in nature.

The point then moves in the field of n virtual points that trail behind it on its trajectory, for equation 1 can be satisfied only by distinct points M_k, save in the very special case of rectilinear motion in which $u = \omega$.

3. The quasiperiodic motions to be studied can be characterized by the number n of virtual active points. It would appear[1] that the permanent motions will be those for which n is constant. The energy of the particle in the field of its anterior n positions and in general every integral invariant of its motion will experience a finite change when the number n changes by unity. This change from one permanent trajectory with n active virtual points to another one with $n + n'$ active virtual points will be accompanied by vibrations defined entirely by classical dynamics and the law of emission that has been adopted.

4. It then seems that one can formulate a hypothesis endowed with the qualities essential to represent the essential properties of the Bohr atom, if we make an appropriate choice of the law of emission.

Hypothesis I. In addition to the velocity of light, the universal medium (aether of space) possesses a propagation velocity very much smaller (of the order of factors of ten in kilometers/sec (?)). Quantum phenomena would appear when the electrons move with a velocity greater than this along quasiperiodic orbits, in such a way that the electron should at every instant be in the field of a finite number of its anterior positions.

The nature and size of the mechanical discontinuities accompanying the change in the integer n will depend on the law of emission from the moving electron and on the nature of the propagated waves (with or without rotation).

One can readily imagine how chemical phenomena can be attached to the actual hypothesis.

5. *Hypothesis II. Let us suppose that in particular the energy of the moving point in the field of one of its anterior position is*

$$\phi = \pm \frac{B^2}{r_k} \tag{3}$$

Eq. 1, which determines this position, can be written

$$\phi \int_{M_k}^{M} \frac{ds}{u} = \frac{B^2}{\omega} \tag{4}$$

The quantity on the left of this equation is an *action,* playing a role in the motion of the point, and equation 4 shows that this action has a constant value; it is Planck's constant h.

This constant comes into play once for each anterior active position of the point, or n times if the trajectory under study contains n anterior active points.

The problems set forth in this note appear to me to warrant an elaborate study (not given here), a study that would have significance from the purely dynamical point of view and from the point of view of chemistry and physics.

[1]This statement is quasi-intuitive. But since it is this that gives birth to the discontinuity analogous to the quantum idea, it is essential that it should hold for the laws of emission for which a complete analytical study is to be made.

2

WAVES AND QUANTA

Louis de Broglie

This article was translated expressly for this Benchmark volume by R. Bruce Lindsay, Brown University, from Acad. Sci. C. R. ***177:***507–510 *(1923), by permission of the publisher, Gauthier-Villars, Paris*

Consider a material particle of proper mass m_o moving with respect to a fixed observer with velocity $v = \beta c$ ($\beta < 1$). According to the principle of the inertia of energy, the particle should possess an intrinsic energy equal to m_0c^2. On the other hand, the quantum principle attributes this intrinsic energy to a simple periodic phenomenon of frequency ν_o such that

$$h\nu_0 = m_0 c^2,$$

c always being the limiting velocity according to the theory of relativity and h Planck's constant. [*Editor's Note:* c is, the velocity of light in free space.]

For the fixed observer a frequency $\nu = (m_0c^2)/(h\sqrt{1-\beta^2})$ will be associated with the total energy of the particle. But if this fixed observer observes the intrinsic periodic phenomenon of the particle, he will see it diminished and will attribute to it a frequency $\nu_1 = \nu_0\sqrt{1-\beta^2}$. For him this phenomenon then varies as

$$\sin 2\pi\nu_1 t.$$

Let us now suppose that at time $t = o$, the moving particle coincides in space with a wave of frequency ν defined above and propagating in the same direction as the above wave with velocity c/β. This wave of velocity greater than c cannot correspond to the transmission of energy; we consider it only as a fictitious wave associated with the motion of the particle.

I say that if at time $t = o$ there is phase agreement between the vectors of the wave and the intrinsic phenomenon of the particle, this phase agreement will persist. For at time t the particle is at a distance from the origin equal to $vt = x$. Its internal or intrinsic motion is then repesented by $\sin 2\pi\nu_1 x/v$.

The wave at this point is represented by

$$\sin 2\pi\nu\left(t - \frac{x\beta}{c}\right) = \sin 2\pi\nu x\left(\frac{1}{v} - \frac{\beta}{c}\right).$$

The two sines are equal, and the agreement in phase is realized if we have

$$\nu_1 = \nu(1-\beta^2),$$

a condition evidently satisfied by the definition of ν and ν_1.

The demonstration of this important result is based uniquely on the principle of restricted relativity and the fact that the quantum relation is the same for the fixed observer as for the observer moving with the particle.

Let us apply this at first to an atom of light. I have shown elsewhere[1] that the atom of light should be considered as a particle of very small mass ($< 10^{-50}$ gram) moving with a velocity very nearly equal to c (though slightly less). We then reach the following statement: *The atom of light, equivalent by reason of its total energy to radiation of frequency υ, is the seat of an intrinsic periodic phenomenon that, viewed by a fixed observer, has at every point in space the same phase as a wave of frequency ν propagating in the same direction with a velocity approximately equal to (though very slightly greater than) the constant called the velocity of light (in free space).*

Let us now pass to the case of an electron following with uniform velocity (slightly less than v) a closed trajectory. At the time $t = o$, the particle is at point O. The associated fictitious wave starting from O and describing the whole trajectory with velocity c/β will catch up with the electron at time t at the point O′ such that $\overline{OO'} = \beta c\tau$.

We then have

$$\tau = \frac{\beta}{c}[\beta c(\tau + T_r)] \qquad \text{or} \qquad \tau = \frac{\beta^2}{1-\beta^2}T_r,$$

where T_r is the period of the electron in its orbit. The intrinsic phase of the electron, when it goes from O to O′, varies by

$$2\pi\nu_1\tau = 2\pi\frac{m_0c^2}{h}T_r\frac{\beta^2}{\sqrt{1-\beta^2}}.$$

It is almost necessary to suppose that the trajectory of the electron is stable only if the fictitious wave passing through O′ catches up with the electron in phase with it: The wave of frequency ν and with velocity c/β should be in resonance on the length of the trajectory. This leads to the condition

$$\frac{m_0\beta^2c^2}{\sqrt{1-\beta^2}}T_r = nh, \qquad \text{(n being integral).}$$

We show that this stability condition is precisely that of the theories of Bohr and Sommerfeld for an orbit described with constant velocity. Let us call p_x, p_y, p_z the rectangular components of the momentum of the electron. The general stability condition as given by Einstein is

$$\int_0^{T_r}(p_x\,dx + p_y\,dy + p_z\,dz) = nh \qquad \text{(n integral).}^2$$

This can be written in the present case

$$\int_0^{T_r}\frac{m_0}{\sqrt{1-\beta^2}}(v_x^2 + v_y^2 + v_z^2)\,dt = \frac{m_0\beta^2c^2}{\sqrt{1-\beta^2}}T_r = nh,$$

In the case of an electron revolving with angular velocity ω in a circle of radius R, we deduce from the above (for sufficiently small velocities) the early formula of Bohr: $m_0\omega R^2 = n \cdot (h)/(2\pi)$.

If the velocity varies along the orbit, we still obtain the Bohr-Einstein formula if β is small. If β takes on large values, the question becomes more complicated and necessitates a special examination.

In pursuing the same path, we have obtained important further results, which will be communicated shortly. We are today in a position to explain phenomena of diffraction and interference, taking account of light quanta.

NOTES

1. *Journal de Physique,* 6th Series, Vol. 3, p. 422, 1922.
2. The case of quasiperiodic motions presents no new difficulty. The necessity of satisfying the condition set forth in the text for an infinity of pseudo-periods leads to Sommerfeld's conditions.

3

COMMENTS ON THE QUANTUM MECHANICS OF FREE ELECTRONS

Walter Elsasser

This article was translated expressly for this Benchmark volume by R. Bruce Lindsay, Brown University, from Naturwissenschaften ***13:****711 (1925), by permission of the publisher, Springer-Verlag, Heidelberg*

Einstein[1] has recently in papers devoted to statistics come in a roundabout way to a physically rather noteworthy result. He attaches considerable probability to the assumption that to every translational motion of a material particle, there is associated a wave field that determines the kinematics of the particle. The hypothesis of the existence of such waves, which had indeed been set forth before Einstein by de Broglie,[2] has been so well supported theoretically by Einstein that is seems fitting to seek an experimental basis for it.

Therefore we here confront the task of demonstrating diffraction and interference in the motion of atoms and electrons. The wavelength to be determined by experimental diffraction phenomena we get, according to de Broglie, from the relation

$$\lambda = \frac{h}{mv} \tag{1}$$

where *mv* denotes the momentum of the particle. Einstein has called attention to the phenomena resulting from this and leading to gas degeneracy at low temperatures. We can attempt to extend this by assuming analogous effects for slow electrons. Because of the small electron mass, we should expect to find in an easily accessible velocity domain strong deviations from ordinary mechanics. The purpose of this note is to call attention to a possible connection of such consequences with certain experiments on the behavior of slow electrons. If we seek, for example, to interpret the remarkable behavior of the free paths of electrons that Ramsauer[3] and after him a number of other investigators have found in terms of the above-mentioned hypothesis, we find that the curves corresponding to this behavior bear a striking resemblance to the curves that one gets in the classical theory for the diffraction of light by small colloidal spheres.[4] According to this, it seemed as if the slow electrons are scattered by the atoms according to laws precisely as would hold for light of the calculated wavelength scattered by spheres of the same radius as the atoms. Naturally, the agreement is only qualitative.

There also appear to be indications of interference in an experiment by Davisson and Kunsman,[5] in which the angular distribution of electrons reflected from a platinum plate was investigated. Several strong maxima were observed, and these maxima moved with increasing electron velocity in a direction to be expected from equation 1 if the maxima were to be considered as diffraction bands formed by an optical grating. If for the grating constants we take those of the platinum crystal grating and treat the problem as one of a plane grating to a first approximation because of the relatively slight penetration of the electrons into the surface of the metal, this rough calculation yields for the wavelength values, which in very rough order of magnitude agree with those computed from equation 1. Since no investigations in well-defined crystal objects have been made, the observed deviations cannot be considered decisive. Further experiments, which are in course of preparation here, are awaited with interest.

I thank Professor J. Franck for various suggestions.

Göttingen (II Physical Institute), July 18, 1925

NOTES

1. A. Einstein, Berliner Akad. 1924, p. 24; 1925, p. 1.
2. Thesis of L. de Broglie, Paris, 1924.
3. Ramsauer, Ann. d. Physik *72,* 345, 1923. For further literature and figures, see Minkowski and Sponer, Ergebnisse der Exakten Naturwissenschaften, Vol. 3.
4. G. Mie, *Annalen der Physik 25,* 377, 1908.
5. Phys. Rev. *22,* 243, 1923.

4

Reprinted from *Nature* **119:**558–560 (1927)

THE SCATTERING OF ELECTRONS BY A SINGLE CRYSTAL OF NICKEL

C. Davisson and L. H. Germer

In a series of experiments now in progress, we are directing a narrow beam of electrons normally against a target cut from a single crystal of nickel, and are measuring the intensity of scattering (number of electrons per unit solid angle with speeds near that of the bombarding electrons) in various directions in front of the target. The experimental arrangement is such that the intensity of scattering can be measured

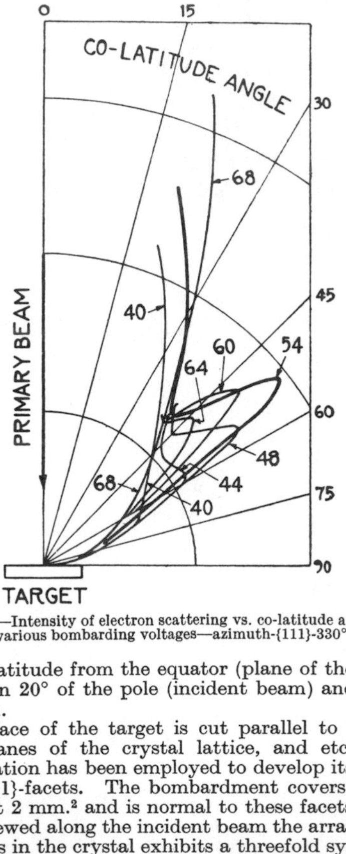

Fig. 1.—Intensity of electron scattering vs. co-latitude angle for various bombarding voltages—azimuth-{111}-330°.

in any latitude from the equator (plane of the target) to within 20° of the pole (incident beam) and in any azimuth.

The face of the target is cut parallel to a set of {111}-planes of the crystal lattice, and etching by vaporisation has been employed to develop its surface into {111}-facets. The bombardment covers an area of about 2 mm.[2] and is normal to these facets.

As viewed along the incident beam the arrangement of atoms in the crystal exhibits a threefold symmetry. Three {100}-normals equally spaced in azimuth emerge from the crystal in latitude 35°, and, midway in azimuth between these, three {111}-normals emerge in latitude 20°. It will be convenient to refer to the azimuth of any one of the {100}-normals as a {100}-azimuth, and to that of any one of the {111}-normals as a {111}-azimuth. A third set of azimuths must also be specified; this bisects the dihedral angle between adjacent {100}- and {111}-azimuths and includes a {110}-normal lying in the plane of the target. There are six such azimuths, and any one of these will be referred to as a {110}-azimuth. It follows from considerations of symmetry that if the intensity of scattering exhibits a dependence upon azimuth as we pass from a {100}-azimuth to the next adjacent {111}-azimuth (60°), the same dependence must be exhibited in the reverse order as we continue on through 60° to the next following {100}-azimuth. Dependence on azimuth must be an even function of period $2\pi/3$.

In general, if bombarding potential and azimuth are fixed and exploration is made in latitude, nothing very striking is observed. The intensity of scattering increases continuously and regularly from zero in the plane of the target to a highest value in co-latitude 20°, the limit of observations. If bombarding potential and co-latitude are fixed and exploration is made in azimuth, a variation in the intensity of scattering of the type to be expected is always observed, but in general this variation is slight, amounting in some cases to not more than a few per cent. of the average intensity. This is the nature of the scattering for bombarding potentials in the range from 15 volts to near 40 volts.

At 40 volts a slight hump appears near 60° in the co-latitude curve for azimuth-{111}. This hump develops rapidly with increasing voltage into a strong spur, at the same time moving slowly upward toward the incident beam. It attains a maximum intensity in co-latitude 50° for a bombarding potential of 54 volts, then decreases in intensity, and disappears in co-latitude 45° at about 66 volts. The growth and decay of this spur are traced in Fig. 1.

A section in azimuth through this spur at its maximum (Fig. 2—Azimuth-330°) shows that it is sharp in azimuth as well as in latitude, and that it forms one of a set of three such spurs, as was to be expected. The width of these spurs both in latitude and in azimuth is almost completely accounted for by the low resolving power of the measuring device. *The spurs are due to beams of scattered electrons which are nearly if not quite as well defined as the primary beam.* The minor peaks occurring in the {100}-azimuth are sections of a similar set of spurs that attains its maximum development in co-latitude 44° for a bombarding potential of 65 volts.

Thirteen sets of beams similar to the one just described have been discovered in an exploration in the principal azimuths covering a voltage range from 15 volts to 200 volts. The data for these are set down on the left in Table I. (columns 1-4). Small corrections have been applied to the observed co-latitude angles to allow for the variation with angle of the 'background scattering,' and for a small angular displacement of the normal to the facets from the incident beam.

If the incident electron beam were replaced by a beam of monochromatic X-rays of adjustable wave-length, very similar phenomena would, of course, be observed. At particular values of wave-length, sets of three or of six diffraction beams would emerge from the incident side of the target. On the right in Table I. (columns 5, 6 and 7) are set down data for the ten sets of X-ray beams of longest wave-length which would occur within the angular range of our observations. Each of these first ten occurs in one of our three principal azimuths.

Several points of correlation will be noted between the two sets of data. Two points of difference will also be noted; the co-latitude angles of the electron beams are not those of the X-ray beams, and the three electron beams listed at the end of the Table appear to have no X-ray analogues.

The first of these differences is systematic and may

be summarised quantitatively in a simple manner. If the crystal were contracted in the direction of the incident beam by a factor 0·7, the X-ray beams would be shifted to the smaller co-latitude angles θ' (column 8), and would then agree in position fairly well with the observed electron beams—the average difference being 1·7°. Associated in this way there is a set of electron beams for each of the first ten sets of X-ray beams occurring in the range of observations, the electron beams for 110 volts alone being unaccounted for.

These results are highly suggestive, of course, of the ideas underlying the theory of wave mechanics, and we naturally inquire if the wave-length of the X-ray beam which we thus associate with a beam of electrons is in fact the h/mv of L. de Broglie. The comparison may be made, as it happens, without assuming a particular correspondence between X-ray and electron beams, and without use of the contraction factor. Quite independently of this factor, the wave-lengths of all possible X-ray beams satisfy the optical grating formula $n\lambda = d \sin \theta$, where d is the distance between lines or rows of atoms in the surface of the crystal—these lines being normal to the azimuth plane of the beam considered. For azimuths {111} and {100}, $d = 2{\cdot}15 \times 10^{-8}$ cm. and for azimuth {110}, $d = 1{\cdot}24 \times 10^{-8}$ cm. We apply this formula to the electron beams without regard to the conditions which determine their distribution in co-latitude angle. The correlation obtained by this procedure between wave-length and electron speed v is set down in the last three columns of Table I.

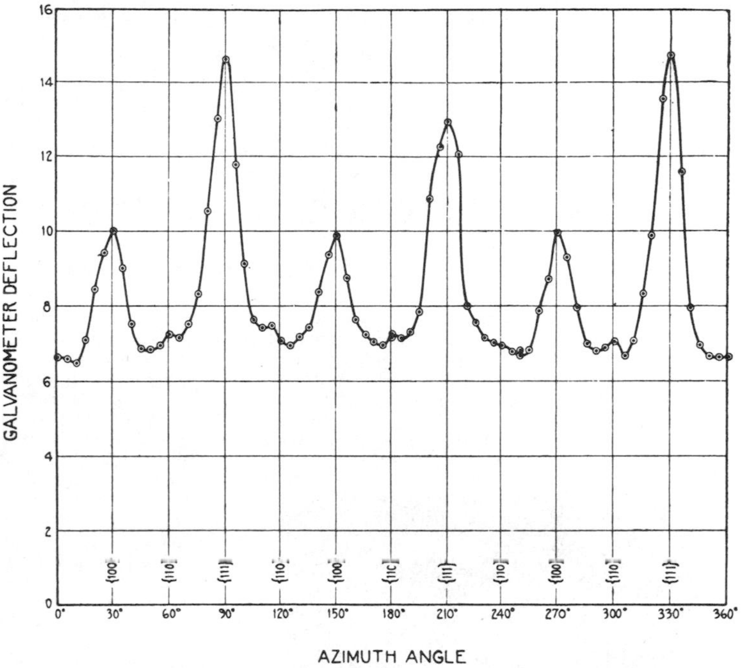

FIG. 2.—Intensity of electron scattering vs. azimuth angle—54 volts, co-latitude 50°.

In considering the computed values of $n(\lambda mv/h)$, listed in the last column, we should perhaps disregard those for the 110-volt beams at the bottom of the Table, as we have had reason already to regard these beams as in some way anomalous. The values for the other beams do, indeed, show a strong bias toward small integers, quite in agreement with the type of phenomenon suggested by the theory of wave mechanics. These integers, one and two, occur just as predicted upon the basis of the correlation between electron beams and X-ray beams obtained by use of the contraction factor. The systematic character of the departures from integers may be significant. We believe, however, that this results from imperfect alignment of the incident beam, or from other structural deficiencies in the apparatus. The greatest departures are for beams lying near the limit of our co-latitude range. The data for these are the least trustworthy.

TABLE I.

Azimuth.	Electron Beams.			X-ray Beams.						
	Bomb. Pot (volts).	Co-lat. θ.	Intensity.	Reflections.	$\lambda \times 10^8$ cm.	Co-lat. θ.	Co-lat. θ'.	$v \times 10^{-8}$ cm./sec.	$n\lambda \times 10^8$ cm.	$n\left\{\frac{\lambda mv}{h}\right\}$.
{111}	54	50°	0·5	{220}	2·03	70·5	52·7	4·36	1·65	0·99
	100	31	0·5	{331}	1·49	44·0	31·6	5·94	1·11	0·91
	174	21	0·9	{442}	1·13	31·6	22·4	7·84	0·77	0·83
	174	55	0·15	{440}	1·01	70·5	52·7	7·84	1·76	2(0·95)
{100}	65	44	0·5	{311}	1·84	59·0	43·2	4·79	1·49	0·98
	126	29	1·0	{422}	1·35	38·9	27·8	6·67	1·04	0·95
	190	20	1·0	{533}	1·04	28·8	20·4	8·19	0·74	0·83
	159	61	0·4	{511}	1·05	77·9	59·0	7·49	1·88	2(0·97)
{110}	138	59	0·07	{420}	1·22	78·5	59·5	6·98	1·06	1·02
	170	46	0·07	{531}	1·04	57·1	41·7	7·75	0·89	0·95
{111}	110	58	0·15	..	..	..	..	6·23	1·82	1·56
{100}	110	58	0·15	..	..	..	..	6·23	1·82	1·56
{110}	110	58	0·25	..	..	..	..	6·23	1·05	0·90

5

Reprinted from *Nature* **119:**890 (1927)

DIFFRACTION OF CATHODE RAYS BY A THIN FILM

G. P. Thomson and A. Reid

IF a fine beam of homogeneous cathode rays is sent nearly normally through a thin celluloid film (of the order 3×10^{-6} cm. thick) and then received on a photographic plate 10 cm. away and parallel to the film, we find that the central spot formed by the undeflected rays is surrounded by rings, recalling in appearance the haloes formed by mist round the sun. A photograph so obtained is reproduced (Fig. 1). If

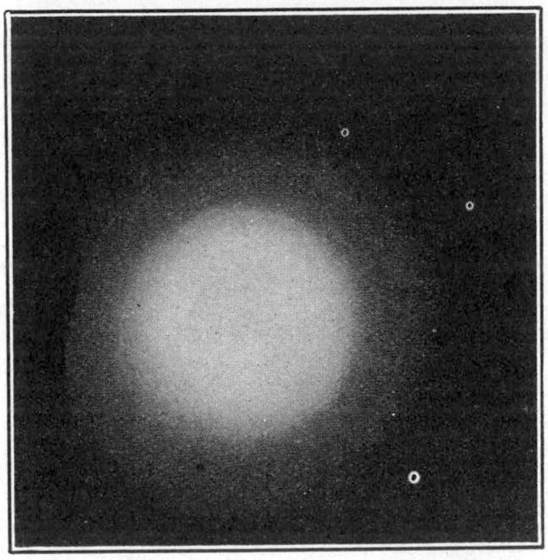

FIG. 1.

the density of the plate is measured by a photometer at a number of points along a radius, and the intensity of the rays at these points found by using the characteristic blackening curve of the plate (see *Phil. Mag.*, vol. 1, p. 963, 1926), the rings appear as humps on the intensity-distance curves. In this way rings can be detected which may not be obvious to direct inspection. With rays of about 13,000 volts two rings have been found inside the obvious one. Traces have been found of a fourth ring in other photographs, but not more than three have been found on any one exposure. This is probably due to the limited range of intensity within which photometric measurements are feasible.

The size of the rings decreases with increasing energy of the rays, the radius of any given ring being roughly inversely proportional to the velocity, but as the rings are rather wide the measurements so far made are not very accurate. The energy of the rays, as measured by their electrostatic deflexion, varied from 3900 volts to 16,500 volts. The rings are sharpest at the higher energies and were indistinguishable at about 2500 volts. In one photograph the radii of the rings were approximately 3, 5, and 6·7 mm. for an energy of 13,800 volts.

It is natural to regard this phenomenon as allied to the effect found by Dymond (NATURE, Sept. 4, 1926, p. 336) for the scattering of electrons in helium, though the angles are of course much smaller than he found. This would be due partly to the greater speed of the rays giving them a smaller wave-length.

Part II

SEMINAL PAPERS IN THE THEORY OF QUANTUM MECHANICS

Editor's Comments on Papers 6, 7, and 8

6 **HEISENBERG**
Excerpts from *On the Quantum Theory Reinterpretation of Kinematical and Mechanical Relations*

7 **SCHRÖDINGER**
Quantization as an Eigenvalue Problem

8 **HEISENBERG**
Excerpt from *On the Intuitive Content of Quantum Kinematics and Mechanics*

The attempt to provide a more powerful and effective method of applying Bohr's fundamental postulates of the quantum theory of atomic structure to the calculation of the energy states of atoms was pursued along two different lines. In connection with efforts made by W. Heisenberg, (1901-1976) in collaboration with H. A. Kramers (1894-1952), to make the Bohr theory work in the explanation of the dispersion of light by atoms, the former finally developed the notion that it was essential to construct a new mechanics based on the quantum concept but in which the fundamental laws would be relations connecting only observable quantities such as frequencies and intensities of spectral lines from atoms. Heisenberg felt that one serious difficulty with Bohr's development of his basic postulates was that his theory involved quantities like the position and velocity of an electron in an atom, which are essentially unobservable. Heisenberg then proceeded to suggest how a radically new quantum mechanics might be constructed without using unobservable quantities. His first paper (1925) in this program, restricted to systems with one degree of freedom, is reproduced here as Paper 6.

Heisenberg's idea was almost immediately taken up by Max Born (1882-1970) and Ernst Pascual Jordan (1902-   ) and generalized into an elaborate theory of quantum mechanics. They observed that to satisfy Heisenberg's fundamental criterion, the quantities to be used in his theory are best represented by matrices, whose properties had indeed been developed in detail by nineteenth-century mathematicians. Hence the Heisenberg, Born, and Jordan theory came to be called matrix mechanics. Born and Jordan worked out the details and showed that the energy of an atomic system is to be represented by a diagonal matrix, the elements of which are the energies of the various quantized states of the atom. They worked out the quantization of the harmonic and anharmonic oscillators and discussed some applications of their theory to electrodynamics. Although not reproduced here, theirs was a very important paper. It is referenced in the bibliography at the end of this commentary.

Erwin Schrödinger (1887-1961) decided to approach the problem of establishing a general quantum mechanics from the standpoint of the wave idea originally proposed by L. de Broglie. He was probably stimulated in this direction by the reflection that though the wave equation in unbounded space can have harmonic solutions for a continuous range of frequencies, once the waves are forced to lie in a bounded medium, the possible frequencies are restricted to a discrete set. The analogy with the discrete energy states of an atom is suggestive. If an appropriate wave equation containing the energy of an atomic system can be set up, the imposition of appropriate boundary conditions might be expected to lead to solutions corresponding to a discrete series of energy values (the eigenvalues). Schrödinger applied his idea to the evaluation of the energy states of the hydrogen atom and obtained results in agreement with those obtained by Bohr. This work marked the beginning of what came to be called wave mechanics; indeed this form of quantum mechanics with the quantized states of an atomic system represented by the so-called eigenfunctions associated with the energy eigenvalues has proved to be the most convenient one in all applications. It was soon shown that there is a very close relation between the matrix mechanics of Heisenberg, Born, and Jordan and the wave mechanics of Schrödinger. In fact, the matrices appropriate for a given quantized atomic system can be calculated from the permitted solution of the corresponding Schrödinger equation.[1] Schrödinger's fundamental 1926 article is reproduced here as Paper 7.

The application of Schrödinger's wave mechanics to the evaluation of the quantized energy states of polyelectronic atoms was initiated and carried through to a large extent by the British physicist Douglas R. Hartree (1897-1958) with the use of the so-called self-consistent field concept, first introduced into the study of the structure of polyelectronic atoms with the use of the earlier Bohr theory. Some relevant material is referenced in the bibliography at the end of the section. Hartree's work was generalized by V. A. Fock (1898-   ) to the calculation of relativistic atomic wave functions and the associated energy levels.

Heisenberg's point of view ultimately made a significant contribution to the development of quantum mechanics, although the detailed calculations of the properties of quantized atomic systems are now carried out almost exclusively in terms of wave mechanics. Although Heisenberg originally based his theoretical considerations on the fundamental assumption that his theory should be expressed in terms of observable quantities only, he ultimately felt it necessary to assign quantum mechanical significance to essentially unobservable quantities. He then found that the price to be paid for this procedure is the famous indeterminacy principle, which in effect says that the measurement of any particular atomic quantity to a definite degree of precision puts a limit on the precision with which its so-called conjugate quantity can be measured simultaneously. Thus the precision with which the position of an electron can be measured is inextricably related to the precision with which its momentum can be measured simultaneously, in the sense that the more precise the determination of the position, the less precise is the simultaneous determination of the momentum, and vice-versa. There also exists a similar indeterminacy relation connecting the energy of an atomic particle or system and the time of observation. The shorter the time of observation, the greater is the spread in the measured energy values. This principle has had considerable use in the estimation of the energy values of atomic systems and a considerable bearing on what may be called the philosophy of quantum mechanics.

Heisenberg's seminal paper on the indeterminacy principle is reproduced here in part as Paper 8.

REFERENCE

1. R. B. Lindsay and H. Margenau, *Foundations of Physics* (Woodbridge, Connecticut: Ox Bow Press, 1981) p. 453ff.

BIBLIOGRAPHY

Born, M., 1926, Quantenmechanik der Stossvorgänge, *Z. Phys.* **38:**803–807. Probability interpretation of quantum mechanics.

Born, M., and E. P. Jordan, 1925, On Quantum Mechanics, *Z. Phys.* **34:**858–888.

De Broglie, L., 1964, *The Current Interpretation of Wave Mechanics: A Critical Study,* Elsevier, Amsterdam.

Dirac, P. A. M., 1967, *The Principles of Quantum Mechanics,* 3rd ed., Clarendon Press, Oxford.

Hartree, D. R., 1957, *The Calculation of Atomic Structures,* Wiley, New York.

Jammer, M., 1966, *The Conceptual Development of Quantum Mechanics,* McGraw-Hill, New York.

Lindsay, R. B., 1924, On the Atomic Models of the Alkali Metals, *J. Math. Phys.* **3:**191–236.

6

ON THE QUANTUM THEORY REINTERPRETATION OF KINEMATICAL AND MECHANICAL RELATIONS

W. Heisenberg

This article was translated expressly for this Benchmark volume by R. Bruce Lindsay, Brown University, from pages 879–889 and 893 of Z. Phys. ***33:****879–893 (1925), by permission of the publisher, Springer-Verlag, New York.*

ABSTRACT

The aim of this work is to establish foundations for a quantum theoretical mechanics, which is based exclusively on relations between quantities that are observable in principle.

It is well known that the formal rules used in the quantum theory to calculate observed quantities (such as, for example, the energy of the hydrogen atom) suffer from the serious objection that they contain as an essential ingredient relations between quantities that in principle are apparently not observable as, for example, the position and period of rotation of the electron in the atom. Moreover, it appears that there is no physical foundation for these rules unless we cling to the hope that the hitherto unobservable quantities will eventually become observable. Such hope might appear to be justified if the rules in question were consistent among themselves and were applicable to a definite domain of quantum theoretical problems. Experiment shows, however, that it is only the hydrogen atom and the Stark effect in this atom that agree with these formal rules of quantum theory. On the other hand, fundamental difficulties already arise in the case of the hydrogen atom subjected to electric and magnetic fields in different directions, and it appears certain that the reaction of atoms to periodically changing fields cannot be described by the rules in question. Finally, the extension of the quantum rules to the handling of atoms with many electrons has proved impossible. [*Editor's Note:* This last extreme statement is not wholly justified, since studies of the stationary states of polyelectronic atoms based on Bohr's general theory of the periodic table of the elements did provide approximate agreement with experiment, as the first attempts with the self-consistent field showed.]

It has become customary to attribute this breakdown of the Bohr quantum theoretical rules, which are indeed characterized essentially by the application of classical mechanics, as a definite deviation from this mechanics. But this way of looking at the matter can hardly be considered as cogent when we reflect that the generally applicable Einstein-Bohr frequency condition is already a complete renunciation of classical mechanics. From the standpoint of wave theory, involving the kinematics at the basis of this mechanics, even in the simplest quantum theoretical problem, classical mechanics cannot be considered as having validity. In this situation, it appears advisable to give up the hope of any ability to observe previously unobservable quantities such as the position and angular velocity of the electron and at the same time to interpret the partial agreement of the quantum conditions with experience as more or less a matter of chance. It would also seem to be advisable to attempt to construct a quantum theoretical mechanics in analogy with classical mechanics, in which we use only relations connecting observable quantities. For the first assumptions of such a quantum theoretical mechanics, we could, besides the frequency rule, use the basic ideas of Kramers' dispersion theory[1] and the further work based on this.[2] In the following we shall attempt to set up some new quantum theoretical relations and use them for the complete solution of some specific problems. We shall restrict our attention to problems with a single degree of freedom.

1). The radiation from a moving electron in classical theory (in the wave region where $\mathfrak{E}$ and $\mathfrak{H}$ vary as $1/r$) is given to a first approximation by the expressions

$$\mathfrak{E} = \frac{e}{r^3 c^2}[\mathfrak{r}[\mathfrak{r}\dot{\mathfrak{v}}]],$$

$$\mathfrak{H} = \frac{e}{r^2 c^2}[\dot{\mathfrak{v}}\mathfrak{r}]$$

where $\mathfrak{E}$ and $\mathfrak{H}$ are the electric and magnetic field intensities, respectively, e is the charge on the electron, $\mathfrak{r}$ is the vector from the origin to the electron, and $\mathfrak{v}$ is the velocity of the electron. In the next approximation, further terms are introduced of the form

$$\frac{e}{r c^3}\dot{\mathfrak{v}}\mathfrak{v},$$

usually referred to as connected with "quadrupole radiation." In a still higher approximation, there are terms of the form

$$\frac{e}{r c^4}\dot{\mathfrak{v}}\mathfrak{v}^2;$$

We may now ask how such higher terms must appear in the quantum theory. Since in the classical theory the higher approximations can be simply calculated if the motion of the electron or its Fourier representation is given, we would expect the same to be true in the quantum theory. This has nothing to do with electrodynamics; it is a purely kinematical matter, a rather important point. We can present it in its simplest form as follows: if we have given a quantum theoretical quantity that is to stand in place of a classical quantity $x(t)$, what quantum theoretical quantity should appear in the place of $[x(t)]^2$?

Before we are able to answer this question, we must recall that in the quantum theory, it was not possible to associate with the electron a point in space as a function of time by means of observable quantities. However, even in the quantum theory, we can associate radiation with the electron. This radiation will be described in the first instance by the frequencies that appear quantum theoretically as functions of two variables in the form

$$\nu(n, n-\alpha) = \frac{1}{h}\{W(n) - W(n-\alpha)\},$$

In the classical theory they have the form

$$\nu(n, \alpha) = \alpha \cdot \nu(n) = \alpha \frac{1}{h}\frac{dW}{dn}.$$

(Here $nh = J$, one of the canonical variables in the classical theory.)

To compare the classical theory with the quantum theory so far as concerns frequency, we can write the combination relations as follows:

Classical

$$\nu(n, \alpha) + \nu(n, \beta) = \nu(n, \alpha + \beta).$$

Quantum

$$\nu(n, n-\alpha) + \nu(n-\alpha, n-\alpha-\beta) = \nu(n, n-\alpha-\beta)$$

or

$$\nu(n-\beta, n-\alpha-\beta) + \nu(n, n-\beta) = \nu(n, n-\alpha-\beta).$$

To describe the radiation, we need in addition to the frequencies to introduce the amplitudes. The latter can be represented as complex vectors, each having six independent parameters. These vectors determine polarization and phase. They are also functions of the two variables n and α. We can write the following expressions for them:

Quantum

$$Re\{\mathfrak{A}(n, n-\alpha)\, e^{i\omega(n, n-\alpha)t}\}. \tag{1}$$

Classical

$$Re\{\mathfrak{A}_\alpha(n)\, e^{i\omega(n)\cdot\alpha t}\}. \tag{2}$$

The phase (contained in $\mathfrak{A}$) does not at first appear to have a meaning in quantum theory since the quantum frequencies are not in general commensurable with their harmonics. We can see at once, however, that the phase does have a meaning in quantum theory analogous to its significance in classical theory. If we now consider a particular quantity $x(t)$ in the classical theory, we can think of it as represented by a totality of quantities of the form

$$\mathfrak{A}_\alpha(n)\, e^{i\omega(n)\cdot\alpha t},$$

which whether the motion is periodic or not can represent $x(t)$ in the form of a series or an integral, as follows:

$$\left.\begin{aligned} x(n,t) &= \sum_{-\infty}^{+\infty}{}_{\alpha}\, \mathfrak{A}_{\alpha}(n)\, e^{i\omega(n).\alpha t} \\ x(n,t) &= \int_{-\infty}^{+\infty} \mathfrak{A}_{\alpha}(n)\, e^{i\omega(n)\alpha t}\, d\alpha. \end{aligned}\right\} \tag{2a}$$

[*Editor's Note:* The sum is, over α.]

Such a combination of the corresponding quantum theoretical quantities does not appear possible without considerable arbitrariness. But we can look upon the totality of the quantities

$$\mathfrak{A}(n, n-\alpha)\, e^{i\omega(n, n-\alpha)t}$$

as representing the quantity $x(t)$ and then ask the question: in what way will the quantity $[x(t)]^2$ be represented? In the classical theory we can write

$$\mathfrak{B}_{\beta}(n)\, e^{i\omega(n)\beta t} = \sum_{-\infty}^{+\infty}{}_{\alpha}\, \mathfrak{A}_{\alpha}\, \mathfrak{A}_{\beta-\alpha}\, e^{i\omega(n)(\alpha+\beta-\alpha)t} \tag{3}$$

or in the integral form

$$= \int_{-\infty}^{+\infty} \mathfrak{A}_{\alpha}\, \mathfrak{A}_{\beta-\alpha}\, e^{i\omega(n)(\alpha+\beta-\alpha)t}\, d\alpha, \tag{4}$$

whereupon we have

$$x(t)^2 = \sum_{-\infty}^{+\infty}{}_{\beta}\, \mathfrak{B}_{\beta}(n)\, e^{i\omega(n)\beta t} \tag{5}$$

or

$$= \int_{-\infty}^{+\infty} \mathfrak{B}_{\beta}(n)\, e^{i\omega(n)\beta t}\, d\beta. \tag{6}$$

In the quantum theory it appears simplest and most natural to replace equations 3 and 4 by the following:

$$\mathfrak{B}(n, n-\beta)\, e^{i\omega(n,n-\beta)t} = \sum_{-\infty}^{+\infty}{}_{\alpha}\, \mathfrak{A}(n, n-\alpha)\, \mathfrak{A}(n-\alpha, n-\beta)\, e^{i\omega(n,n-\beta)t} \tag{7}$$

or

$$= \int_{-\infty}^{+\infty} d\alpha\, \mathfrak{A}(n, n-\alpha)\, \mathfrak{A}(n-\alpha, n-\beta)\, e^{i\omega(n,n-\beta)t}; \tag{8}$$

Indeed the combination relation for the frequency practically demands this representation. If we make the assumptions in equations 7 and 8, we see that the phases of the quantum theoretical $\mathfrak{A}$ have just as great a meaning as they do in the classical theory. However, the time origin and hence a phase constant common to all $\mathfrak{A}$ are arbitrary and without physical meaning. Yet the phases of the individual $\mathfrak{A}$s are included in the quantities $\mathfrak{B}$.[3] A geometrical interpretation of quantum theoretical phase relations in analogy with classical theory appears in the first instance to be hardly possible.

So far as the quantum theoretical representation of the quantity $[x(t)]^3$ is concerned, we ought to be able to write

$$\mathfrak{C}(n,\gamma) = \sum_{-\infty}^{+\infty}\sum_{-\infty}^{+\infty}{}_{\alpha,\beta}\, \mathfrak{A}_\alpha(n)\,\mathfrak{A}_\beta(n)\,\mathfrak{A}_{\gamma-\alpha-\beta}(n). \tag{9}$$

for the classical case and

$$\mathfrak{C}(n,n-\gamma) = \sum_{-\infty}^{+\infty}\sum_{-\infty}^{+\infty}{}_{\alpha,\beta}\, \mathfrak{A}(n,n-\alpha)\,\mathfrak{A}(n-\alpha,n-\alpha-\beta)\,\mathfrak{A}(n-\alpha-\beta,n-\gamma) \tag{10}$$

for the quantum theoretical case, along with the corresponding appropriate integrals.

In similar fashion all quantities of the form $[x(t)]^n$ can have their quantum analogs found, and hence any function of $[x(t)]$ that can be represented by a series expansion in powers of $x(t)$ can be represented quantum theoretically.

There is, however, a difficulty about the quantum representation of a product like $x(t)y(t)$. Let $x(t)$ be characterized by $\mathfrak{A}$ and $y(t)$ by $\mathfrak{B}$; then the representation of $x(t)y(t)$ will be

Classical

$$\mathfrak{C}_\beta(n) = \sum_{-\infty}^{+\infty}{}_\alpha\, \mathfrak{A}_\alpha(n)\,\mathfrak{B}_{\beta-\alpha}(n).$$

Quantum

$$\mathfrak{C}(n,n-\beta) = \sum_{-\infty}^{+\infty}{}_\alpha\, \mathfrak{A}(n,n-\alpha)\,\mathfrak{B}(n-\alpha,n-\beta).$$

Classically $x(t)y(t)$ always equals $y(t)x(t)$, but in the quantum case this is not generally so. With respect to combinations of the form

$$v(t)\,\dot{v}(t)$$

in the quantum case we replace $v\dot{v}$ by

$$\frac{v\dot{v}+\dot{v}v}{2}$$

in order to have $v\dot{v}$ come out as the derivative of $v^2/2$. In similar fashion, we can form average values in the quantum case, though they will doubtless be more hypothetical in character than formulas 7 and 8.

Aside from the above-mentioned difficulties, formulas like 7 and 8 should be satisfactory to express the interaction of the electrons in an atom through their characteristic amplitudes.

2). Following the foregoing considerations concerning the kinematics of the quantum theory, we take up next the mechanical problem of the determination of the $\mathfrak{A}$, v, w, etc. from the given forces on the system. In the earlier (Bohr) formulation of the theory, the procedure was as follows:

1) Integration of the equation of motion

$$\ddot{x} + f(x) = 0. \tag{11}$$

2) Determination of the constants appropriate for periodic motions by means of the quantum conditions

$$\oint p\, dq = \oint m \dot{x}\, dx = J (= nh). \tag{12}$$

If we now attempt to construct a quantum theoretical mechanics that is to be as close an analog to the classical variety as possible, it is plausible to take over the equation of motion (equation 11) directly into the quantum theory, in which it is necessary only (in order not to give up the fundamental principle of employing only *observable* quantities) to replace the classical quantities $\ddot{x}$, x(t), etc. in paragraph 1 by their quantum mechanical analogs.

In classical theory, it is possible to seek the solution of equation 11 by representing x as a Fourier series or integral with undetermined coefficients and frequencies. To be sure, in this case we wind up in general with infinitely many equations in infinitely many unknowns or with integral equations that only in special cases lead to simple recursion formulas for the quantities $\mathfrak{A}$. For the present, however, we are forced into this type of solution in the quantum case since, as has been emphasized above, no one of the functions x(*n*,*t*) leads to a direct quantum theoretical analog.

This leads to the result that the quantum theoretical solution of equation 11 can be carried out in the first instance only in the simplest cases. Before taking up such simple examples, let us see how the quantum theoretical determination of constants in accordance with equation 12 is to be found. We assume that the classical motion is periodic, that is,

$$x = \sum_{-\infty}^{+\infty}{}^{\alpha}\, a_\alpha(n)\, e^{i\alpha\omega_n t}; \tag{13}$$

whence

$$m\dot{x} = m \sum_{-\infty}^{+\infty} a_\alpha(n)\,.\, i\alpha\,\omega_n\, e^{i\alpha\omega_n t}$$

and

$$\oint m\dot{x}\, dx = \oint m\dot{x}^2\, dt = 2\pi m \sum_{-\infty}^{+\infty}{}^{\alpha}\, a_\alpha(n)\, a_{-\alpha}(n)\, \alpha^2\omega_n.$$

Since $a_{-\alpha}(n) = \overline{a_\alpha(n)}$ (x being taken as real), it follows that

$$\oint m\dot{x}^2\, dt = 2\pi m \sum_{-\infty}^{+\infty}{}^{\alpha}\, |a_\alpha(n)|^2\, \alpha^2\omega_n. \tag{14}$$

In the previous version of the quantum theory, it has been customary to put this phase integral equal to an integral multiple of *h*, that is, *nh*. However, such a condition is inserted into mechanical calculations only in a very formal fashion. Even from the standpoint of the correspondence principle, it seems arbitrary. For on the basis of this principle the *J*s are integral multiples of *h* only up to an arbitrary constant. In place of equation 14, there really should appear

$$\frac{d}{dn}(nh) = \frac{d}{dn}\cdot\oint m\dot{x}^2\,dt,$$

or

$$h = 2\pi m \cdot \sum_{-\infty}^{+\infty}{}^{\alpha}\, \alpha \frac{d}{dn}(\alpha\,\omega_n \cdot |a_\alpha|^2). \qquad (15)$$

Even such a condition gives the a_α only to within an arbitrary constant. This indeterminacy led in the earlier quantum theory (Bohr) to the difficulty of half quantum numbers.

If we now try to find the quantum mechanical analog to equations 14 and 15 for the case of observable quantities, we run into the problem of uniqueness.

Equation 15 does indeed possess a simple quantum theoretical transformation connected with the Kramers dispersion theory[4]:

$$h = 4\pi m \sum_{0}^{\infty}{}^{\alpha} \{|a(n,n+\alpha)|^2\,\omega(n,n+\alpha) - |a(n,n-\alpha)|^2\,\omega(n,n-\alpha)\}, \quad (16)$$

This relation is satisfactory for the unique determination of *a*. For the undetermined constants hidden in the quantities *a* are themselves fixed by the condition that the state defined is a normal state from which there is no radiation. If we denote this normal state by n_0, we must accordingly have

$$a(n_0, n_0 - \alpha) = 0 \text{ (für } \alpha > 0)$$

The problem of half-integral quantum numbers should therefore not arise in a quantum theoretical mechanics, which employs only relations connecting observable quantities.

Equations 11 and 16 taken together, if they can be solved, provide a complete determination not only of the frequencies and energies but also the quantum theoretical transition probabilities. The complete mathematical calculation, however, can for the present be carried out only for the simplest cases. A particular complication arises in the case of many systems, as for example, the hydrogen atom: that the solutions correspond in part to periodic and in part to aperiodic motions. The consequence is that the quantum theoretical series in equations 7 and 8 and equation 16 always break down into a sum and an integral. From the quantum theoretical point of view, a separation into periodic and aperiodic motions cannot in general be carried through.

In spite of this, we may perhaps consider equations 11 and 16 as at least in principle satisfactory solutions of the mechanical problem, if it can be shown that this solution agrees with or is at any rate not in contradiction to the previously well-established quantum relations. This means that a small perturbation of a mechanical problem gives rise to additional terms in the energy or the frequency that correspond to the expressions found by Kramers and Born, in contrast to the expressions derived from the classical theory. We must investigate further whether in general equation 11 in the proposed quantum theoretical mechanics has an energy integral

$$m\frac{\dot{x}^2}{2} + U(x) = \text{const}$$

and also whether the energy so derived satisfies the condition

$$\Delta W = h \cdot \nu$$

(where W is the energy and ν is the frequency).
Also

$$\nu = \frac{\partial W}{\partial J}$$

where J is the action (see equation 12).

A general answer to these questions is required in order to be able to verify the inner consistency of the foregoing quantum theoretical investigations and to lead to a quantum mechanics operating with observable quantities only. Aside from a general relation between the Kramers dispersion theory and equations 11 and 16, we can answer the questions just raised only in very special cases, solvable by simple recursion.

The general relation between the Kramers dispersion theory and our equations 11 and 16 resides in the fact that from equation 11 (that is, quantum theoretical analog), it follows that just as in the classical theory an oscillating electron behaves with respect to light of much shorter wavelengths than all normal modes of the system as a free electron. This result also follows from the Kramers theory, if we also take equation 16 into account. Kramers indeed finds for the moment induced by the wave $E \cdot \cos 2\pi\nu t$

$$M = e^2 E \cos 2\pi\nu t \cdot \frac{2}{h} \sum_0^\infty{}_\alpha \left\{ \frac{|a(n, n+\alpha)|^2 \nu(n, n+\alpha)}{\nu^2(n, n+\alpha) - \nu^2} - \frac{|a(n, n-\alpha)|^2 \nu(n, n-\alpha)}{\nu^2(n, n-\alpha) - \nu^2} \right\},$$

Accordingly, for $\nu >> \nu(n, n+\alpha)$, we have

$$M = -\frac{2Ee^2 \cos 2\pi\nu t}{\nu^2 . h} \sum_0^\infty{}_\alpha \{ |a(n, n+\alpha)|^2 \nu(n, n+\alpha) - |a(n, n-\alpha)|^2 \nu(n, n-\alpha) \},$$

In view of equation 16, this takes the form

$$M = -\frac{e^2 E \cos 2\pi\nu t}{\nu^2 \cdot 4\pi^2 m}.$$

3). As the simplest example we now discuss the anharmonic oscillator, for which the equation of motion is

$$\ddot{x} + \omega_0^2 x + \lambda x^2 = 0. \tag{17}$$

In the classical theory, this equation can be satisfied by

$$x = \lambda a_0 + a_1 \cos \omega t + \lambda a_2 \cos 2\omega t + \lambda^2 a_3 \cos 3\omega t + \cdots \lambda^{\tau-1} a_\tau \cos \tau\omega t,$$

in which the *a*s are power series in λ, beginning with a term free of λ. For a quantum theoretical analog, we represent *x* by an aggregate of terms of the form

$$\lambda a(n,n); \quad a(n,n-1)\cos\omega(n,n-1)t; \quad \lambda a(n,n-2)\cos\omega(n,n-2)t;$$
$$\ldots \lambda^{\tau-1} a(n,n-\tau)\cos\omega(n,n-\tau)t\ldots$$

Using equations 3 and 4 or 7 and 8, we obtain the following recursion formulas for the determination of *a* and ω

Classical theory

$$\left.\begin{aligned}
\omega_0^2 a_0(n) + \frac{a_1^2(n)}{2} &= 0;\\
-\omega^2 + \omega_0^2 &= 0;\\
(-4\omega^2 + \omega_0^2)a_2(n) + \frac{a_1^2}{2} &= 0;\\
(-9\omega^2 + \omega_0^2)a_3(n) + a_1 a_2 &= 0;\\
\ldots\ldots\ldots\ldots\ldots\ldots &
\end{aligned}\right\} \tag{18}$$

Quantum theoretical

$$\left.\begin{aligned}
\omega_0^2 a_0(n) + \frac{a^2(n+1,n) + a^2(n,n-1)}{4} &= 0;\\
-\omega^2(n,n-1) + \omega_0^2 &= 0;\\
(-\omega^2(n,n-2) + \omega_0^2)a(n,n-2) + \frac{a(n,n-1)a(n-1,n-2)}{2} &= 0;\\
(-\omega^2(n,n-3) + \omega_0^2)a(n,n-3) \qquad\qquad &\\
+\frac{a(n,n-1)a(n-1,n-3)}{2} + \frac{a(n,n-2)a(n-2,n-3)}{2} &= 0;\\
\ldots\ldots\ldots\ldots\ldots\ldots\ldots\ldots\ldots &
\end{aligned}\right\} \tag{19}$$

To these equations we must add the quantum conditions

Classical (J = nh)

$$1 = 2\pi m \frac{d}{dJ} \sum_{-\infty}^{+\infty} \tau^2 \frac{|a_\tau|^2 \omega}{4}.$$

Quantum

$$h = \pi m \sum_{0}^{\infty} [|a(n+\tau, n)|^2 \omega(n+\tau, n) - |a(n, n-\tau)|^2 \omega(n, n-\tau)].$$

To a first approximation this yields

$$\left.\begin{aligned} a_1^2(n) &= \frac{(n + \text{const})\, h}{\pi m \omega_0} \quad \text{(classical)} \\ a^2(n, n-1) &= \frac{(n + \text{const})\, h}{\pi m \omega_0} \quad \text{(quantum)} \end{aligned}\right\} \qquad (20)$$

The constants in equation 20 can be determined quantum theoretically by the condition that $a(n_0, n_0-1)$ must be zero in the normal state. If we number the n so that n is equal to zero in the normal state—that is, $n_0 = 0$—it follows that

$$a^2(n, n-1) = \frac{n h}{\pi m \omega_0}.$$

From the recursion equations 18, it then follows that in the classical theory (to the first approximation in λ)a_τ will be of the form $\kappa(\tau) n^{\tau/2}$, where $\kappa(\tau)$ represents a factor independent of n. In the quantum theoretical case we have from equation 19

$$a(n, n-\tau) = \varkappa(\tau) \sqrt{\frac{n!}{(n-\tau)!}}, \qquad (21)$$

in which $\kappa(\tau)$ represents the same proportionality factor, independent of n. For large values of n, the quantum theoretical value of a_τ goes over asymptotically to the classical value.

For the energy, it is natural to investigate the energy equation

$$\frac{m \dot{x}^2}{2} + m \omega_0^2 \frac{x^2}{2} + \frac{m \lambda}{3} x^3 = W$$

which to the approximation carried out here is also constant in the quantum theoretical case. From equations 19, 20, and 21, the energy comes out to be

Classical case

$$W = \frac{n h \omega_0}{2\pi}. \qquad (22)$$

Quantum case (from (7) and (8)

$$W = \frac{(n + \frac{1}{2})\, h\, \omega_0}{2\,\pi} \tag{23}$$

(cut to quantities of the order of λ^2.)

We see that even for the case of the harmonic oscillator the energy is not given by the classical value in equation 22 but has the form in equation 23.

[*Editor's Note:* In the remainder of the paper, not reproduced here, the author extends the theory of the anharmonic oscillator to a higher approximation and obtains agreement with the earlier dispersion analysis of Born and Kramers (cf. reference 1). He also works out the case of the simple rotator (an electron moving in a circular orbit about the nucleus) and shows that he is led to results in agreement with band spectra observations without the artificial introduction of half-quantum numbers. The author closes his paper with the following paragraph.]

Whether it can be considered satisfactory to use a method for the determination of quantum theoretical data based on relations between observable quantities, in line with the proposal made here, or whether this method represents too rough and approximate an attack on the obviously very complicated problem of a quantum theoretical mechanics can be decided only by a deeper mathematical investigation of the rather superficial methods used in this early approach.

Göttingen, Institute of Theoretical Physics

NOTES

1. H. A. Kramers, *Nature, 113,* 673, 1924.
2. M. Born, *Zs. f. Physik, 26,* 379, 1924. H. A. Kramers and W. Heisenberg, *Zs. f. Physik, 31,* 681, 1925. M. Born and P. Jordan, *Zs. f. Physik* (in press).
3. Cf. H. A. Kramers and W. Heisenberg, op. cit. In this paper, the phases are essentially included in the expression for the individual scattering moments.
4. This relation has already been given on the basis of dispersion considerations by W. Kuhn, *Zs. f. Phys. 33,* 408, 1925 and Thomas, *Nature, 13,* 1925.

7

QUANTIZATION AS AN EIGENVALUE PROBLEM

E. Schrödinger

This article was translated expressly for this Benchmark volume by R. Bruce Lindsay, Brown University, from Ann. Phys. ***79:***361–376 *(1926), by permission of the publisher, Johann Ambrosius Barth, Leipzig*

In this communication, I should like to show in the simplest case of the nonrelativistic and unperturbed hydrogen atom that the usual quantization prescription can be replaced by another condition, which contains no word about "whole numbers." On the contrary, the integrality arises in the same natural way as the integral character of the number of nodes in a vibrating string. The new treatment is generalizable and touches very deeply the true nature of the quantum conditions.

The usual form of the quantum condition is closely related to the Hamiltonian partial differential equation

$$H\left(q, \frac{\partial S}{\partial q}\right) = E \; . \tag{1}$$

We shall seek a solution of this equation, representable as a sum of functions, each of which is a function of only one of the independent variables q.

We now introduce for S a new unknown function ψ of such a nature that ψ will appear as a product of functions of the individual coordinates. Thus we set

$$S = K \lg \psi \, . \tag{2}$$

The constant K must be introduced for reasons of dimensionality. It has the dimensions of action. We then obtain

$$H\left(q, \frac{K}{\psi}\frac{\partial \psi}{\partial q}\right) = E \; . \tag{1'}$$

We do not seek a solution of equation 1′. Rather we present the following condition: Neglecting change of mass, equation 1′ can always be made to take the form: a quadratic function of ψ and its first derivative is equal to zero. (Consideration of mass change can be introduced if only the single electron problem is considered.) We look for those real functions ψ, which are single valued, finite, and twice continuously differentiable in the whole configuration space, which make the integral of the above-mentioned quadratic form, extended over the whole configuration space, an extremal. This variation problem replaces the usual quantum conditions.[1]

We take for H the Hamiltonian function for the Keplerian motion and then proceed to show that the condition cited can be fulfilled for all positive values of E, but only for a discrete set of negative values. This means that the variation problem in question has a discrete and continuous eigenvalue spectrum. The discrete spectrum corresponds to the

Balmer terms, whereas the continuous spectrum corresponds to the energies of the hyperbolic (unbound) orbits. In order to secure numerical agreement with experiment, we must set K equal to $h/2\pi$, where h is Planck's constant.

Since the choice of coordinates is immaterial in the setting up of the variation equations, we shall choose rectangular coordinates. In our case (with e and m the charge and mass of the electron, respectively) equation 1′ becomes

$$\left(\frac{\partial \psi}{\partial x}\right)^2 + \left(\frac{\partial \psi}{\partial y}\right)^2 + \left(\frac{\partial \psi}{\partial z}\right)^2 - \frac{2m}{K^2}\left(E + \frac{e^2}{r}\right)\psi^2 = 0 \, . \qquad (1'')$$
$$r = \sqrt{x^2 + y^2 + z^2} \, .$$

The variation problem takes the form

$$\left\{ \begin{aligned} \delta J = \delta \iiint dx\,dy\,dz \Bigg[\left(\frac{\partial \psi}{\partial x}\right)^2 + \left(\frac{\partial \psi}{\partial y}\right)^2 + \left(\frac{\partial \psi}{\partial z}\right)^2 \\ - \frac{2m}{K^2}\left(E + \frac{e^2}{r}\right)\psi^2\Bigg] = 0 \, , \end{aligned} \right. \qquad (3)$$

The integral is extended over all space. From this we find in the usual way

$$\left\{ \begin{aligned} \tfrac{1}{2}\delta J = \int df\, \delta\psi \frac{\partial \psi}{\partial n} - \iiint dx\,dy\,dz\, \delta\psi \Bigg[\Delta \psi + \\ + \frac{2m}{K^2}\left(E + \frac{e^2}{r}\right)\psi\Bigg] = 0 \, . \end{aligned} \right. \qquad (4)$$

In the first place, we must have

$$\Delta \psi + \frac{2m}{K^2}\left(E + \frac{e^2}{r}\right)\psi = 0 \qquad (5)$$

In the second place, the integral taken over the infinitely distant closed surface must also vanish. Thus we have

$$\int df\, \delta\psi \frac{\partial \psi}{\partial n} = 0 \, . \qquad (6)$$

Because of this last requirement, the variation problem must be supplemented by a requirement concerning the behavior of $\delta\psi$ at infinity—that is, that the *continuous* eigenvalue spectrum really exists. More about this later.

The solution of equation 5 can be carried through in spherical coordinates r, θ, ϕ, by treating ψ as a product of functions of r, θ, and ϕ separately. The method is sufficiently well known. For the dependence of ψ on the angles θ and ϕ, we get a spherical harmonic. For the dependence on r (the function in question being designated by χ), we obtain the differential equation

$$\frac{d^2\chi}{dr^2} + \frac{2}{r}\frac{d\chi}{dr} + \left(\frac{2mE}{K^2} + \frac{2me^2}{K^2 r} - \frac{n(n+1)}{r^2}\right)\chi = 0 \, . \qquad (7)$$
$$n = 0,\ 1,\ 2,\ 3\ \ldots .$$

The restriction of n to integral values is necessary in order that the dependence on the coordinates θ and ϕ shall be single-valued.

We require solutions of equation 7, which remain finite for all nonnegative real values of r. Eq 7 has two singularities in the r plane: at $r = 0$ and $r = \infty$, the second of which is an essential singularity of all integrals.[2] On the other hand, the first is not an essential singularity. These two singularities form the boundary points of our interval along the r axis. In such a case, we know that the requirement of finiteness of the function χ at the boundary points is equivalent to a boundary condition. The equation in general has no integral at all that remains finite at both boundary points. Such an integral exists for only certain special values of the constants occurring in the equation. These special values are the ones we seek to obtain.

The state of affairs just set forth is the springboard for the whole investigation.

We consider in the first place the singular point $r = 0$. The fundamental equation determining the behavior of the integral at this point is

$$\varrho(\varrho - 1) + 2\varrho - n(n + 1) = 0 \tag{8}$$

[*Editor's Note:* The author does not define the variable ρ here. It turns out that this is the exponent of r in the dependence of the solution of equation 7 on r.]

This equation has the roots

$$\varrho_1 = n, \qquad \varrho_2 = -(n + 1) \ . \tag{8'}$$

The two canonical integrals at this singular point belong accordingly to the exponents n and $-(n + 1)$. Since n is not negative, we can use only the first case. The corresponding integral will be represented by a power series, whose first term is r^n. (The other integral, which has no interest for us, can contain a logarithm because of the integral number difference between the exponents in the two cases.) Since the next singular point does not occur until infinity, the power series in question converges uniformly and represents a complete transcendental function. We can therefore say:

The solution we seek (to an arbitrary constant factor) is a uniquely determined complete transcendental function, which vanishes at $r = 0$ as r^n.

We now have to investigate the behavior of this function at $r = \infty$ on the positive real axis. For this we simplify equation 7 by the substitution

$$\chi = r^{\alpha} U \ , \tag{9}$$

in which α is chosen in such a way that the term in $1/r^2$ disappears. We can readily see that α must have one of the two values n or $-(n + 1)$. Equation 7 then takes the form

$$\frac{d^2 U}{d r^2} + \frac{2(\alpha + 1)}{r} \frac{d U}{d r} + \frac{2m}{K^2}\left(E + \frac{e^2}{r}\right) U = 0 \ . \tag{7'}$$

The integrals for this at $r = 0$ have the exponents 0 and $-2\alpha - 1$. For the first α value, $\alpha = n$, the first of these integrals is a complete transcendental function, and for the second α value, the second of these integrals is a complete transcendental function and, according to equation 9, leads to the desired solution, which is indeed single valued. Hence we lose nothing be restricting our attention to one of the α values. We choose

$$\alpha = n \, . \tag{10}$$

Our solution U corresponds, then, at $r = 0$ to the exponent zero. Mathematicians call equation 7′ Laplace's equation. The general form is

$$U'' + \left(\delta_0 + \frac{\delta_1}{r}\right) U' + \left(\varepsilon_0 + \frac{\varepsilon_1}{r}\right) U = 0 \, . \tag{7''}$$

[*Editor's Note:* here $U' = du/dr$, etc.]

In our case the constants have the values

$$\delta_0 = 0 \, , \quad \delta_1 = 2(\alpha + 1) \, , \quad \varepsilon_0 = \frac{2\,m\,E}{K^2} \, , \quad \varepsilon_1 = \frac{2\,m\,e^2}{K^2} \, . \tag{11}$$

This type of equation is comparatively easy to handle, since the so-called Laplace transformation, which in general leads to another equation of the second order here gives one of the first order, solvable by quadrature. This permits a repesentation of the solutions of equation 7′′ itself by means of integrals in the complex plane. We present here only the final result.[3] The integral

$$U = \int_L e^{zr} (z - c_1)^{\alpha_1 - 1} (z - c_2)^{\alpha_2 - 1} \, dz \tag{12}$$

is a solution of equation 7′′ for the integration path L, for which

$$\int_L \frac{d}{dz} \left[e^{zr} (z - c_1)^{\alpha_1} (z - c_2)^{\alpha_2} \right] dz = 0 \, . \tag{13}$$

The constants c_1, c_2, α_1, α_2 have the following values: c_1 and c_2 are the roots of the quadratic equation

$$z^2 + \delta_0 \, z + \varepsilon_0 = 0 \tag{14}$$

and

$$\alpha_1 = \frac{\varepsilon_1 + \delta_1 \, c_1}{c_1 - c_2} \, , \qquad \alpha_2 = \frac{\varepsilon_1 + \delta_1 \, c_2}{c_2 - c_1} \, . \tag{14'}$$

For the particular case of equation 7′, we have from equations 11 and 10

$$\begin{cases} c_1 = + \sqrt{\dfrac{-2\,m\,E}{K^2}} \, , \quad c_2 = - \sqrt{\dfrac{-2\,m\,E}{K^2}} \; ; \\[2ex] \alpha_1 = \dfrac{m\,e^2}{K\sqrt{-2\,m\,E}} + n + 1 \, , \quad \alpha_2 = - \dfrac{m\,e^2}{K\sqrt{-2\,m\,E}} + n + 1. \end{cases} \tag{14''}$$

The integral representation in equation 12 permits us not only to follow the asymptotic behavior of the totality of solutions as r approaches infinity in any specific fashion but also allows us to derive this behavior for a specific solution, which is always a more difficult task.

We shall now in the first place exclude the solution for the case in which α_1 and α_2 are real integers. This case arises simultaneously for both quantities and only when

$$\frac{m e^2}{K \underset{+}{\sqrt{-2mE}}} = \text{a real integer} \tag{15}$$

We accordingly assume that equation 15 is not satisfied.

The behavior of the totality of solutions for a certain mode of approach of r to infinity (we always restrict ourselves to approach along the real axis) will then be characterized by the behavior[4] of the two linear independent solutions obtained from the following two special cases of the integration path L and which we call U_1 and U_2, respectively. In both cases, z comes from infinity and returns there by the same route and indeed in such a direction that

$$\lim_{z=\infty} e^{zr} = 0 \;, \tag{16}$$

that is, the real part of zr becomes infinitely negative. In this way the condition in equation 13 is satisfied. In the one case (solution U_1), the point c_1 is revolved around once, and in the other case (solution U_2), the point $z = c_2$ is revolved around once.

These two solutions for very large real positive values of r are represented asymptotically (in the Poincaré sense) by

$$\begin{cases} U_1 \sim e^{c_1 r} r^{-\alpha_1} (-1)^{\alpha_1} (e^{2\pi i \alpha_1} - 1) \Gamma(\alpha_1)(c_1 - c_2)^{\alpha_2 - 1} \,, \\ U_2 \sim e^{c_2 r} r^{-\alpha_2} (-1)^{\alpha_2} (e^{2\pi i \alpha_2} - 1) \Gamma(\alpha_2)(c_2 - c_1)^{\alpha_1 - 1} \,, \end{cases} \tag{17}$$

in which we here satisfy ourselves with the first member of the asymptotic series in integral negative powers of r.

We have now to distinguish between the two cases $E > 0$ and $E < 0$.

1. $E > 0$. In the first place, the fact that equation 15 does not hold is at once guaranteed, since this quantity is then pure imaginary. Further, because of equation 14'', c_1 and c_2 are also pure imaginary. The exponential functions in equation 17 are accordingly finite periodic functions, since r is real. From equation 14'', the values of α_1 and α_2 show that U_1 and U_2 both go to zero as $r \to \infty$ as r^{-n-1}. The same must therefore hold for our whole completely transcendental U (whose behavior we are investigating), as they are always made up linearly from U_1 and U_2. Further, we learn from equations 9 and 10 that the function χ—that is, the complete transcendental solution of the original equation (7)—always goes to zero as $1/r$, since it arises from U by multiplication by r^n. We can express all this as follows:

The Euler differential equation (5) of our variation problem for every positive E has solutions that are continuous, finite, and single-valued and go to zero at infinity (under steady oscillations) as $1/r$. Of the surface condition (equation 6) we must speak later.

2. $E < 0$. In this case that equation 15 may hold is not automatically excluded, yet tentatively we adhere to its stipulated exclusion. Then from equations 14'' and 17, U_1 for $r = \infty$ grows beyond all limits for $E < 0$. On the other hand U_2 vanishes exponentially. Our complete transcendental function U (and the same is true for χ) will remain finite then and only then if U is identical with U_2 except for a numerical factor. This is, however, not the case. We see this as follows: If in equation 12 we choose for the integral path L a closed circuit about both points c_1 and c_2, a circuit that (because the sum $\alpha_1 + \alpha_2$ is integral) is then really closed on the Riemann surface of the integrand, so that automatically condition 13 is satisfied, it can readily be shown that the integral (equation 12) represents our complete transcendental function U. It then permits an expansion in a series with positive powers of r, which in any case converges for sufficiently small r and hence satisfies the differential equation (7'), which must coincide with that for U. Accordingly U is represented by equation 12 if L is a closed circuit about both points c_1 and c_2. This closed circuit can, however, be so distorted that it can appear as additively combined from the two earlier considered integration paths belonging to U_1 and U_2, an additive combination with nonvanishing factors such as 1 and $e^{2\pi i \alpha_1}$. Hence, U cannot coincide with U_2 but must also contain U_1.

Hence our complete transcendental function U, which is the only one among the solutions of equation (7') which comes into consideration in the solution of our problem, does not remain finite for large r under the assumptions laid down. Pending the completeness investigation, that is, the demonstration that our procedure permits us to find all linearly independent solutions of the problem, we may accordingly state:

For negative values of E that do not satisfy condition 15, our variation problem has no solution.

We have now only to investigate the discrete array of negative E values for which equation 15 is satisfied. For this α_1 and α_2 are accordingly integral. Of the two integration paths, which produced earlier the fundamental system U_1 and U_2, the first must surely be changed to provide for nonvanishing. For $\alpha_1 - 1$ is certainly positive, and the point c_1 is accordingly neither a branch point nor a pole of the integrand but an ordinary zero. The point c_2 can also be a regular point if $\alpha_2 - 1$ is also nonnegative. In every case, however, we can readily find two suitable integration paths and carry out the integration along them in closed form with known functions, so that the behavior of the solutions is completely clear.

Let us take

$$\frac{m e^2}{K \sqrt{-2mE}} = l ; \quad l = 1, 2, 3, 4 \ldots . \tag{15'}$$

Then from equation 14'' we have

$$\alpha_1 - 1 = l + n , \quad \alpha_2 - 1 = -l + n . \tag{14'''}$$

We must now distinguish between the two cases: $l \leq n$ and $l > n$. Let us take first

a) $l \leq n$. Then c_2 and c_1 lose any singular character and thereby gain the ability to function as initial and end points by virtue of the fulfillment of condition 13. A third point suitable for this purpose is the point at negative infinity along the real axis. Every path between two of these three points yields a solution, and of these three solutions any two are linearly independent, as can readily be verified by expressing the integrals in closed form. In particular, the complete transcendental solution is given by the integration path from c_1 to c_2. The fact that this integral remains regular at $r = 0$ can be readily recognized without computing it. This is mentioned since the actual working out of the solution is likely to cloud over this result. On the other hand, the solution shows that the integral grows without limit for positively increasing r. One of the two other integrals remains finite for large r but becomes infinite for $r = 0$.

Hence, for $l \leq n$ we obtain no solution to our problem.

b) $l > n$. Here according to equation 14''', c_1 is a zero point, whereas c_2 is at least a pole of first order of the integrand. Two independent integrals then result. The one comes from the path from $z = -\infty$ to the origin, being careful to avoid the pole; the other comes from the residue at the pole. The latter is the complete transcendental function. We give the calculated value, already multiplied by r^n, whereby according to equations 9 and 10 we obtain the solution χ of the original equation (7). (The multiplicative constant is arbitrarily adjusted.) Thus we find:

$$\chi = f\left(r \frac{\sqrt{-2mE}}{K}\right); \quad f(x) = x^n e^{-x} \sum_{k=0}^{l-n-1} \frac{(-2x)^k}{k!} \binom{l+n}{l-n-1-k}. \tag{18}$$

We recognize that this is really a usable solution since it remains finite for all real, nonnegative r. The surface condition (6) is satisfied through the vanishing of the solution at $r = \infty$. For negative E we summarize the results as follows:

For negative E our variation problem has solutions when and only when E satisfies the condition in equation 15. The integer n, which gives the order of the spherical harmonic appearing in the solution, may take on only values smaller than l. The part of the solution dependent on r is given by equation 18.

By enumeration of the constants in the spherical harmonics (namely $2n + 1$), we find further:

The solution we have obtained, for a permitted (n, l) combination, contains exactly 2n + 1 arbitrary constants. For a given value of l, there are accordingly l^2 arbitrary constants.

We have therefore satisfied in main outlines the original requirements for the eigenvalue spectrum of our variation problem. However, some gaps remain.

In the first place, there is the question of the completeness of the whole system of eigenfunctions. This will not be considered in this article. Through considerations to be provided elsewhere, we can feel sure that we have not overlooked any eigenvalues.

In the second place, we must remember that the eigenfunctions obtained for positive E do not without further investigation solve the variation problem in the form originally set up, since they go to zero at infinity only as $1/r$ and that therefore $\partial\psi/\partial r$ goes to zero on the surface of a large sphere as $1/r^2$. The surface integral (equation 6) therefore remains at infinity of the order of $\delta\psi$. If one wishes really to obtain the continuous spectrum, it is necessary to insert still another condition: that $\delta\psi$ vanishes at infinity or at least that it approaches a constant value, independent of the direction in which infinity is approached. In the latter case the spherical harmonic function will cause the surface integral to vanish.

The condition (15) yields

$$-E_l = \frac{me^4}{2K^2 l^2}. \tag{19}$$

If we assign to the constant K, which had to be introduced on dimensional grounds, the value

$$K = \frac{h}{2\pi}. \tag{20}$$

the above energy values are those of Bohr, corresponding to the Balmer terms. We then get

$$-E_l = \frac{2\pi^2 me^4}{h^2 l^2}. \tag{19'}$$

Our l is then the principal quantum number, and $n + 1$ is analogous to the azimuthal quantum number. The further splitting of this number in the closer determination of the spherical harmonics can be considered analogous to the splitting of the azimuthal quantum number into an "equatorial" and a "polar" quantum number. These numbers determine here the system of nodal lines on the sphere. Moreover the "radial" quantum number, $l - n - 1$, exactly determines the number of "nodal spheres." For one can readily be convinced that the function $f(x)$ in equation 18 has precisely $l - n - 1$ positive real roots. The positive values of E correspond to the continuum of hyperbolic orbits, to which we may plausibly assign the radial quantum number infinity. Corresponding to this, the relevant functions of the solution move out to infinity, executing continual oscillations.

It is of interest that the region within which the functions in equation 18 are appreciably different from zero, and within which they carry out their oscillations, is of the general order of magnitude of the major axis of the associated ellipse in Bohr's calculations. The factor that makes the argument of the function f dimensionless is obviously the reciprocal of a length. This length is

$$\frac{K}{\sqrt{-2mE}} = \frac{K^2 l}{me^2} = \frac{h^2 l}{4\pi^2 me^2} = \frac{a_l}{l}, \tag{21}$$

in which a_l is the half major axis of the elliptical orbit associated with l. (These equations follow from equation 19 with the use of the relation $E_l =$

$-e^2/2a_l$.) The quantity (equation 21) has the order of magnitude of the root domain for small values of *l* and *n*. For then it may be assumed that the roots of $\dot{f}(x)$ are of the order of magnitude unity. This will naturally no longer be the case if the coefficients of the polynomial are large numbers. We do not go here into the more precise evaluation of the roots but believe that the above claim can be substantiated as reasonably accurate.

There is a strong suggestion to relate the function ψ to an oscillation phenomenon in the atom. The reality of electron orbits is doubted by many today, and perhaps greater reality can be attributed to such oscillations. It was my original intent to develop the new presentation of the quantum prescription in this more picturesque fashion. However, I finally decided to use the preceding neutral mathematical formulation, since it has permitted me to present the essential features in a clearer way.

In my opinion, the essential thing is that in the quantum prescription, the mysterious requirement of integral numbers no longer appears; the mystery is, so to speak, carried back a step further. It has its basis in the finiteness and single-valuedness of a certain function of space.

I should prefer now not to go into greater detail in the discussion of the representation possibilities of the oscillation idea before somewhat more complicated cases have been studied and evaluated with success on the basis of the new conception. It is not out of the question that these will prove to be a mere stereotype of the standard quantum theory. For example, if one carries out the relativistic Kepler problem in the method above presented, it is noteworthy that half-integral quantum numbers appear.

However, we may permit ourselves a few observations on the oscillation representation. In the first place, I must not leave unmentioned the fact that I owe the stimulus to these considerations to the ingenious points of view of Louis de Broglie[5] and reflection on the spatial distribution of the "phase waves," associated with which he has shown there is always an integer measured along the path corresponding to a period or quasiperiod of the electron. The principal difference in our methods is that de Broglie is thinking of continuous waves, whereas if we provide an oscillatory representation for our formulation, we are led to stationary waves. I have recently showed[6] that we can base Einstein's gas theory on the consideration of such stationary waves by applying the dispersion laws of the de Broglie phase waves. [*Editor's note:* What the author is thinking of is probably the quantity k in the oscillator differential equation, viz. $m\ddot{\xi} = -k\xi$, with $k/m = 4\pi^2 f^2$, in which f is the frequency.] The above application of such considerations to the atom might be considered as a generalization of the gas model treatment.

If we interpret the individual functions (equation 18) multiplied by a spherical harmonic of order *n* as the description of characteristic oscillations, the quantity *E* must have something to do with the frequency of the oscillation in question. Now in oscillational problems we are accustomed to find the so-called parameter (usually referred to as λ) proportional to the square of the frequency. However, we encounter the difficulty that such a result in our presentation would lead for negative *E* values to imaginary frequencies, and in the second place in the quantum theory we are accustomed to have the energy proportional to the frequency itself and not to its square.

The contradiction may be resolved as follows: For the parameter E in the variation problem (equation 5), there is no provisional rational zero level set, particularly since the continuous function ψ appears multiplied with a function of r, which can be changed by a constant through a corresponding alteration in the zero level of E. Consequently the expectation associated with classical oscillation theory has to be corrected to the extent that not E itself but E increased by a certain constant is expected to be proportional to the square of the frequency. Suppose we let this constant be *very large* in comparison with the magnitude of all negative E values [which are limited indeed by condition 15]. Then in the first place, the frequencies will be real, and in the second place, our E values, since they correspond to only relatively small frequency differences, will actually be very approximately proportional to these frequency differences. That is really all that the "natural feeling" of the quantum theorist is entitled to demand as long as the zero level of the energy is not fixed.

The conception that the frequency of the oscillation process should be given by

$$\nu = C' \sqrt{C + E} = C' \sqrt{C} + \frac{C'}{2\sqrt{C}} E + \ldots \qquad (22)$$

in which C is a constant very large compared with all E, has another valuable advantage. It leads to an understanding of the Bohr frequency condition. According to the latter, the emission frequencies are proportional to the differences in E and accordingly from equation 22 are also proportional to the differences of the characteristic frequencies of the hypothetical oscillation processes. Indeed the characteristic oscillation frequencies are all very large compared with the emission frequencies and are themselves rather close together. The emission frequencies are deep "difference tones" of the much higher-frequency characteristic oscillations. It is indeed understandable that in the passage of the energy from one normal oscillation to another, some phenomenon should appear (I mean the light wave) whose frequency is the frequency difference of the characteristic oscillations. One needs only to look at it from this point of view: the light wave is causally connected with the beats necessarily occurring during the transition at any point in space, and the light frequency is determined by the frequency with which the intensity maximum of the beat process repeats itself.

The doubt may arise that the above conclusion rests on the relation (22) in its approximate form (through expansion of the square root), whereby the Bohr frequency condition would necessarily assume the character of an approximation formula. This is, however, only an apparent circumstance and is completely avoided if we develop the relativistic theory, through which for the first time a deeper understanding can be made available. The large additive constant C is, naturally, closely connected with the rest energy mc^2 of the electron. Moreover, the apparently repeated appearance of the constant h in the frequency condition in independent fashion [already introduced by the assumption in equation 20] is cleared up by the relativity theory. Unfortunately, attempts to carry this through in objection-free fashion at this time have encountered the sort of difficulties already mentioned above.

It is hardly necessary to emphasize the attractiveness of the view that in a quantum transition, the energy passes from one oscillator to another compared with the representation of such a transition by a "jumping" electron. The alteration in the form of oscillation can take place continuously in space and time; it can readily last as long as the emission process has been shown to last by experiment (for example, the canal rays research of W. Wien). Moreover, if during such a transition the atom is exposed for a relatively short time to an electric field, the characteristic (eigen) frequencies get out of tune as well as the beat frequencies, and this happens only so long as the field is on. This well-established experimental result has made it very difficult to achieve a satisfactory explanation. We recall here the discussion about the investigations of Bohr, Kramers, and Slater.

In the satisfaction over the ability of mankind to approach close to an understanding of all these things, we must not forget that the point of view that the atom is vibrating even if it does not radiate, and in the form of a single natural mode of oscillation, if such a representation must be adhered to, it must differ widely from the natural, commonly accepted picture of an oscillating system, for a macroscopic system does not behave in this fashion. In general, it involves what may be called a potpourri of its natural vibrations. However, we must not attach ourselves to this point of view too hastily. Even a potpourri of natural vibrations in the individual atom would not provide a serious difficulty insofar as no other beat frequencies were to appear than those whose emission the atom is able to bring about. Moreover, the simultaneous actual emission of many spectral lines by the same atom does not contradict experience. We can also conceive that in the normal state and even in certain metastable states, the atom may oscillate with a single characteristic frequency, and it therefore does not radiate since no beats are produced. Excitation might then consist of a simultaneous production of one or more characteristic frequencies, whereby beats are produced, thus bringing about light emission.

Under all circumstances, I am inclined to believe that the eigenfunctions belonging to the same frequency are all excited simultaneously. Multiplicity of eigenvalues then corresponds to degeneracy in the language of the earlier quantum theory of atomic structure. The reduction of the quantization of degenerate systems then ought to correspond to the arbitrary distribution of energy among the eigenfunction belonging to a single eigenvalue.

ADDENDUM MADE IN THE PROOF FEBRUARY 28, 1926

For the case of the classical mechanics of conservative systems, the variation problem can be formulated better than previously with explicit reference to the Hamiltonian partial differential equation. Let *T(q,p)* be the kinetic energy as a function of coordinates and momenta, *V* the potential energy, $d\tau$ the volume element in configuration space "rationally measured," that is, not simply the product $dq_1 dq_2 dq_3 \ldots dq_n$ but this divided by the square root of the discriminant of the quadratic form *T(q,p)* (See Gibbs, "Statistical Mechanics"). Then we require that ψ shall make the Hamiltonian integral

$$\int d\tau \left\{ K^2 T\left(q, \frac{\partial \psi}{\partial q}\right) + \psi^2 V \right\} \tag{23}$$

stationary under the normalizing condition

$$\int \psi^2 d\tau = 1. \tag{24}$$

The eigenvalues of this variation problem, as is well known, are the stationary values of the integral (equation 23) and thus provide, in accordance with our theory, the quantum energy levels.

We may remark concerning equation 14'' that in the quantities α_2 we have essentially the well-known Sommerfeld expression

$$-\frac{B}{\sqrt{A}} + \sqrt{C}$$

(cf. A. Sommerfeld: *Atombau und Spektrallinen,* 4th edition, p. 775).

Zürich, Physical Institute of the University, January 27, 1926

NOTES

1. I am well aware that this formulation is not unique.
2. For guidance in the handling of equation 7, I am deeply indebted to Hermann Weyl. For the following proposition for which no proof is given, I refer the reader to L. Schlesinger, *Differential Gleichungers* (Sammlung Schubert, No. 13. Göschen, 1900, especially chapters 3 and 5.)
3. Cf. L. Schlesinger, loc. cit. The relevant theory is due to H. Poincaré and J. Horn.
4. If equation 15 is satisfied, at least one of the two paths of integration described in the text is not usable since it leads to a vanishing result.
5. L. de Broglie, Ann. de Physique (10) **3,** 22, 1925 (Thesis, Paris, 1924)
6. To appear soon in *Physikalische Zeitschrift.*

8

ON THE INTUITIVE CONTENT OF QUANTUM KINEMATICS AND MECHANICS

W. Heisenberg

This article was translated expressly for this Benchmark volume by R. Bruce Lindsay, Brown University, from pages 172–179 of Z. Phys. ***43:****172–198 (1927), by permission of the publisher, Springer-Verlag, New York*

ABSTRACT

In this work, we first formulate precise definitions of the words "position," "velocity," "energy," etc. (for example, of an electron), words that also retain validity in quantum mechanics. It will be shown that canonically conjugate quantities can be determined simultaneously only with a characteristic indeterminacy (paragraph 1). This indeterminacy is the essential reason for the appearance of statistical connections in quantum mechanics. It can be mathematically formulated by means of the Dirac-Jordan theory (paragraph 2). Proceeding from the fundamental theorems of this theory, we can show how macroscopic phenomena can be understood in terms of quantum mechanics (paragraph 3). Several special thought experiments are discussed to illustrate the theory.

We believe we understand a physical theory intuitively if we can grasp qualitatively in all simple cases the experimental consequences of the theory and if we have recognized that the application of the theory never contains internal contradictions. For example, we believe we can understand intuitively the Einstein representation of closed three-dimensional space since the experimental consequences of this representation are conceivable by us in noncontradictory fashion. To be sure, these consequences contradict our customary space-time conceptions. However, we can convince ourselves that the possibility of the applications of the customary space-time concept to very large spaces follow neither from our laws of thinking nor from experience.

So far the intuitive meaning of quantum mechanics has been full of contradictions, which express themselves in the struggle of opinions about continuity and discontinuity in theory and the difference between corpuscles and waves. One might already conclude from this that an interpretation of quantum mechanics using our usual kinematic and mechanical concepts is not possible. Quantum mechanics arose indeed from the attempt to break with those customary kinematic concepts and to replace them with relations between experimentally given numbers. Since this appears to have succeeded, the mathematical schema of quantum mechanics would appear to need no revision. Nor will we need to contemplate revision of the space-time geometry for small spaces and short times, since by choice of sufficiently heavy masses the quantum mechanical laws can approximate the classical ones sufficiently closely and also when one handles small spaces and times. However, it appears to follow from the fundamental equations of quantum mechanics that a revision of the kinematic and mechanical concepts is necessary. If a particular mass *m* is given in our usual way of looking at things, to speak of the position and velocity of the center of mass of the mass *m* has a simply understood sense. In quantum mechanics, however, a relation

$$pq - qp = \frac{h}{2\pi i}$$

exists connecting mass, position, and velocity. We therefore have good reason to be cautious about the uncritial application of the words "position" and "velocity." If we admit that discontinuities are somewhat typical for small spaces and short times, a breakdown of the very concepts of "position" and "velocity" appears distinctly plausible. If, for example, we think of the one-dimensional motion of a material point in a continuum theory, we can draw a curve *x(t)* for the path of the particle (more precisely its center of mass) (Fig. 1), and the tangent

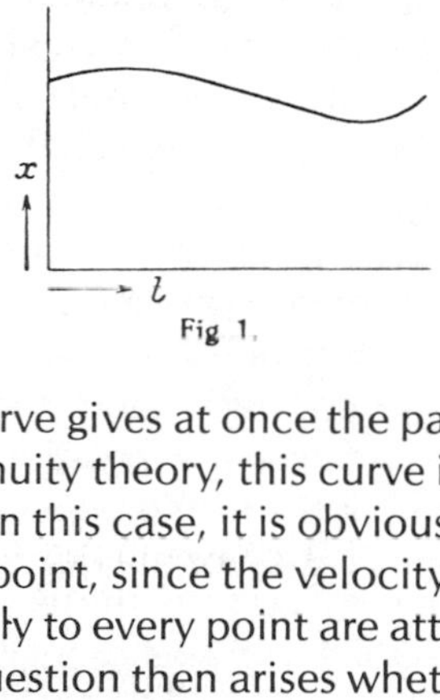

Fig 1.

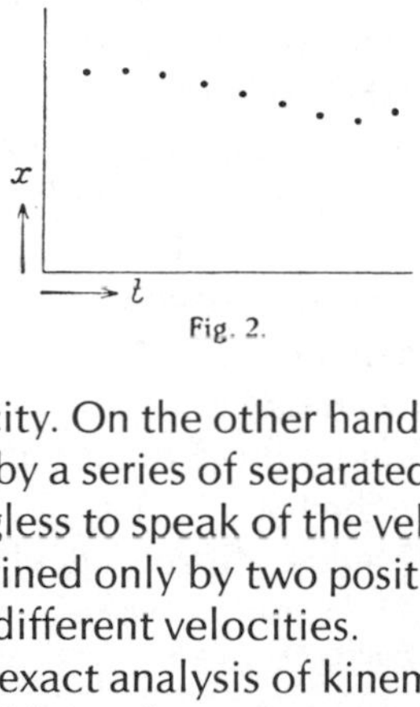

Fig. 2.

to the curve gives at once the particle velocity. On the other hand, on the discontinuity theory, this curve is replaced by a series of separated points (Fig. 2). In this case, it is obviously meaningless to speak of the velocity at a definite point, since the velocity can be defined only by two positions and conversely to every point are attached two different velocities.

The question then arises whether a more exact analysis of kinematic and mechanical concepts may not make it possible to clear up the contradictions existing up to now in the intuitive meaning of quantum mechanics and thus to arrive at an intuitive understanding of the quantum mechanical relations.[1]

I. THE CONCEPTS OF POSITION, PATH, VELOCITY, AND ENERGY

In order to be able to follow the quantum mechanical behavior of any object, we must know the mass of the object and the forces of interaction between it and other objects and fields that happen to be present. Only then can the Hamiltonian function of the quantum mechanical system be set up. [The following considerations in general refer to nonrelativistic quantum mechanics, since the laws of quantum theoretical electrodynamics are still very incompletely known. Just very recently however, great advances in this domain have been made by P. Dirac (Proc. Roy. Soc., (A) **114,** 243, 1927 and later papers).]

Further statements about the "form" of the object are unnecessary. It is most convenient to denote the totality of the interaction forces with the word "form".

If we wish to be clear as to what we shall mean by the words "position of the object"—for example, that of an electron relative to a given reference system—we must provide certain experiments with whose help we believe we can measure the "position of the electron." Otherwise the words have no meaning. There is indeed no lack of such experiments by which in principle the "position of the electron" can be determined with arbitrary accuracy. For example, we illuminate the electron and observe it under a microscope. The greatest attainable accuracy of the position determination is given essentially by the wavelength of the light used. In principle one would construct a gamma-ray microscope and with this carry out the determination of the position of the electron as precisely as one could wish. However, in this determination there is a side effect: the Compton effect. Every observation of the scattered light coming from the electron implies a photoelectric effect (in the eye, in the photographic plate, in the photo cell). This can be interpreted as meaning that a light quantum hits the electron and having been reflected or diffracted by the electron and further deviated by the lens of the microscope produces the photo effect. At the instant of the fixing of the position—that is, the instant at which the light quantum is deflected by the electron—the electron changes its momentum discontinuously. This change is the greater the shorter the wavelength of the light employed—that is, the more precise the position determination. At the moment at which the position of the electron is known, its momentum can be known only within the limits corresponding to that discontinuous change. Accordingly, the more precisely the position is determined, the less precisely the momentum is known, and conversely. Here we see a direct intuitive, physical exemplification of the quantum mechanical relation

$$pq - qp = \frac{h}{2\pi i}.$$

Let q_1 be the precision with which the value q is known (q_1 may be taken to be the mean error in q); accordingly, here the wavelength of the light, p_1 the precision with which the value of p is determinable; accordingly, here the discontinuous change in p through the Compton effect. Then from the elementary formulas of that effect, p_1 and q_1 are related by the expression)

$$p_1 q_1 \sim h. \qquad (1)$$

It will later be shown that the relation (1) can be mathematically deduced from the exchange relation

$$pq - qp = \frac{h}{2\pi i}$$

It should be pointed out that equation 1 is the exact expression for the facts that one sought to describe through the subdivisions of phase space into cells of magnitude h.

For the determination of the position of the electron, we can also try other experiments, such as collisions. A precise measurement of the position demands collisions with very fast-moving particles, since for slow electrons, the diffraction effects, which according to Einstein are a consequence of the de Broglie waves, are an obstacle to a precise determination of position. (See, for example, the Ramsauer effect.) In a precise measurement of position, the momentum of the electron accordingly again varies discontinuously and a simple estimation of the indeterminacies using the de Broglie wave formulas again leads to equation 1.

The above discussion seems to have defined the concept "position of the electron" sufficiently clearly. A word about the "size" of the electron may perhaps be added. If two very fast-moving particles hit the electron one after the other in an interval of time Δt, the positions of the electron defined by the two particles lie within a spatial range Δl. From the laws observed in the behavior of α particles, we conclude that Δl can be as small as quantities of the order of 10^{-12}cm, if Δt is chosen sufficiently small and the particles move sufficiently fast. This is the meaning involved in the statement that the electron is a corpuscle whose radius is not greater than 10^{-12}cm.

We now take up the concept "path of the electron." By "path" we understand a sequence of space points (in a given reference system) that the electron assumes one after another as "positions." Since we already know what we are to understand by "position at a definite time," no new difficulties arise here. Nevertheless, it is easy to see that the often-used expression "the 1*S* orbit of the electron in the hydrogen atom" has no meaning from the point of view developed here. In order to measure this "1*S*" orbit, the atom must be illuminated with light whose wavelength is considerably shorter than 10^{-8}cm. However, a single light quantum of such light is sufficient to eject the electron from its "orbit," so that only a single space point of such an orbit can be defined. Hence the word "orbit" has no rational sense here. Even without a knowledge of the new theories, this simply follows from experimental possibilities.

On the other hand, the measurement of position under consideration can be carried out on many atoms in the 1*S* state. (Atoms in a given "stationary" state in principle can be isolated by the Stern-Gerlach experiment). Accordingly, for a definite state of the atom, for example, the 1*S*, there must be a probability function for the position of the electrons that corresponds to the average of the classical orbit over all phases and is determinable by measurements with arbitrary precision. According to Born[2] this function is given by $\psi_{1S}(q)\ \overline{\psi}_{1S}(q)$, where $\psi_{1S}(q)$ means the Schrödinger wave function belonging to the 1*S* state. With Dirac[2] and Jordan[2] I might put it this way with respect to later investigations: The probability is given by $S(1S,q)S(1S,q)$, where $S(1S,q)$ means that column of the transformation matrix $S(E,q)$ of E by q which belongs to $E = E_{1S}$ (E means energy).

The fact that in quantum theory there can be given only the probability function of the electron position in a given state, for example, the 1*S* state, allows us, following Born and Jordan, to see in quantum theory a characteristic statistical property foreign to classical theory. However, we may, if we wish, also say, following Dirac, that the statistics is introduced by our experiments. For obviously, even in the classical theory, only the probability of a definite electron position is available so long as we do not know the phases of the atom. The difference between classical mechanics and quantum mechanics consists rather in the fact that classically we can always think of the phase as determined by a previous experiment. However, in reality this is impossible, since every experiment for the determination of phase disturbs and changes the atom. In a given stationary "state" of the atom, the phases are in principle undetermined, which one can see as direct consequences of the well-known equations

$$\boldsymbol{Et} - \boldsymbol{tE} = \frac{h}{2\pi i} \quad \text{or} \quad \boldsymbol{Jw} - \boldsymbol{wJ} = \frac{h}{2\pi i}$$

where $\boldsymbol{J}$ is an action variable and $\boldsymbol{w}$ is an angle variable.

The word "velocity" of an object can readily be defined by experiment, if force-free motions are in question. For example, we can illuminate the object with red light and determine the velocity of the particle by means of the Doppler effect of the scattered light. The determination of the velocity will be the more precise the longer the wavelength of the light being used, since then the change in velocity of the particle for each incident light quantum by the Compton effect will be the smaller. The determination of position will be correspondingly imprecise corresponding to equation 1. If it is desired to measure the velocity of an electron in the atom at a given instant, we will allow the charge on the nucleus and the forces due to the other electrons to vanish suddenly at this instant so that the motion from then on takes place force-free, and we can then carry out the program given above. Again we can convince ourselves that a function $p(t)$ for a given state of the atom, for example, 1*S*, cannot be defined. On the other hand, there is again a probability function of p in this state, which according to Dirac and Jordan has the value $S(1S,p)\bar{S}(1S,p)$. $S(1S,p)$ again denotes that column of the transformation matrix $S(E,p)$ of $\boldsymbol{E}$ according to $\boldsymbol{p}$, which belongs to $E = E_{1S}$.

Finally we must pay attention to the experiments that permit us to measure the energy or the values of the action variables J. Such experiments are particularly important since it is only with their help that we can define what we mean when we speak of the discontinuous change in energy and of the Js. The Franck-Hertz collision investigations permit us, because of the validity of the law of conservation of energy in quantum theory, to carry out the measurement of the energy of the atom by measuring the energy of electrons moving in a straight line. In principle this measurement can be carried out with arbitrary precision if we only refrain from the simultaneous measurement of the position of the electron (that is neglecting the phase) (recall the determination of p above), corresponding to the relation $\boldsymbol{Et} - \boldsymbol{tE} = h/2\pi i$. The Stern-Gerlach experiment permits the determination of the magnetic or average electric moment of the atom, and accordingly the measurement of quantities that depend only on the action variables J. The phases remain in principle undetermined. Just as it is meaningless to speak of the frequency of a light wave at a single instant, it is equally meaningless to speak of the energy of an atom at a given instant. This corresponds to the fact in the Stern-Gerlach experiment that the precision of the measurement of energy is the less the shorter the time interval in which the atoms remain under the influence of the deviating force.[3] An upper limit for the deviating force is given by the fact that the potential energy of this deviating force may vary within the bundle of rays only by amounts materially smaller than the energy differences between the stationary states, if a determination of the energy of the stationary states is to be possible. Let E_1 be an energy amount that satisfies this condition (E_1 gives at once the precision of that energy measurement). Accordingly E_1/d is the maximum value of the deviating force, if d denotes the breadth of the bundle of rays (measurable by the width of the diaphragm used to make the bundle). The angular deviation of the atomic beam is then $E_1 t_1/dp$, where t_1 denotes the time interval during which the atoms are under the influence of the deviating force and p is the momentum of the atoms in the direction of the beam. This deviation must be at least of the same order of magnitude as the naturally produced broadening of the beam produced by diffraction due to the diaphragm, through which a measurement is possible. The angular deviation produced by diffraction is approximately λ/d, if λ is the de Broglie wavelength. Accordingly

$$\frac{\lambda}{d} \sim \frac{E_1 t_1}{dp} \quad \text{or since} \quad \lambda = \frac{h}{p}, \tag{2}$$

$$E_1 t_1 \sim h.$$

This equation corresponds to equation 1 and shows how a precise energy determination can be attained only with a corresponding indeterminacy in the time.

[*Editor's Note:* In the remaining sections of this paper the author first attaches to equations 1 and 2 the name "indeterminacy relations," a nomenclature that has persisted. He then shows how these relations can be formally deduced from the Born-Jordan statistical theory of quantum mechanics. As indicated in the abstract, the author discusses at length the transition from the micromechanics of the quantum theory to the macromechanics of ordinary large-scale experience. Finally, there is discussion of several "thought" experiments along the lines of the philosophical views of Niels Bohr. We do not reproduce this part of the paper here.]

NOTES

1. The work presented here has arisen out of the efforts and desires that other scientists have expressed even before the origin of quantum mechanics. I recall particularly Bohr's work on the fundamental postulates of the quantum theory (cf., for example, Zs. fur Physik, **13,** 117, 1923) and Einstein's discussions on the connection between wave field and light quanta. The problems involved here

have been discussed most recently and the relevant questions answered in part by W. Pauli (Quantum Theory in Handbuch der Physik, Vol. 23, cited in what follows as 1.c). The formulation of these problems by Pauli has not been materially altered by quantum mechanics. It is a particular pleasure at this place to thank Herr Pauli for the great stimulation I have received from my discussions with him, both in person and through correspondence. They have proved an essential influence on the present work.

2. The statistical intepretation of the de Broglie waves was first formulated by A. Einstein (Sitzungsberichte d. preuss. Akad. d. Wiss. 1925, page 3). This statistical element in quantum mechanics plays an essential role in M. Born, W. Heisenberg and P. Jordan, Quantum Mechanics II, (Zs. f. Physik 35, 557, 1926) with particular reference to Chap. 4, Para. 3, and P. Jordan (Zs. f. Physik, *37,* 376, 1926); it was mathematically analyzed in a fundamental work by M. Born and applied to the interpretation of collision phenomena (Zs. F. Physik *38,* 803, 1926). The foundation of the probability theory through the transformation theory of matrixes is found in the articles: W. Heisenberg (Zs. F Physik, *40,* 661, 1926), P. Jordan (Zs. F. Physik, *40,* 661, 1926), W. Pauli, (Note in Zs. F. Physik, 41, 81, 1927), P. A. M. Dirac (Proc. Roy. Soc. (Al), *113,* 621, 1926), P. Jordan, (Zs. f. Physik, *40,* 809, 1926). The statistical side of quantum mechanics is discussed in a general way by P. Jordan (Naturwissenschaften, *15,* 105, 1927) and M. Born (Naturwissenschaften, *15,* 238, 1927).
3. Cf. on this W. Pauli, 1.c. p. 61.

Part III

BEGINNINGS OF QUANTUM STATISTICS

Editor's Comments on Papers 9 Through 14

The classical theory of statistics, with its application to the kinetic theory of gases, was developed in the second half of the nineteenth century by such figures as James Clerk Maxwell (1831–1879), Ludwig Boltzmann (1844–1906), Rudolf Clausius (1822–1888), August Krönig (1822–1879), and J. Willard Gibbs (1839–1903).[1]

The early classical statistics was reasonably successful when applied to ideal gases under normal conditions of temperature and pressure; however, it led to difficulty in the case of gases under extreme conditions. Moreover, when the attempt was made to apply it to the electron gas in a metal in order to provide an understanding of thermal and electrical conduction in metals, certain difficulties were encountered.[2] In particular, when H. A. Lorentz (1853–1928) tried to improve the simple theory of Paul Drude (1863–1906) by bringing it more in consonance with the Maxwell-Boltzmann theory, it was found that the Wiedemann-Franz law constant connecting the electrical and thermal conductivities of the metal turned out to be in serious disagreement with the experimental value. Moreover, the theory predicted that the free electrons in the metal should make a contribution to the specific heat of the same order of magnitude as that due to the atoms, but this was in complete disagreement with experiment.

In addition, the classical theory was not seen to apply successfully the theory of radiation to a photon gas. With the advent of quantum mechanics, it was natural to try to see how it would modify classical statistics. Such a trial was first made in epoch-making fashion by the Indian physicist Satyendranath Bose (1894–1974). The essential feature of Bose's idea is that in the distribution of photons over energy cells in phase space, the statistical probability is the same no matter how many photons lie in any cell. With this hypothesis as a base, Bose was able to give in a 1924 paper (Paper 9) a derivation of Planck's radiation law different from all previous derivations. Einstein (1879–1955) immediately recognized the fundamental significance of Bose's work and himself applied it to the quantum statistics of a monatomic ideal gas. He showed that this statistics might be of importance in the case of a gas under extreme conditions, for example, near the critical point (1924). This led to the introduction of what has come to be called the Bose-Einstein statistics. Further study showed this statistics holds for all assemblies of particles for which the quantum mechanical wave functions are symmetrical in character.[3] Bose's fundamental paper was written while he was still a student in India. (He studied in France and Germany during 1925-26, then spent the rest of his life in India.)

In 1925 Wolfgang Pauli (1900–1958) published a paper based on the original Bohr orbital quantum theory of atomic structure in which by examining the application of the theory to atomic spectra, he arrived at a result now known as the Pauli exclusion principle. This says, in effect, that there can never be two or more equivalent electrons in the atom for which the values in strong fields of all quantum numbers are identical. This principle was taken over into quantum mechanics where it has led to the conclusion that the atomic wave functions for elementary charged particles like electrons and protons must be antisymmetric in character; that is, for an assembly of such charged particles, the interchange of the coordinates of two systems in the assembly produces a change in sign without changing the absolute value of the wave function.[3] We reproduce the first part of Pauli's article, leading up to the statement of his exclusion principle, as Paper 10.

In papers published in 1926, the Italian physicist Enrico Fermi (1901–1954) and the British physicist Paul A. M. Dirac, (1902-) showed that on the basis of Pauli's exclusion principle, the statistics for elementary charged particles must be different from classical statistics and the Bosé Einstein statistics. We reproduce the articles of Fermi and Dirac as Papers 11 and 12, respectively.

The Fermi-Dirac statistics, as now conventionally known, has had wide application in the modern quantum mechanics of atomic systems. In particular, the German physicist Arnold Sommerfeld (1868–1951) showed in 1927 that the Fermi-Dirac statistics, when applied to the electron gas in a metal, is able to clear up the difficulties encountered by classical statistics in this case. The trouble over specific heat is removed by the fact that the electron gas in a metal at normal temperature is degenerate to the extent that the contribution of the electrons to the specific heat as computed on the Fermi-Dirac statistics is negligible as compared with that of the atoms of the metal. Moreover, with the use of this form of statistics, the constant in the Wiedemann-Franz law turns out to be in complete agreement with the experimental value. Sommerfeld also proceeded to show that other thermal and electrical properties of metals can be described successfully with the use of the Fermi-Dirac statistics, though the full understanding of the details has had to await the use of methods more powerful than those he used. We reproduce a translation of Sommerfeld's 1927 general article as Paper 13.

The early work on quantum mechanics and quantum statistics dealt with atomic systems in isolation from external influences and in particular from radiation fields. No attempt was made to discuss the process of radiation from such systems or that of the absorption of radiation from the surroundings. Dirac very early realized the significance of this omission and set about trying to remedy it. In 1927, shortly after the formulation of both the matrix mechanics of Heisenberg, Born, and Jordan and Schrödinger's wave mechanics, as well as the Bose-Einstein statistics, Dirac applied the latter to an assembly of light quanta (photons) interacting with an ordinary atom and was able to deduce Einstein's law for the emission and absorption of radiation.[4]

The fundamental paper of Dirac marks the beginning of concern for the development of quantum electrodynamics, also referred to as quantum field theory. This subject has presented considerable difficulty over the years, leading to the existence of embarrassing infinities in the mathematical development. It has received great attention during the past forty-five years. The details are too elaborate to justify the inclusion of further articles dealing specifically with it. Fortunately, there exists an excellent collection of relevant papers on the subject by Julian Schwinger.[5] We reproduce Dirac's 1927 article as Paper 14.

NOTES AND REFERENCES

1. For the classical theory of physical statistics see R. Lindsay, ed., *Early Concepts of Energy in Atomic Physics,* Benchmark Papers on Energy, vol. 7 (Stroudsburg, Pa.: Dowden, Hutchinson & Ross, 1979), pp. 172–187. For the work of Boltzmann, see S. G. Brush, trans., *Lectures*

on Gas Theory (Berkeley, Calif.: University of California Press, 1964). For Gibbs, see J. W. Gibbs, *Elementary Principles in Statistical Mechanics* (New Haven, Conn.: Yale University Press, 1902).

2. P. Drude, "On the Electron Theory of Metals," *Annalen der Physik,* ser. 4, **1**:567-613, 1900. See also the translation of that article in R. Lindsay, ed., *Early Concepts of Energy in Atomic Physics,* Benchmark Papers on Energy, vol. 7 (Stroudsburg, Pa.: Dowden, Hutchinson & Ross, 1979), pp. 356-366.
3. R. B. Lindsay and H. Margenau, *Foundations of Physics,* reprint ed., (Woodbridge, Conn.: Ox Bow Press, 1981), pp. 488ff. See also A. Einstein, *Quantentheorie des einatomigen idealen gases* (Sitzungsberichte der Preussischen Akademie der Wissenschaften, 1924), pp. 261-267.
4. For a translation of Einstein's fundamental 1916 paper with the derivation of Planck's radiation law, see R. Lindsay, ed., *Early Concepts of Energy in Physics,* Benchmark Papers on Energy, vol. 7 (Stroudsburg, Pa.: Dowden, Hutchinson & Ross, 1979), pp. 293-297.
5. J. Schwinger, ed., *Selected Papers on Quantum Electrodynamics* (New York: Dover, 1968). For a helpful commentary on the subject, see S. Weinberg, "The Search for Unity—Notes for a History of Quantum Field Theory," *Daedalus,* Fall 1977, pp. 17-35. This also contains an extensive bibliography covering the whole subject.

9

PLANCK'S LAW AND THE HYPOTHESIS OF LIGHT QUANTA

Satyendranath Bose

This article was translated expressly for this Benchmark volume by R. Bruce Lindsay, Brown University, from Z. Phys. ***26:****178–181 (1924), by permission of the publisher, Springer-Verlag, New York*

ABSTRACT

The phase space of a light quantum occupying a given volume is divided into cells of volume h^3. The number of possible distributions of the light quanta of macroscopically defined radiation among these cells leads to the entropy and hence all the thermodynamic properties of the radiation.

Planck's formula for the distribution of energy in black-body radiation initiated the quantum theory, which has seen much development in the last twenty years and has borne fruit in all branches of physics. Since Planck's publication in the year 1901, many derivations of the radiation law have been proposed. It is well known that the fundamental assumptions of the quantum theory are incompatible with the laws of classical electrodynamics. All previous derivations make use of the relation

$$\varrho_\nu \, d\nu = \frac{8\pi\nu^2 \, d\nu}{c^3} E,$$

that is, the relation between the radiation density ρ_ν and the average energy E of an oscillator. And they make assumptions about the number of degrees of freedom of the aether of space and how they appear in the above equation (the first factor on the right-hand side). However, this factor can be derived only from classical theory. This is the unsatisfactory point in all derivations, and it is not surprising that many efforts have been made to give a derivation that is free from this logical defect.

A particularly elegant derivation has been given by Einstein. He has recognized the logical inadequacy of all previous derivations and has attempted to deduce the formula independently of the classical theory. Proceeding from very simple assumptions concerning the exchange of energy between molecules and radiation field, he finds the relation

$$\varrho_\nu = \frac{\alpha_{mn}}{e^{\frac{\varepsilon_m - \varepsilon_n}{kT}} - 1}.$$

However, in order to bring this formula into agreement with Planck's formula, he must make use of Wien's displacement law and Bohr's correspondence principle. Wien's law is based on the classical theory, and

Bohr's correspondence principle assumes that the quantum theory agrees with classical theory in certain limiting cases.

In all cases, it seems to me that the derivations are not sufficiently justified logically. On the other hand, the hypothesis of light quanta in conjunction with statistical mechanics (as it has been adjusted to the needs of quantum theory by Planck) seems to me to be sufficient for the derivation of the law independently of classical theory. In what follows, I shall give a brief sketch of the method.

Let the radiation be confined in volume V and let its total energy be E. Let there be various kinds of quanta to the number N_s and energy $h\nu_s$ (s going from zero to ∞). The total energy is then

$$E = \sum_s N_s h \nu_s = V \int \varrho_\nu d\nu. \tag{1}$$

The solution of the problem then demands the determination of N_s, which in turn determine ρ_ν. If we can provide the probability for each distribution characterized by given N_s, the final solution will be determined by the condition that this probability will be a maximum subject to the auxiliary condition (1). We now seek this probability.

The quantum (photon) has a momentum of magnitude $h\nu_s/c$ in the direction of its motion. The instantaneous state of the quantum is characterized by its coordinates x, y, z and the conjugate momenta p_x, p_y, p_z. These six quantities can be considered as the point coordinates in a six-dimensional space, in which we have the condition

$$p_x^2 + p_y^2 + p_z^2 = \frac{h^2 \nu^2}{c^2},$$

In accordance with this, the phase point in question is forced to remain on a cylindrical surface determined by the frequency of the quantum. In this sense, to the frequency domain $d\nu_s$ there is associated the phase space

$$\int dx\, dy\, dz\, dp_x\, dp_y\, dp_z = V \cdot 4\pi \left(\frac{h\nu}{c}\right)^2 \frac{h\, d\nu}{c} = 4\pi \frac{h^3 \nu^2}{c^3} V d\nu.$$

If we divide the whole phase volume into cells of magnitude h^3, there are accordingly $4\pi V \nu^2/c^3\, d\nu$ cells in the frequency interval $d\nu$. Nothing definite can be said about the nature of this division. However, the total number of cells must be looked upon as the number of possible arrangements of a quantum in the given volume. In order to take account of the fact of polarization, however, it is essential to multiply this number by 2, so that the number of cells assigned to $d\nu$ becomes $8\pi V\nu^2 d\nu/c^3$.

It is now a simple matter to calculate the thermodynamic probability of a macroscopically defined state. Let N^s be the number of quanta belonging in the frequency interval $d\nu^s$. In how many ways can these be distributed over the cells belonging to $d\nu^s$? Let $p_o{}^s$ be the number of vacant cells, $p_1{}^s$ be the number containing one quantum, p_2^s the number containing two quanta, etc. The number of possible distributions is then

$$\frac{A^s!}{p_0^s!\, p_1^s! \ldots}, \quad \text{where} \quad A^s = \frac{8\pi\nu^2}{c^3} d\nu^s,$$

and where

$$N^s = 0.p_0^s + 1.p_1^s + 2.p_2^s + \ldots$$

[*Editor's Note:* This equation has been corrected from the original.]

is the number of quanta belonging to $d\nu^s$.

The probability of the state defined by all the p_2^s is then

$$W = \prod_s \frac{A^s!}{p_0^s!\, p_1^s! \ldots}$$

[*Editor's Note:* This equation has been corrected from the original.]

Keeping in mind that we can consider the p_r^s as large numbers, we have

$$\lg W = \sum_s A^s \lg A^s - \sum_s \sum_r p_r^s \lg p_r^s,$$

where

$$A^s = \sum_r p_r^s.$$

The expression for lg W is to be a maximum under the auxiliary conditions

$$E = \sum_s N^s h\nu^s; \quad N^s = \sum_r r p_r^s.$$

Carrying out the necessary variation leads to the conditions

$$\sum_s \sum_r \delta p_r^s (1 + \lg p_r^s) = 0, \qquad \sum_s \delta N^s h\nu^s = 0$$

$$\sum_r \delta p_r^s = 0 \qquad \delta N^s = \sum_r r\,\delta p_r^s.$$

From this it follows that

$$\sum_s \sum_r \delta p_r^s (1 + \lg p_r^s + \lambda^s) + \frac{1}{\beta} \sum_s h\nu^s \sum_r r\,\delta p_r^s = 0.$$

From this one gets at first

$$p_r^s = B^s e^{-\frac{rh\nu^s}{\beta}}.$$

Since, however,

$$A^s = \sum_r B^s e^{-\frac{r h \nu^s}{\beta}} = B^s \left(1 - e^{-\frac{h\nu^s}{\beta}}\right)^{-1},$$

we get

$$B_s = A^s \left(1 - e^{-\frac{h\nu^s}{\beta}}\right).$$

We have further

$$N^s = \sum_r r p_r^s = \sum_r r A^s \left(1 - e^{-\frac{h\nu^s}{\beta}}\right) e^{-\frac{r h \nu^s}{\beta}}$$

$$= \frac{A^s e^{-\frac{h\nu^s}{\beta}}}{1 - e^{-\frac{h\nu^s}{\beta}}}.$$

Taking account of the value of A^s found above, we have accordingly

$$E = \sum_s \frac{8\pi h \nu^{s^3} d\nu^s}{c^3} V \frac{e^{-\frac{h\nu^s}{\beta}}}{1 - e^{-\frac{h\nu^s}{\beta}}}.$$

We also further find

$$S = k\left[\frac{E}{\beta} - \sum_s A^s \lg\left(1 - e^{\frac{h\nu^s}{\beta}}\right)\right],$$

Since

$$\frac{\partial S}{\partial E} = \frac{1}{T}$$

it follows that

$$\beta = kT$$

If we insert this into the above equation for *E*, we get

$$E = \sum_s \frac{8\pi h \nu^{s^3}}{c^3} V \frac{1}{e^{\frac{h\nu^s}{kT}} - 1} d\nu^s,$$

which is equivalent to Planck's formula.
[Note by the translator (of the original into German): A. Einstein. In my opinion Bose's derivation of Planck's formula signifies an important advance. The method used here also leads to the quantum theory of an ideal gas, as I shall show in another place.]

10

ON THE CONNECTION BETWEEN THE CLOSING OF THE ELECTRON GROUPS IN THE ATOM AND THE COMPLEX STRUCTURE OF SPECTRA

W. Pauli

This article was translated expressly for this Benchmark volume by R. Bruce Lindsay, Brown University, from pages 765–776 of Z. Phys. ***31****:765–783 (1925), by permission of the publisher, Springer-Verlag, New York*

ABSTRACT

With respect to the Millikan-Landé discovery of the representation of the alkali doublet by relativistic formulas and on the basis of results obtained in a previous work, the idea is proposed that in this doublet and its anomalous Zeeman effect, a not classically describable ambiguity of the quantum theory properties of the valence electron is revealed even though the closed noble gas configuration of the atomic core is not considered in this connection in the form of a body momentum or as the seat of the magneto-mechanical anomaly of the atom. The attempt is then made to apply as far as possible this idea as a working hypothesis to other atoms besides the alkalis, in spite of the difficulties confronting it. It then becomes apparent that in contrast to the usual conception, this hypothesis makes it possible, in the case of a strong external magnetic field in which one can neglect the coupling forces between the atom core and the valence electron, to assign to these two partial systems with respect to the number of their stationary states as well as the values of their quantum numbers and their magnetic energy no other properties than those ascribed to the free atom core and the valence electron respectively in the alkalis. On the basis of this result, we can arrive at a general classification of every electron in the atom through a principal quantum number n and two auxiliary quantum numbers k_1 and k_2, to which in the presence of an external field still another quantum number, m, must be added. In conjunction with a recent work by E. C. Stoner, this classification leads to a general quantum theoretical foundation for the closure of electron groups in the atom.

1. PERMANANCE OF THE QUANTUM NUMBERS (SYNTHESIS PRINCIPLE) IN THE COMPLEX STRUCTURE AND THE ZEEMAN EFFECT.

In an earlier work[1] it was emphasized that the usual representation considering the inner closed electron shells in the atom in the form of a core momentum and as the essential seat of the magneto-mechanical anomalies in the complex structure of the optical spectra and the anomalous Zeeman effect leads to various serious difficulties. There is a strong inducement to replace this representation by another, namely that the doublet structure of the alkali spectra, as well as their anomalous Zeeman effect, is caused by an ambiguity in the quantum theoretical properties of the valence electron, which is not describable in classical terms. This conception finds support in the discovery by Millikan and Landé that the optical doublets of the alkalis are analogous to the relativity doublets in X-ray spectra, and that their magnitude is determined by a relativistic formula.

Following this line of thought, we adopt the method of Bohr and Coster for X-ray spectra and assign to the stationary states associated with the emission of the alkali spectra two auxiliary quantum numbers k_1 and k_2 in addition to the principal quantum number n. The first, k_1 (usually designated simply as k) has the values 1, 2, 3, . . . , for the *s, p, d,* . . . , terms, respectively, and always changes by unity in permitted transition processes; it determines the magnitude of the central field interaction between the valence electron and the atom core. The second auxiliary quantum number k_2 is equal to $k_1 - 1$ and k_1, respectively, for the two terms of the doublet (for example, p_1 and p_2). In transition processes, it changes by ± 1 or zero and determines the magnitude of the relativity correction (modified according to Landé with respect to the penetration of the valence electron into the core region). If, following Sommerfeld, we define the total momentum quantum number j of the atom as the maximum value of the quantum number m_1 determining (for the stationary state in question) the momentum component parallel to an external field (usually referred to simply as m), then for the alkalis we have to set $j = k_2 - \frac{1}{2}$. The number of stationary states in the magnetic field for given k_1 and k_2 is $2j + 1 = 2k_2$. The number of these states for both doublet terms with given k_1 is equal to $2(2k_1 - 1)$.

If we now consider the case of strong fields (the Paschen-Back effect), besides k_1 and the above-mentioned quantum number m, we can introduce a magnetic quantum number m_2 in place of k_2, providing directly the energy of the atom in the magnetic field—that is, the component of the magnetic moment of the valence electron parallel to the field. Corresponding to the two terms of the doublet, this has the values $m_1 + \frac{1}{2}$ and $m_1 - \frac{1}{2}$. In the doublet structure of the alkali spectra, the "anomaly of the relativity correction" comes to the fore; its magnitude is controlled by another quantum number just as in the case of the magnitude of the central force interaction energy of the valence electron and the atom core. Similarly, in the deviations of the Zeeman type from the normal Lorentz triplet the "magneto-mechanical anomaly" analogous to the just-mentioned anomaly makes its appearance (for the magnitude of the magnetic moment of the valence electron, another quantum number is controlling, as also for the moment of momentum.). The appearance of half (effective) integral quantum numbers and the value $g = 2$ for the splitting factor in the *s*-term for the alkalis, which is formally connected with this, is apparently closely associated with the twofold character of the term "level." However, we shall not seek here further theoretical analogies of this situation, but in the following considerations shall treat the Zeeman effect of the alkalis as a fact of experience.

Without initially troubling ourselves over the difficulties facing the point of view in question (which will be considered in detail shortly), we now seek to extend this formal classification of the valence electron by means of the four quantum numbers n, k_1, k_2, and m, to atoms more complicated than the alkalis. Then it turns out that on the basis of this classification (in contrast to the usual conception), we can retain completely the permanence of the quantum numbers (the synthesis principle) even for the complex structure of spectra and the anomalous Zeeman effect. This principle, enunciated by Bohr, says that in the attachment of an additional electron to an atom (appropriately charged as a whole), the quantum numbers of the electrons already bound to the atom retain the same values taken on by them in the associated stationary state of the free atom core.

Let us first consider the alkaline earths. The spectrum for these consists of a singlet and a triplet system. In an external magnetic field $1. (2k_1 - 1)$ stationary states for the first system correspond to quantum states with a definite value of the quantum number k_1 of the valence electron. At the same time $3. (2k_1 - 1)$ stationary states in the latter system have the similar correspondence. Up to now, this has been interpreted to mean that in strong fields $2k_1 - 1$ arrangements are associated with the valence electron, whereas the atom core can take on one arrangement in the first case and three in the latter case. The number of these arrangements is apparently different from the number two for the arrangements of the free atom core in the field (similar to the alkali *s*-term). Bohr characterized this state of affairs as a "constraint," which is not analogous to the action of external force fields.[2] However, we can now simply interpret the total number of states of the atom, $4(2k_1 - 1)$, by saying that (both before and after) there are two states for the atom core in the presence of the field and $2(2k_1 - 1)$ states for the valence electron, as in the case of the alkalis.

More generally in accordance with a branching formula devised by Heisenberg and Landé,[3] a stationary state of the atom core with N states gives rise in the field to two term systems through the addition of a further electron, to which a total of $(N + 1)(2k_1 - 1)$ or $(N - 1)(2k_1 - 1)$ states, respectively, correspond in the presence of the field for a particular value of the quantum number k_1 of the added electron. In accordance with our interpretation, these $2N(2k_1 - 1)$ states of the atom as a whole in the presence of strong fields come about through N states of the atom core and $2(2k_1 - 1)$ states of the valence electron. Therefore, in the assumed quantum theoretical classification of the electrons, the manifold structure of the terms demanded by the branching rule appears simply as a consequence of the synthesis principle. According to the point of view proposed here, Bohr's "constraint" finds its meaning not in a breaking down of the permanence of the quantum numbers in the coupling of the series electron to the atomic core but only in the characteristic ambiguity of the quantum theoretical properties of the individual electrons in the stationary states of the atom.

Nevertheless, on this interpretation of the synthesis principle, we can calculate not only the number of stationary states but also the energy values in the case of strong fields (that is, their parts proportional to the field intensity) additively from those of the free atomic core and those of the valence electron, the latter of which are obtainable from the spectra of the alkalies. In this case indeed the total components $\overline{m}_1$ of the moment of momentum of the atom parallel to the field (measured in units of $h/2\pi$) and also the components $\overline{m}_2$ of the magnetic moment of the atom in the same direction (measured in Bohr magnetons) are equal, respectively, to the sum of the quantum numbers m_1 and m_2, respectively, for the individual electron. Thus

$$\overline{m}_1 = \sum m_1, \qquad \overline{m}_2 = \sum m_2. \tag{1}$$

Here the individual electrons independently of each other run through all values belonging to the momentum quantum numbers k_1 and k_2 of these electrons in the particular stationary state of the atom. (Here m_2oh, where o = the Larmor frequency, is the part of the energy of the atom proportional to the field intensity.)

As an example let us consider the two s-terms (Singlet S-term and Triplet S-term) of the alkaline earths. It is at first sufficient to take account of only the two valence electrons, since the contribution of the remaining electrons to the sums in equation 1 vanishes. In this case, for each of these two valence electrons in accordance with our general conclusion (independent of the other electrons) we are to take the values $m_1 = -\frac{1}{2}$, $m_2 = -1$ and $m_1 = \frac{1}{2}$, $m_2 = 1$ of the s term of the alkalis. Hence in accordance with equation 1 we obtain the following values of the quantum numbers $\overline{m}_1$ and $\overline{m}_2$ of the whole atom.

$$\begin{aligned} \overline{m}_1 &= -\tfrac{1}{2}-\tfrac{1}{2}, \quad -\tfrac{1}{2}+\tfrac{1}{2}, \quad \tfrac{1}{2}-\tfrac{1}{2}, \quad \tfrac{1}{2}+\tfrac{1}{2} \\ \overline{m}_2 &= -1-1, \quad -1+1, \quad 1-1, \quad 1+1 \end{aligned}$$

or

$\overline{m}_1$	-1	0	1
$\overline{m}_2$	-2	0,0	2

(These correspond to a term with $j = 0$ and a term with $j = 1$ in weak fields.)[4] In order to obtain the *p, d,* . . . , terms for the alkaline earths, we must for the unchanged contribution of the first valence electron (*S*-term) and correspondingly for the second electron insert in equation 1 the m_1 and m_2 values of the *p, d,* . . . , terms of the alkalis.

In general the prescription (1) leads directly to a procedure given by Landé[5] for the calculation of the energy values in strong fields, from which this author has shown that it also gives correct results in more complicated cases. According to Landé, this procedure provides the Zeeman terms of the neon spectrum (at first in the case of strong fields), if in the atomic core one assumes an effective electron in a *p*-term (instead as above in an *s*-term)[6] and lets the valence electron again go through the *s, p, d, f,* . . . , terms.

This result leads to the demand that in general every electron in the atom is to be characterized not only by the principal quantum number n but also by the two auxiliary quantum numbers k_1 and k_2, whether in the presence of several equivalent electrons or in closed electron groups. Further (even in the cases just mentioned), we can imagine a magnetic field strong enough so that we can assign to every electron independently of the other electrons present the two quantum numbers m_1 and m_2 in addition to the quantum numbers n and k_1 (wherein these last determine the contribution of this electron to the magnetic energy of the atom). The connection between k_2 and m_2 for given k_1 and m_1 is to be provided from the alkali spectra.

Before we apply this quantum theoretical classification of the electrons in the atom to the problem of the closing of electron groups, we must consider more closely the difficulties encountered by the proposed interpretation of the complex structure and the anomalous Zeeman effect, as well as the limitations of its significance.

In the first place, this interpretation does not immediately satisfy the separate more-or-less self-evident appearance of the various term systems (for example, in the case of the alkaline earths, the simple and triplet systems), which holds also for the position of the terms of these systems, as well as for the Landé interval rule. Surely one should not assume two different causes for the energy differences of the triplet levels in the case of the alkaline earths, nor for the anomaly of the relativity correction of the valence electron or the dependence of the interchange energy of this with the atom core on the orientation of these two systems.

A still more serious difficulty is provided by the connection of the proposed interpretation with the correspondence principle, which forms an indispensable means for the explanation of the selection rules for the quantum numbers k_1, j, and m, as well as for the polarization of the Zeeman components. According to this principle, it is not demanded that in a definite stationary state, a uniquely prescribed path is to be assigned in the sense of ordinary kinematics. Rather, the totality of the stationary states of the atom must correspond to a class of orbits with a definite type of periodicity properties. Thus in our case, the well-known selection and polarization rules according to the correspondence principle demand the type of motion of a central force orbit with the superposed precession of the orbital plane about a definite axis. To this is added a precession about an axis through the nucleus in the field direction in the case of weak external magnetic fields. The previously assumed dynamical explanation of this type of motion of the valence electron resting on the assumption of the deviation of the forces exerted by the atom core on this electron from central symmetry does not appear to be compatible with the representation of the alkali doublet (and hence accordingly the magnitude of the corresponding precession frequency) by means of relativistic formulas. There is a corresponding difficulty with the types of motion in the case of strong fields.

Accordingly we meet here the difficult problem of providing a physical interpretation for the existence of the type of motion of the valence electron demanded by the correspondence principle independently of its previously assumed special dynamical significance, which is indeed hard to justify. The question as to the magnitude of the term values in the Zeeman effect (particularly for the alkali spectra) appears to be most closely connected with this problem.

To the extent that this problem is not yet solved, the conception of the complex structure and the anomalous Zeeman effect proposed here certainly cannot be considered as a sufficient physical foundation for an explanation of these phenomena, even if the latter are in many respects better than by the usually accepted interpretation. It is not out of the question that in the future a merger of these two interpretations may be successful. In the present state of the question, it is of interest to follow out the first interpretation in its consequences as far as possible. In this sense, it may be worthwhile in the following paragraphs to apply its fundamental standpoint, at least tentatively, to the problem of the closing of electron groups in the atom, without regard to the difficulties confronting it. We shall thereby draw conclusions concerning the number of possible stationary states of the atom in the presence of several equivalent electrons, but not about the position and arrangement of term values.

2. ON A GENERAL QUANTUM THEORETICAL RULE FOR THE POSSIBILITY OF THE EXISTENCE OF EQUIVALENT ELECTRONS IN THE ATOM.

It is well known that the appearance of several equivalent electrons in an atom (that is, in the sense of their having the same quantum numbers and the same binding energies) is possible only under special circumstances, and these are very closely connected with the regularities of the complex structure of the spectra. For example, in the alkaline earths, the normal state, in which both valence electrons are equivalent, corresponds to a singlet s-term, whereas in those stationary states of the atom belonging to the triplet system, the valence electrons are never bound in equivalent fashion, since the largest triplet s-term possesses a principal quantum number greater by unity than that for the normal state. As a second example, consider the spectrum of neon. This consists of two groups of terms with different series limits corresponding to different states of the atom core. The first group, belonging to the transition of an electron with quantum numbers $k_1 = 2$, $k_2 = 1$ from the atom core, can be considered as being made up of a singlet and a triplet system, whereas the second group, belonging to the transition of an electron with $k_1 = k_2 = 2$ from the atom core, can be designated as a triplet plus quintet system. The ultraviolet resonance lines of neon have not yet been observed, but there can be scarcely any doubt that the normal state of the neon atom with respect to its combinations with the known excited states of the atom must be regarded as a *p* term. And indeed according to the unique definiteness and the diamagnetic behavior of the noble gas configuration, there can be only one such term, and indeed[7] with the value $j = 0$. Since the only *p*-terms with $j = 0$ are the lowest triplet terms *p* of both groups, we can also conclude that for neon for the the value 2 of the principal quantum number, these two triplet terms alone exist and moreover are identical for both term groups.

In general, we can expect that for those values of the quantum numbers n and k_1 for which electrons already are present in the atom, certain multiplet terms of the spectra will fall out or coincide. The question arises what quantum theoretical rules govern this behavior of the terms.

As is already clear from the example of the neon spectrum, this question is closely bound up with the problem of the closing of the electron groups in the atom that condition the lengths 2, 8, 18, 32, . . . , of the periods in the natural system of the elements. This closing implies that an n-quantized electron group is unable to hold more than $2n^2$ electrons, either by emission or absorption of radiation, or by other external influences.

It is well known that in his theory of the natural system, Bohr has introduced a subdivision of these electron groups into subgroups. His theory contains a unified summary of spectroscopic and chemical data and in particular a quantum theoretical foundation for the appearance of chemically similar elements, such as the iron and platinum groups and the rare earths in the later periods of the system. By characterizing every electron in the stationary states of the atom with reference to the stationary states of a central motion by means of a symbol $n_k (k \leq n)$, he obtains in general for an electron group with principal quantum number n, n subgroups. In this way Bohr was led to the scheme of the atomic structure of the noble gases reproduced in Table 1. He himself emphasized, however, that the assumed equality of the number of electrons in the different subgroups of a main group is hypothetical to a high degree, and that a complete and satisfactory theoretical explanation of the closing of the electron groups in the atom and in particular an explanation of the period lengths 2, 8, 18, 32, . . . , in the natural system could not be given at that time.[8]

More recently an essential advance has been achieved with respect to the problem of the closing of electron groups in atoms by the observations of E. C. Stoner.[9] This author proposes a scheme for the atomic structure of the noble gases, which in contrast to that of Bohr permits no opening of a

Table 1. Original Bohr Scheme for the Configurations of the Nable Gases

Element	*Atomic Number*	*Number of n_k electrons*														
		1_1	2_1	2_2	3_1	3_2	3_3	4_1	4_2	4_3	4_4	5_1	5_2	5_3	6_1	6_2
Helium	2	2	–	–	–	–	–	–	–	–	–	–	–	–	–	–
Neon	10	2	4	4	–	–	–	–	–	–	–	–	–	–	–	–
Argon	18	2	4	4	4	4	–	–	–	–	–	–	–	–	–	–
Krypton	36	2	4	4	6	6	6	4	4	–	–	–	–	–	–	–
Xenon	54	2	4	4	6	6	6	6	6	6	–	4	4	–	–	–
Emanation (Radon)	86	2	4	4	6	6	6	8	8	8	8	6	6	6	4	4

closed subgroup by the addition of further electrons of the same principal group. In this scheme, the number of electrons in a closed subgroup is to depend only on the value of n—that is, on the presence of further subgroups of the same principal group. This in itself involves a great simplification, which can also be supported by a variety of experimental data. Thus for $k = 1$, there must be assumed to be two electrons in the closed state of the corresponding subgroup. Similarly for $k = 2$, we must assume six such electrons, and for $k = 3$ ten such electrons. In general, for a given k value there will be $2(2k - 1)$ electrons in the closed state of the corresponding subgroup in order to remain in agreement with the empirically known number of electrons in the atoms of the noble gases.

Stoner remarked further that this assignment of numbers of electrons agrees with the assignment of the number of stationary states of the alkali atoms in an external field for given values of k. He carries the analogy to the stationary states of the spectra of the alkalis still further by assuming a further subdivision of the subgroups corresponding to the complex structure of these spectra (and the X-ray spectra as well). This subdivision is into two partial subgroups characterized by the two numbers k_1 and k_2 in which k_1 is identical with Bohr's number k and $k_2 = k_1 - 1$ or $k_2 = k_1$ respectively (though for $k_1 - 1$, in consideration of the simplicity of the s-term, we have only $k_2 = 1$). There are $2k_2$ stationary states into which a stationary state of the alkali atom with definite values of the quantum numbers k_1 and k_2 breaks down in the presence of the external field. Corresponding to this state, Stoner assumes $2k_2$ electrons in the closed partial subgroup belonging to the quantum numbers n, k, and k_2. The scheme at which Stoner arrives for the atomic structure of the noble gases is represented in Table 2.

We can now generalize this representation of Stoner and make it more precise if we apply the conception of the complex structure of spectra and the anomalous Zeeman effect to the case of the presence of equivalent electrons in the atom. Then, supported by the possibility of the maintenance of the permanence of the quantum numbers, we were led to characterize every electron in the atom not only by the principal quantum number n but also by the auxiliary quantum numbers k_1 and k_2. In strong magnetic fields, a momentum quantum number m_1 had to be introduced for every electron, and in addition to this, besides k_1 and m_1 in place of k_2, another quantum number m_2 arising from the magnetic moment was introduced. In the first place, we see that the application of the two quantum numbers k_1 and k_2 to every electron is in very good agreement with Stoner's subdivision of Bohr's subgroups.[10] From a consideration of the case of strong magnetic fields, we can then base Stoner's result, whereby the number of electrons in a closed subgroup is the same as the number of corresponding terms of the Zeeman effect of the alkali spectra, on the following more general rule concerning the appearance of equivalent electrons in the atom:

There can never be two or more equivalent electrons in the atom for which the values in strong fields of all quantum numbers n, k_1, k_2, m_1 (or

Table 2. Stoner's scheme for the Nable Gas Configurations

Element	Atomic Number	Number of n_k electrons														
		1_1	2_1	2_2, (1+2)	3_1	3_2, (1+2)	3_3, (2+3)	4_1	4_2, (1+2)	4_3, (2+3)	4_4, (3+4)	5_1	5_2, (1+2)	5_3, (2+3)	6_1	6_2, (1+2)
Helium	2	2	–	–	–	–	–	–	–	–	–	–	–	–	–	–
Neon	10	2	2	2+4	–	–	–	–	–	–	–	–	–	–	–	–
Argon	18	2	2	2+4	2	2+4	–	–	–	–	–	–	–	–	–	–
Krypton	36	2	2	2+4	2	2+4	4+6	2	2+4	–	–	–	–	–	–	–
Xenon	54	2	2	2+4	2	2+4	4+6	2	2+4	4+6	–	2	2+4	–	–	–
Emanation (Radon)	86	2	2	2+4	2	2+4	4+6	2	2+4	4+6	6+8	2	2+4	4+6	2	2+4

what amounts to the same thing) n, k_1, m_1, m_2 *are identical. If there exists an electron in the atom for which these quantum numbers have definite values in an external field, this state if fully occupied.*

We should keep in mind that on this rule, the principal quantum number n enters in essential fashion. It is obvious that general nonequivalent electrons can exist in the atom whose values of the quantum numbers k_1, k_2, m_1 are identical, for which, however, the principal quantum number n is different.

We are unable to give a more precise foundation for this rule. However, it appears to be a very natural one. As mentioned, it relates in the first place to the case of strong fields. On thermodynamic grounds (invariance of the statistical weights in adiabatic transformations of the system,)[11] however, the number of stationary states of the atom for given values of the number k_1 and k_2 for the individual electrons and the values of $\overline{m}_1 = \Sigma m_1$ for the whole atom in strong and weak fields alike must be in agreement. Hence, in the latter case we can also make predictions about the number of stationary states and their associated values of j (which belong to the different values of k_1 and k_2 for given number of the equivalent electrons). Hence the number of realization possibilities of the different unclosed electron shells can be deduced, and the questions raised at the beginning of this section concerning the absence or coincidence of certain multiplet terms in the spectra can be uniquely answered in every single case. Such terms involve values of the principal quantum number for which several equivalent electrons are present in the atom. It must be confessed, however, that here we are able to say something about only the number of the terms and the values of their quantum numbers but not about their magnitudes and interval relationships.

[*Editor's note:* The author devotes the remainder of the paper to a study of the consequences of the rule he has laid down and the comparison with experimental data. In particular he points out that Stoner's results are an immediate consequence of his principle. He then goes on to a discussion of the application to multiplet structure. We shall not reproduce this part of the paper.]

NOTES

1. Zs. F. Physik, **31,** 373, 1925. At the conclusion of this paper, reference is made to the present work.
2. N. Bohr, Ann. der Physik, 71, 228, 1923, in particular p. 276.
3. Zs. fur Physik, 25, 279, 1924. We do not consider here the question of the limits of the validity of the rule and in particular the theoretical meaning of the so-called excluded terms. For the interpretation of the latter in the sense of the branching rule, we must probably assume that the atomic core is different in the bound and free states. In any case, the combination rule holding for these terms, which differs from the usual one, must be carefully considered.
4. We see here that two different terms (with respect to the parts of the energy independent of field intensity) must be assigned to the two cases $m_1 = -½$ for

the first, and $m_1 = ½$ for the second electron on the one hand and $m_1 = +½$ for the first and $m_1 = -½$ for the second electron on the other hand. This is perhaps a kind of incompleteness of the classification being carried out here. However, we shall see later that through the equivalence of the inner and outer valence electrons, both terms are in fact identical.

5. Ann. der Physik, **76,** 273, 1925; see in particular para. 2.
6. The replacement of the seven shell (atomic core of neon) by a single electron proposed here will find theoretical justification in the following paragraph.
7. As has already been noted, the value of *j* here and in what follows is always defined as the maximum value of the quantum number m_1.
8. Cf. N. Bohr, *Drei Aufsätze uher Spektren and Atomban,* 2. Auflage, Braunschweig, 1924, Appendix
9. Phil. Mag. **48,** 719, 1924. The preface to the new edition of Sommerfeld's book "Atombau and Spektrallimen" has called attention to this important work.
10. That this subdivision and the question of the number of electrons in the partial subgroups also has meaning for closed electron groups is confirmed by the Millikan-Landé result on the relativity doublets in X-ray spectra. These numbers enter clearly the expression for the energy of the whole group as a function of the order number as factors of the Moseley-Sommerfeld expressions formed with definite values of the screening constants (determined by k_1) and the relativity correction (determined by k_2).
11. This invariance is independent of the validity of classical mechanics in the transformation.

11

ON THE QUANTIZATION OF THE IDEAL MONATOMIC GAS

Enrico Fermi

This article was translated expressly for this Benchmark volume by R. Bruce Lindsay, Brown University, from Z. Phys. ***36:****902–912 (1926), by permission of the publisher, Springer-Verlag, New York*

ABSTRACT

If the Nernst heat theorem is also valid for the ideal gas, we must assume that at low temperatures the ideal gas laws must deviate from the classical laws. The cause of this degeneracy is to be sought in the quantization of molecular motions. In all theories of degeneracy, more-or-less arbitrary assumptions are always made concerning the statistical behavior of molecules or concerning their quantizations. In the present work, we use only the assumption first set forth by Pauli and used as the fundamental assumption justified by countless facts of spectroscopy: in any system we can never have two equivalent elements whose quantum numbers are the same. The equation of state and the internal energy of an ideal gas will be derived with the use of this hypothesis. The entropy value for high temperatures is found to agree with that of Stern and Tetrode.

In classical thermodynamics the molecular heat at constant volume is set equal to

$$c = \tfrac{3}{2} k T$$

[*Editor's Note:* k is Boltzmann's gas constant.] If, however, we desire to be able to apply the Nernst heat theorem to an ideal gas, we must look upon this expression merely as an approximation that holds at high temperatures, since c must vanish in the limit of $T = 0$. We are therefore forced to assume that the motion of the molecules of an ideal gas is quantized. This quantization makes itself evident at low temperatures through certain degeneracy phenomena, so that both the specific heat and the equation of state will deviate from their classical expressions.

The purpose of this paper is to present a method for the quantization of an ideal gas that in our judgment is as independent as possible of arbitrary assumptions concerning the statistical behavior of gas molecules.

Recently there have been many attempts to establish theoretically the equation of state of an ideal gas.[1] The equations of state obtained by the various authors differ among themselves as well as from ours, and from the classical equations of state $pV = NkT$ through terms that become of important magnitude only at very low temperatures and great pressure. Unfortunately the deviations of real gases from the ideal are just the largest under these circumstances, so that the degeneracy phenomena, which are not insignificant in themselves, have not previously been able to be observed. It is not impossible, however, that a deeper knowledge of gaseous equations of state will permit one to separate the degeneracy from the other deviations from the equation $pV = NkT$ so that an experimental decision may be possible between the various degeneracy theories.

To apply the quantum rules to the motion of the molecules of our ideal gas, we can proceed in various ways, but the end result is always the same. For example, we can think of the molecules as confined in a vessel of parallelepiped form with elastically reflecting walls. Then the motion of the molecules flying to and fro between the walls becomes conditionally periodic and can hence be quantized. More generally, we can think of the molecules as exposed to an external field of force of such a character that their motion becomes conditionally periodic. The assumption that the gas is ideal permits us to neglect the mechanical actions of the molecules on each other so that their mechanical motion is carried out only under the influence of the external force. It is evident, however, that the quantization of the molecular motion carried out under the assumption that the molecules are completely independent of each other is not sufficient to account for the expected degeneracy. We see this best in the example of the molecules enclosed in a vessel whose linear dimensions are permitted to grow larger, in which case the quantized energy values of each individual molecule become more densely packed together, so that for vessels of macroscopic dimensions, every influence of the discontinuous character of the energy values practically disappears. Besides the volume of the vessel, this influence depends on the choice of the number of molecules in the vessel in such a way that the density remains constant.

Calculation[2] shows that we get a degeneracy of the expected order of magnitude only if we choose the vessel so small that on the average it contains only one molecule.

We therefore strongly feel that for the quantization of an ideal gas, a supplement to the Sommerfeld quantum conditions is necessary.

Recently W. Pauli,[3] in commenting on a work of E. C. Stoner,[4] has set forth the rule that if an atom contains an electron whose quantum numbers (exclusive of the magnetic quantum numbers) have definite values, there can exist in this atom no other electron whose path is characterized by the same numbers. In other words, a quantum orbit (in an external magnetic field) is already completely occupied by a single electron.

[*Editor's Note:* This is the statement that later became known as the Pauli exclusion principle. (See Paper 10).]

Since Pauli's rule has proved itself extremely fruitful in the interpretation of spectroscopic phenomena,[5] we shall attempt to see whether it is not also of use in the problem of the quantization of an ideal gas.

We shall show that this indeed is the case and that the application of Pauli's rule permits us to set up a completely persuasive theory of the degeneracy of an ideal gas.

In the following we shall therefore assume that at the most, a single molecule with given quantum numbers can be present in our gas. As quantum numbers, we have to consider not only those that determine the internal motions of the molecule but also those that determine its translational motion.

In the first place, we must insert our molecules in a suitable external field so that their motion will be conditionally periodic. This can happen in infinitely many ways. Since, however, the result does not depend on the choice of field of force, we shall subject the molecules to a central elastic attraction directed to a fixed point O (the origin of the coordinate system) so that every molecule constitutes a harmonic oscillator. This central force serves to keep the gas in the neighborhood of O. The gas density will decrease with the distance from O and will vanish at infinite distance. Let ν be the natural frequency of the oscillators. Then the force acting on the molecules is given by

$$4\pi^2 \nu^2 m r$$

where *m* is the mass of the molecule and *r* is the distance from 0. The potential energy of the force of attraction is then

$$u = 2\pi^2 \nu^2 m r^2. \tag{1}$$

Let the quantum numbers of the oscillator formed by a single molecule be designated by s_1, s_2, and s_3.

The total energy of this molecule is given by the expression

$$w = h\nu(s_1 + s_2 + s_3) = h\nu s \tag{2}$$

The total energy can therefore be an arbitrary integral multiple of $h\nu$. The value $sh\nu$ can, however, be realized in many ways. Each possibility corresponds to a solution of the equation

$$s = s_1 + s_2 + s_3, \tag{3}$$

in which s_1, s_2, s_3, can assume the values 0, 1, 2, 3, . . . Equation 3 has

$$Q_s = \frac{(s+1)(s+2)}{2} \tag{4}$$

solutions. Energy zero can therefore be realized in only one way. Energy $h\nu$ can be realized in three ways, energy $2h\nu$ in six ways, and so on. A molecule with the energy $sh\nu$ will for simplicity be denoted as an S-molecule.

In accordance with our assumptions, in the whole quantity of gas there can be at most Q_s "s-molecules," accordingly at most one molecule with energy zero, at most three with energy $h\nu$, at most six with energy $2h\nu$, and so on.

In order to grasp clearly the consequences of this situation, we shall consider the extreme case in which the absolute temperature of the gas is zero. Let N be the number of the molecules. At the absolute zero, the gas must be in the state of lowest energy. If now there were no restriction on the number of molecules of a given energy, each molecule would then be in the state of energy zero ($s_1 = s_2 = s_3 = 0$). However, from the preceding, there can be at most one molecule with energy zero. Therefore if $N = 1$, then that single molecule at absolute zero would have the energy zero. If $N = 4$, then one molecule will be in the state of energy zero, and the three others will occupy the three places with energy $h\nu$. If $N = 10$, there will still be one molecule in the state of energy zero, three others will be in the state of energy $h\nu$, and the six remaining will be in the six places corresponding to energy $2h\nu$, and so on.

At absolute zero, therefore, the molecules of the gas have a kind of shell structure having a certain analogy to the shell arrangements of the electrons in a polyelectronic atom.

We shall now inquire how a certain quantity of energy

$$W = Eh\nu \tag{5}$$

(E integral) is distributed among the N molecules.

Let N_s be the number of molecules that lie in a state with energy $sh\nu$. According to our assumption

$$N_s \leq Q_s. \tag{6}$$

We have further the equations

$$\sum N_s = N, \tag{7}$$

and

$$\sum s N_s = E, \tag{8}$$

expressing the fact that the total number of molecules is fixed at N and the total energy is $Eh\nu$.

We shall now calculate the number of arrangments P of the N molecules such that N_0 occupy states with energy zero, N_1 are in states of energy $h\nu$, N_2 in states with energy $2h\nu$, and so on. We shall look upon two arrangements as equivalent if the states occupied by the molecules are the same. Two arrangements that differ only by a permutation of the molecules among their states are therefore looked upon as the same arrangement. If we were to consider two such arrangements as different, we should have to multiply P with the constant N! We can easily see, however, that this would have no influence on what follows. The number of arrangements of N_s molecules among the Q_s states of energy $sh\nu$ is given by

$$\binom{Q_s}{N_s}$$

[*Editor's Note:* $\binom{Q_s}{N_s} = \frac{Q_s!}{N_s!\,(Q_s - N_s)!}$]

Hence for *P* we get the expression

$$P = \binom{Q_0}{N_0}\binom{Q_1}{N_1}\binom{Q_2}{N_2}\cdots = \prod \binom{Q_s}{N_s}. \tag{9}$$

We then get the most probable value of N_s by making *P* a maximum subject to the conditions 7 and 8. By use of Stirling's theorem for *N*! we can write to a sufficiently good approximation

$$\log P = \sum \log \binom{Q_s}{N_s} = -\sum \left(N_s \log \frac{N_s}{Q_s - N_s} + Q_s \log \frac{Q_s - N_s}{Q_s}\right. . \tag{10}$$

We therefore look for the values of N_s that satisfy equations 7 and 8 and for which log *P* is a maximum. We find

$$\alpha e^{-\beta s} = \frac{N_s}{Q_s - N_s},$$

in which α and β are constants. This equation can be written

$$N_s = Q_s \frac{\alpha e^{-\beta s}}{1 + \alpha e^{-\beta s}}. \tag{11}$$

The values of α and β can be determined from equations 7 and 8. Or we can also think of α and β as given. Then equations 7 and 8 determine the total number and total energy of the molecules. We thus find

$$\left.\begin{aligned} N &= \sum_0^\infty Q_s \frac{\alpha e^{-\beta s}}{1 + \alpha e^{-\beta s}}, \\ \frac{W}{h\nu} = E &= \sum_0^\infty s\, Q_s \frac{\alpha e^{-\beta s}}{1 + \alpha e^{-\beta s}}, \end{aligned}\right\} \tag{12}$$

The absolute temperature *T* of the gas is a function of *N* and *E* or alternatively of α and β. This function can be determined by two methods, though both lead to the same result. For example, we could follow the Boltzmann principle and write the entropy S

$$S = k \log P$$

and then evaluate the temperature by means of the formula

$$T = \frac{dW}{dS}$$

However, like all other methods based on the Boltzmann principle, this has the disadvantage that for its application, we need a more-or-less arbitrary supplementary assumption for the state probability. We therefore prefer to proceed as follows. We observe that the density of the gas is a function of the distance *r* from the origin *O* and vanishes for infinite distance. For infinitely large *r*, consequently, the degeneracy phenomena will cease, and the statistics of the gas will revert to the classical variety. In particular for $r = \infty$, the mean kinetic energy of the molecule must become $3kT/2$, and the velocity distribution must reduce to that of Maxwell.We can accordingly determine the temperature from the distribution of velocities in the region of infinitely small density, and since the whole mass of gas is at constant temperature, we shall therefore know the temperature for the domain of high density also. For this determination, we will make use of a gas thermometer with an infinitely dilute ideal gas.

First, we must calculate the density of the molecules with kinetric energy between L and $L + dL$ at distance *r*. The total energy of these molecules will, in accordance with equation 1, lie between

$$L + 2\pi^2\nu^2 m r^2 \quad \text{and} \quad L + 2\pi^2\nu^2 m r^2 + dL$$

But the total energy of a molecule is equal to $sh\nu$. For our molecules, therefore, *s* must lie between *s* and $s + ds$, where

$$s = \frac{L}{h\nu} + \frac{2\pi^2\nu m}{h} r^2, \quad ds = \frac{dL}{h\nu}. \tag{13}$$

Let us consider now a molecule whose motion is characterized by the quantum numbers s_1, s_2, s_3. Its coordinates *x, y, z,* are then given as functions of the time by

$$\left.\begin{aligned} x = \sqrt{Hs_1}\cos(2\pi\nu t - \alpha_1), \quad y = \sqrt{Hs_2}\cos(2\pi\nu t - \alpha_2), \\ z = \sqrt{Hs_3}\cos(2\pi\nu t - \alpha_3) \end{aligned}\right\} \tag{14}$$

Here we have set

$$H = \frac{h}{2\pi^2\nu m} \tag{15}$$

The quantities $\alpha_1, \alpha_2, \alpha_3$ are phase constants that with equal probability can take on any arbitrary system of values. From this and from equations 14, it follows that $x \leq \sqrt{Hs_1}$, $y \leq \sqrt{Hs_2}$, and $z \leq \sqrt{Hs_3}$ and that the probability that *x, y, z* lie between the limits *x* and $x + dx$, *y* and $y + dy$, *z* and $z + dz$ takes the following form

$$\frac{dx\,dy\,dz}{\pi^3\sqrt{(Hs_1 - x^2)(Hs_2 - y^2)(Hs_3 - z^2)}}.$$

If we do not know the individual values of s_1, s_2, s_3, but only their sum, this probability takes the form

$$\frac{1}{Q_s}\frac{dx\,dy\,dz}{\pi^3}\sum\frac{1}{\sqrt{(Hs_1 - x^2)(Hs_2 - y^2)(Hs_3 - z^2)}} \tag{16}$$

The sum is to be taken over all integral solutions of the equation 3, which satisfy the inequalities

$$H s_1 \geqslant x^2, \quad H s_2 \geqslant y^2, \quad H s_3 \geqslant z^2$$

If we multiply the probability (16) by the number N_s of the s molecules, we get the number of s molecules contained in the volume element *dx dy dz*. Taking account of equation 11, we find therefore that the density of s molecules at the point *x, y, z* is given by

$$n_s = \frac{\alpha e^{-\beta s}}{1 + \alpha e^{-\beta s}} \frac{1}{\pi^3} \sum \frac{1}{\sqrt{(H s_1 - x^2)(H s_2 - y^2)(H s_3 - z^2)}}$$

For sufficiently large *s*, we can replace the sum by a double integral, and after carrying out the integration find

$$n_s = \frac{2}{\pi^2 H^2} \frac{\alpha e^{-\beta s}}{1 + \alpha e^{-\beta s}} \sqrt{H s - r^2}.$$

With the case of equations 13 and 15 we now find that the density of molecules with kinetic energy between *L* and *L* + *dL* at the point *x, y, z* may be expressed as follows:

$$n(L)\, dL = n_s\, ds = \frac{2\pi (2m)^{3/2}}{h^3} \sqrt{L}\, dL \frac{\alpha e^{-\frac{2\pi^2 \nu m \beta r^2}{h}} e^{-\frac{\beta L}{h\nu}}}{1 + \alpha e^{-\frac{2\pi^2 \nu m \beta r^2}{h}} e^{-\frac{\beta L}{h\nu}}}. \tag{17}$$

This formula should be compared with the classical expression for the Maxwell distribution law

$$n^*(L)\, dL = K \sqrt{L}\, dL\, e^{-L/kT}. \tag{17'}$$

We see then that in the limit as $r \to \infty$, equation 17 goes over into equation 17′ if we set

$$\beta = \frac{h\nu}{kT} \tag{18}$$

We can now write equation 17 in the following way

$$n(L)\, dL = \frac{(2\pi)(2m)^{3/2}}{h^3} \sqrt{L}\, dL \frac{A e^{-L/kT}}{1 + A e^{-L/kT}}, \tag{19}$$

where

$$A = \alpha e^{-\frac{2\pi^2 \nu^2 m r^2}{kT}}. \tag{20}$$

The total density of the molecules at distance *r* then becomes

$$n = \int_0^\infty n(L)\, dL = \frac{(2\pi m k T)^{3/2}}{h^3} F(A), \tag{21}$$

where

$$F(A) = \frac{2}{\sqrt{\pi}} \int_0^\infty \frac{A \sqrt{x} e^{-x}\, dx}{1 + A e^{-x}}. \tag{22}$$

The mean kinetic energy of the molecules at distance *r* is

$$\bar{L} = \frac{1}{n}\int_0^\infty L\, n(L)\, dL = \frac{3}{2} kT \frac{G(A)}{F(A)}, \tag{23}$$

where

$$G(A) = \frac{4}{3\sqrt{\pi}} \int_0^\infty \frac{A x^{3/2} e^{-x}\, dx}{1 + A e^{-x}}. \tag{24}$$

We can determine *A* as a function of density and temperature by means of equation 21. If the value so found is inserted in equations 19 and 23, we get the velocity distribution and the mean kinetic energy of the molecules as a function of density and temperature.

To set up the equation of state, we apply the virial theorem. According to the latter, the pressure is given by

$$p = \frac{2}{3} n \bar{L} = nkT \frac{G(A)}{F(A)} \tag{25}$$

The value of *A* is again to be taken from equation 12 as a function of density and temperature.

Before we go further, we shall mention some mathematical properties of the functions *F(A)* and *G(A)*.

For $A \leq 1$, we can represent both functions by the convergent series

$$\left.\begin{aligned} F(A) &= A - \frac{A^2}{2^{3/2}} + \frac{A^3}{3^{3/2}} - \cdots, \\ G(A) &= A - \frac{A^2}{2^{5/2}} + \frac{A^3}{3^{5/2}} - \cdots \end{aligned}\right\} \tag{26}$$

For large *A* we have the asymptotic expression

$$\left.\begin{aligned} F(A) &= \frac{4}{3\sqrt{\pi}} (\log A)^{3/2} \left[1 + \frac{\pi^2}{8(\log A)^2} + \cdots\right], \\ G(A) &= \frac{8}{15\sqrt{\pi}} (\log A)^{5/2} \left[1 + \frac{5\pi^2}{8(\log A)^2} + \cdots\right]. \end{aligned}\right\} \tag{27}$$

We have the further relation

$$\frac{d\,G(A)}{F(A)} = d \log A. \tag{28}$$

We must introduce still another function *P(Θ)* defined by the relation

$$P(\Theta) = \Theta \frac{G(A)}{F(A)}, \quad F(A) = \frac{1}{\Theta^{3/2}} \tag{29}$$

For very large and very small *Θ*, respectively, the function *P(θ)* is given by the approximation formulas

$$\left.\begin{aligned} P(\Theta) &= \Theta \left\{1 + \frac{1}{2^{5/2}\Theta^{3/2}} + \cdots\right\} \\ P(\Theta) &= \frac{3^{2/3}\pi^{1/3}}{5 \cdot 2^{1/3}} \left\{1 + \frac{5 \cdot 2^{2/3}\pi^{4/3}}{3^{7/3}} \Theta^2 + \cdots\right\} \end{aligned}\right\} \tag{30}$$

Using equations 29, 28, and 27, we see further that

$$\int_0^\Theta \frac{d\,P(\Theta)}{\Theta} = \frac{5}{3}\frac{G(A)}{F(A)} - \frac{2}{3}\log A. \tag{31}$$

We are now in a position to eliminate the parameter A from the state equation 15 and equation 23. We then find the pressure and the mean kinetic energy of the molecules as explicit functions of density and temperature.

$$p = \frac{h^2 n^{5/3}}{2\pi m} P\left(\frac{2\pi m k T}{h^2 n^{2/3}}\right), \tag{32}$$

$$\overline{L} = \frac{3}{2}\frac{h^2 n^{2/3}}{2\pi m} P\left(\frac{2\pi m k T}{h^2 n^{2/3}}\right). \tag{33}$$

In the limiting case of weak degeneracy (T large and n small), the equation of state assumes the following form

$$p = n k T\left\{1 + \frac{1}{16}\frac{h^3 n}{(\pi m k T)^{3/2}} + \cdots\right\}. \tag{34}$$

The pressure is accordingly greater than that for the classical state equation ($p = nkT$). For an ideal gas with the atomic weight of helium, at $T = 5°$ and a pressure of 10 atmospheres, the difference is about 15%.

In the limiting case of large degeneracy, equations 32 and 33 take, respectively, the from

$$p = \frac{1}{20}\left(\frac{6}{\pi}\right)^{2/3}\frac{h^2 n^{5/3}}{m} + \frac{2^{4/3}\pi^{8/3}}{3^{5/3}}\frac{m n^{1/3} k^2 T^2}{h^2} + \cdots \tag{35}$$

$$\overline{L} = \frac{3}{40}\left(\frac{6}{\pi}\right)^{2/3}\frac{h^2 n^{2/3}}{m} + \frac{2^{1/3}\pi^{8/3}}{3^{2/3}}\frac{m k^2 T^2}{h^2 n^{2/3}} + \cdots \tag{36}$$

We conclude from these equations that as a result of the degeneracy, there is a zero-point pressure and a zero-point energy.

From equation 36, we can also calculate the specific heat at low temperatures. We find

$$c_v = \frac{d\overline{L}}{dT} = \frac{2^{4/3}\pi^{8/3}}{3^{2/3}}\frac{m k^2 T}{h^2 n^{2/3}} + \cdots \tag{37}$$

It follows that the specific heat at absolute zero temperature vanishes, and indeed that at very low temperatures, it is proportional to the absolute temperature.

Finally we wish to show that our theory leads to the Stern-Tetrode value for the absolute entropy of the gas. By application of equation 33, we find that

$$S = n\int_0^T \frac{d\overline{L}}{T} = \frac{3}{2} n k \int_0^\Theta \frac{P'(\Theta)\,d\Theta}{\Theta}.$$

Equation 31 then yields

$$S = n\,k \left\{\frac{5}{2}\,\frac{G(A)}{F(A)} - \log A\right\}, \tag{38}$$

where the value of A is again to be taken from equation 21. By application of equation 26, we find for high temperatures

$$A = \frac{n\,h^3}{(2\,\pi\,m\,k\,T)^{3/2}}, \qquad \frac{G(A)}{F(A)} = 1.$$

Then from equation 38, there follows

$$S = n\,k \left\{\log \frac{(2\,\pi\,m\,k\,T)^{3/2}}{n\,h^3} + \frac{5}{2}\right\}$$
$$= n\,k \left\{\frac{3}{2}\log T - \log n + \log \frac{(2\,\pi\,m)^{3/2}\,k^{3/2}\,e^{5/2}}{h^3}\right\},$$

in agreement with the entropy value of Stern and Tetrode.

NOTES

1. Cf., for example, A. Einstein, Berliner Berichte, 1924, p. 261; 1925, p. 318. Also M. Planck, in the same journal, 1925, p. 49. Our method is related to that of Einstein insofar as the assumption of the statistical independence of the molecules is given up in both methods, although the sort of dependence we have assumed is wholly different from that of Einstein. The end result for the deviations from the classical equations of state is quite different in the two cases.
2. E. Fermi "Nuovo Cimento," 1, 145 (1924)
3. W. Pauli, Jr., Zs. für Physik, 31, 765, 1925.
4. E. C. Stoner, Phil. Mag. 48, 719, 1924.
5. Cf, for example, B. F. Hund, Zs. f. Physik, 33, 345, 1925.

12

Reprinted from *Roy. Soc. (London) Proc.*, ser. A, **112:**661–677 (1926)

On the Theory of Quantum Mechanics.

By P. A. M. DIRAC, St. John's College, Cambridge.

(Communicated by R. H. Fowler, F.R.S.—Received August 26, 1926.)

§ 1. *Introduction and Summary.*

The new mechanics of the atom introduced by Heisenberg* may be based on the assumption that the variables that describe a dynamical system do not obey the commutative law of multiplication, but satisfy instead certain quantum conditions. One can build up a theory without knowing anything about the dynamical variables except the algebraic laws that they are subject to, and can show that they may be represented by matrices whenever a set of uniformising variables for the dynamical system exists.† It may be shown, however (see § 3), that there is no set of uniformising variables for a system containing more than one electron, so that the theory cannot progress very far on these lines.

A new development of the theory has recently been given by Schrödinger.‡ Starting from the idea that an atomic system cannot be represented by a trajectory, *i.e.*, by a point moving through the co-ordinate space, but must be represented by a wave in this space, Schrödinger obtains from a variation principle a differential equation which the wave function ψ must satisfy. This differential equation turns out to be very closely connected with the Hamiltonian equation which specifies the system, namely, if

$$H(q_r, p_r) - W = 0$$

is the Hamiltonian equation of the system, where the q_r, p_r are canonical variables, then the wave equation for ψ is

$$\left\{H\left(q_r, ih\frac{\partial}{\partial q_r}\right) - W\right\}\psi = 0, \tag{1}$$

where h is $(2\pi)^{-1}$ times the usual Planck's constant. Each momentum p_r in H is replaced by the operator $ih\,\partial/\partial q_r$, and is supposed to operate on all that exists on its right-hand side in the term in which it occurs. Schrödinger takes the values of the parameter W for which there exists a ψ satisfying (1) that is

* See various papers by Born, Heisenberg and Jordan, 'Zeits. f. Phys.,' vol. 33 onwards.

† 'Roy. Soc. Proc.,' A, vol. 110, p. 561 (1926).

‡ See various papers in the 'Ann. d. Phys.,' beginning with vol. 79, p. 361 (1926).

continuous, single-valued and bounded throughout the whole of q-space to be the energy levels of the system, and shows that when the general solution of (1) is known, matrices to represent the p_r and q_r may easily be obtained, satisfying all the conditions that they have to satisfy according to Heisenberg's matrix mechanics, and consistent with the energy levels previously found. The mathematical equivalence of the theories is thus established.

In the present paper, Schrödinger's theory is considered in § 2 from a slightly more general point of view, in which the time t and its conjugate momentum $-W$ are treated from the beginning on the same footing as the other variables. A more general method, requiring only elementary symbolic algebra, of obtaining matrix representations of the dynamical variables is given.

In § 3 the problem is considered of a system containing several similar particles, such as an atom with several electrons. If the positions of two of the electrons are interchanged, the new state of the atom is physically indistinguishable from the original one. In such a case one would expect only symmetrical functions of the co-ordinates of all the electrons to be capable of being represented by matrices. It is found that this allows one to obtain two solutions of the problem satisfying all the necessary conditions, and the theory is incapable of deciding which is the correct one. One of the solutions leads to Pauli's principle that not more than one electron can be in any given orbit, and the other, when applied to the analogous problem of the ideal gas, leads to the Einstein-Bose statistical mechanics.

The effect of an arbitrarily varying perturbation on an atomic system is worked out in § 5 with the help of a new assumption. The theory is applied to the absorption and stimulated emission of radiation by an atom. A generalisation of the description of the phenomena by Einstein's B coefficients is obtained, in which the phases play their proper parts. This method cannot be applied to spontaneous emission.

§ 2. *General Theory.*

According to the new point of view introduced by Schrödinger, we no longer leave unspecified the nature of the dynamical variables that describe an atomic system, but count the q's and t as ordinary mathematical variables (this being permissible since they commute with one another) and take the p's and W to be the differential operators

$$p_r = -ih\frac{\partial}{\partial q_r}, \qquad -W = -ih\frac{\partial}{\partial t}. \tag{2}$$

Whenever a p_r or W occurs in a term of an equation, it must be considered as meaning the corresponding differential operator operating on all that occurs on its *right-hand side in the term in question.* Thus, by carrying out the operations, one can reduce any function of the p's, q's, W and t to a function of the q's and t only.

The relations (2) require two obvious modifications to be made in the algebra governing the dynamical variables. Firstly, only rational integral functions of the p's and W have a meaning, and, secondly, *one can multiply up an equation by a factor (integral in the p's and W) on the left-hand side, but one cannot, in general, multiply up by factor on the right-hand side.* Thus, if one is given the equation $a = b$, one can infer from it that $\mathrm{X}a = \mathrm{X}b$, where X is arbitrary, but one cannot in general infer that $a\mathrm{X} = b\mathrm{X}$.

There are, however, certain equations $a = b$ for which it is true that $a\mathrm{X} = b\mathrm{X}$ for any X, and these equations we call identities. The quantum conditions

$$q_r p_s - p_s q_r = ih\delta_{rs}, \qquad p_r p_s - p_s p_r = 0,$$

with the similar relations involving $-\mathrm{W}$ and t, are identities, as it can easily be verified (and has been verified by Schrödinger) that the relations

$$(q_r p_s - p_s q_r)\,\mathrm{X} = ih\delta_{rs}\mathrm{X},$$

etc., hold for any X. These relations form the main justification for the assumptions (2).

If $a = b$ is an identity, we can deduce, since $a\mathrm{X} = b\mathrm{X}$ and $\mathrm{X}a = \mathrm{X}b$, that

$$a\mathrm{X} - \mathrm{X}a = b\mathrm{X} - \mathrm{X}b,$$

or

$$[a, \mathrm{X}] = [b, \mathrm{X}].$$

Thus we can equate the Poisson bracket of either side of an identity with an arbitrary quantity, and so our quantum identity is the analogue of an identity on the classical theory. We assume the general equation $xy - yx = ih\,[x, y]$ and the equations of motion of a dynamical system to be identities.

A dynamical system is specified by a Hamiltonian equation between the variables

$$\mathrm{H}(q_r, p_r, t) - \mathrm{W} = 0, \tag{3}$$

or more generally

$$\mathrm{F}(q_r, p_r, t, \mathrm{W}) = 0, \tag{4}$$

and the equations of motion are

$$dx/ds = [x, \mathrm{F}],$$

where x is any function of the dynamical variables, and s is a variable which depends on the form in which (4) is written, and, in particular, is just t if (4) is written in the form (3). On the new theory we consider the equation

$$\mathrm{F}\psi = 0, \tag{5}$$

which, if we take ψ to be a function of the q's and t only, is an ordinary differential equation for ψ. From the general solution of this differential equation the matrices that form the solution of the mechanical problem may be very easily obtained.

Since (5) is linear in ψ, its general solution is of the form

$$\psi = \Sigma c_n \psi_n, \tag{6}$$

where the c_n's are arbitrary constants and the ψ_n's are a set of independent solutions, which may be called eigenfunctions. Only solutions that are continuous, single-valued and bounded throughout the whole domain of the q's and t are recognised by the theory. Instead of a discreet set of eigenfunctions ψ_n there may be a continuous set $\psi(\alpha)$, depending on a parameter α, and satisfying the differential equation for all values of α in a certain range, in which case the sum in (6) must be replaced by an integral $\int c_\alpha \psi(\alpha)\, d\alpha$,* or both a discreet set and a continuous set may occur together. For definiteness, however, we shall write down explicitly only the discreet sum in the following work.

We shall now show that any constant of integration of the dynamical system (either a first integral or a second integral) can be represented by a matrix whose elements are constants, there being one row and column of the matrix corresponding to each eigenfunction ψ_n. Let a be a constant of integration of the system, *i.e.*, a function of the dynamical variables such that $[a, \mathrm{F}] = 0$ identically. We have the relation

$$\mathrm{F}a = a\mathrm{F},$$

which, being an identity, we can multiply by ψ_n on the right-hand side. We thus obtain

$$\mathrm{F}a\psi_n = a\mathrm{F}\psi_n = 0,$$

since $\mathrm{F}\psi_n = 0$ (although not identically). Hence $a\psi_n$ is a solution of the differential equation (5), so that it can be expanded in the form (6), *i.e.*,

$$a\psi_n = \Sigma_m \psi_m a_{mn},$$

where the a_{mn}'s are constants. We take the quantities a_{mn} to be the elements of the matrix that represents a. The matrix rule of multiplication evidently holds, since, if b is another constant of integration of the system, we have

$$ab\psi_n = a\Sigma_m \psi_m b_{mn} = \Sigma_{mk} \psi_k a_{km} b_{mn},$$

* The general solution may contain quantities, such as ψ_a and $\partial\psi_a/\partial a$, which satisfy the differential equation (5), but which cannot strictly be put in the form $\int c_a \psi_a da$, although they may be regarded as the limits of series of quantities which are of this form.

and also

$$ab\psi_n = \Sigma_k \psi_k (ab)_{kn},$$

so that

$$(ab)_{kn} = \Sigma_m a_{km} b_{mn}.$$

As an example of a constant of integration of the dynamical system, we may take the value $x(t_0)$ that an arbitrary function x of the p's, q's, W and t has at a specified time $t = t_0$. The matrix that represents $x(t_0)$ will consist of elements each of which is a function of t_0. Writing t for t_0, we see that an arbitrary function of the dynamical variables, $x(t)$, or simply x, can be represented by a matrix whose elements are functions of t only.

The matrix representation we have obtained is not unique, since any set of independent eigenfunctions ψ_n will do. To obtain the matrices of Heisenberg's original quantum mechanics, we must choose the ψ_n's in a particular way. We can always, by a linear transformation, obtain a set of ψ_n's which makes the matrix representing any given constant of integration of the dynamical system a diagonal matrix. Suppose now that the Hamiltonian F does not contain the time explicitly, so that W is a constant of the system, and is the energy, and we choose the ψ_n's so as to make the matrix representing W a diagonal matrix, *i.e.*, so as to make

$$W\psi_n = W_n \psi_n, \tag{7}$$

where W_n is a numerical constant. Let x be any function of the dynamical variables *that does not involve the time explicitly,* and put

$$x\psi_n = \Sigma_m x_{mn} \psi_m,$$

where the x_{mn}'s are functions of the time only. We shall now show that the x_{mn}'s are of the form

$$x_{mn} = a_{mn} e^{i(W_m - W_n)t/h}, \tag{8}$$

where the a_{mn}'s are constants, as on Heisenberg's theory. We have

$$\begin{aligned} Wx\psi_n &= \Sigma_m W x_{mn} \psi_m \\ &= \Sigma_m (Wx_{mn} - x_{mn}W)\, \psi_m + \Sigma_m x_{mn} W \psi_m \\ &= \Sigma_m ih\dot{x}_{mn} \psi_m + \Sigma_m x_{mn} W_m \psi_m. \end{aligned} \tag{9}$$

Also, since x does not contain t explicitly,

$$\begin{aligned} Wx\psi_n &= xW\psi_n = xW_n\psi_n = W_n x \psi_n \\ &= W_n \Sigma_m x_{mn} \psi_m. \end{aligned} \tag{10}$$

Equating the coefficients of ψ_m in (9) and (10), we obtain

$$ih\dot{x}_{mn} = x_{mn}(W_n - W_m),$$

which shows that x_{mn} is of the form (8).

We have thus shown that with the ψ_n's chosen in this way the matrices satisfy all the conditions of Heisenberg's matrix mechanics, except the condition that the matrices that represent real quantities are Hermitic (*i.e.*, have their *mn* and *nm* elements conjugate imaginaries). There does not seem to be any simple general proof that this is the case, as the proof would have to make use of the fact that the ψ_n's are bounded. It is easy to prove the particular case that the matrix representing W is Hermitic, *i.e.*, that the W_n's are real, since from (7) ψ_n must be of the form

$$\psi_n = u_n e^{-iW_n t/h},$$

where u_n is independent of t, and if W_n contains an imaginary part, ψ_n would not remain bounded as t becomes infinite. In general, the matrices representing real quantities could be Hermitic only if the arbitrary numerical constants by which the ψ_n's may be multiplied are chosen in a particular way.

We may regard an eigenfunction ψ_n as being associated with definite numerical values for some of the constants of integration of the system. Thus, if we find constants of integration $a, b, \ldots$ such that

$$a\psi_n = a_n\psi_n, \qquad b\psi_n = b_n\psi_n, \ldots \tag{11}$$

where $a_n, b_n, \ldots$ are numerical constants, we can say that ψ_n represents a state of the system in which $a, b, \ldots$ have the numerical values $a_n, b_n, \ldots$ (Note that $a, b, \ldots$ must commute for (11) to be possible.) In this way we can have eigenfunctions representing stationary states of an atomic system with definite values for the energy, angular momentum, and other constants of integration.

It should be noticed that the choice of the time t as the variable that occurs in the elements of the matrices representing variable quantities is quite arbitrary, and any function of t and the q's that increases steadily would do. To determine accurately the radiation emitted by the system in the direction of the x-axis, one would have to use $(t - x/c)$ instead of t.* It is probable that the representation of a constant of integration of the system by a matrix of constant elements is more fundamental than the representation of a variable quantity by a matrix whose elements are functions of some variable such as t or $(t - x/c)$. It would appear to be possible to build up an electromagnetic theory in which the potentials of the field at a specified point x_0, y_0, z_0, t_0 in space-time are represented by matrices of constant elements that are functions of x_0, y_0, z_0, t_0.

§ 3. *Systems containing Several Similar Particles.*

In Heisenberg's matrix mechanics it is assumed that the elements of the matrices that represent the dynamical variables determine the frequencies and

* 'Roy. Soc. Proc.,' A, vol. 111, p. 405 (1926).

intensities of the components of radiation emitted. The theory thus enables one to calculate just those quantities that are of physical importance, and gives no information about quantities such as orbital frequencies that one can never hope to measure experimentally. We should expect this very satisfactory characteristic to persist in all future developments of the theory.

Consider now a system that contains two or more similar particles, say, for definiteness, an atom with two electrons. Denote by (mn) that state of the atom in which one electron is in an orbit labelled m, and the other in the orbit n. The question arises whether the two states (mn) and (nm), which are physically indistinguishable as they differ only by the interchange of the two electrons, are to be counted as two different states or as only one state, *i.e.*, do they give rise to two rows and columns in the matrices or to only one? If the first alternative is right, then the theory would enable one to calculate the intensities due to the two transitions $(mn) \rightarrow (m'n')$ and $(mn) \rightarrow (n'm')$ separately, as the amplitude corresponding to either would be given by a definite element in the matrix representing the total polarisation. The two transitions are, however, physically indistinguishable, and only the sum of the intensities for the two together could be determined experimentally. Hence, in order to keep the essential characteristic of the theory that it shall enable one to calculate only observable quantities, one must adopt the second alternative that (mn) and (nm) count as only one state.

This alternative, though, also leads to difficulties. The symmetry between the two electrons requires that the amplitude associated with the transition $(mn) \rightarrow (m'n')$ of x_1, a co-ordinate of one of the electrons, shall equal the amplitude associated with the transition $(nm) \rightarrow (n'm')$ of x_2, the corresponding co-ordinate of the other electron, *i.e.*,

$$x_1(mn\,;\, m'n') = x_2(nm\,;\, n'm'). \tag{12}$$

If we now count (mn) and (nm) as both defining the same row and column of the matrices, and similarly for $(m'n')$ and $(n'm')$, equation (12) shows that each element of the matrix x_1 equals the corresponding element of the matrix x_2, so that we should have the matrix equation

$$x_1 = x_2.$$

This relation is obviously impossible, as, amongst other things, it is inconsistent with the quantum conditions. We must infer that unsymmetrical functions of the co-ordinates (and momenta) of the two electrons cannot be represented by matrices. Symmetrical functions, such as the total polarisation of the atom, can be considered to be represented by matrices without inconsistency, and these matrices are by themselves sufficient to determine all the physical properties of the system.

One consequence of these considerations is that the theory of uniformising variables introduced by the author can no longer apply. This is because, corresponding to any transition $(mn) \rightarrow (m'n')$, there would be a term $e^{i(aw)}$ in the Fourier expansions, and we should require there to be a unique state, $(m''n'')$, say, such that the same term $e^{i(aw)}$ corresponds to the transition $(m'n') \rightarrow (m''n'')$, and $e^{2i(aw)}$ corresponds to $(mn) \rightarrow (m''n'')$. If the m's and n's are quantum numbers, and we take the case of one quantum number per electron for definiteness, we should have to have

$$m'' - m' = m' - m, \qquad n'' - n' = n' - n.$$

Since, however, the state $(m'n')$ may equally well be called the state $(n'm')$, we may equally well take

$$m'' - n' = n' - m, \qquad n'' - m' = m' - n,$$

which would give a different state $(m''n'')$. There is thus no unique state $(m''n'')$ that the theory of uniformising variables demands.

If we neglect the interaction between the two electrons, then we can obtain the eigenfunctions for the whole atom simply by multiplying the eigenfunctions for one electron when it exists alone in the atom by the eigenfunctions for the other electron alone, and taking the same time variable for each.* Thus if $\psi_n(x, y, z, t)$ is the eigenfunction for a single electron in the orbit n, then the eigenfunction for the whole atom in the state (mn) is

$$\psi_m(x_1, y_1, z_1, t)\,\psi_n(x_2, y_2, z_2, t) = \psi_m(1)\,\psi_n(2),$$

say, where x_1, y_1, z_1 and x_2, y_2, z_2 are the co-ordinates of the two electrons, and $\psi(r)$ means $\psi(x_r, y_r, z_r, t)$. The eigenfunction $\psi_m(2)\,\psi_n(1)$, however, also corresponds to the same state of the atom if we count the (mn) and (nm) states as identical. But two independent eigenfunctions must give rise to two rows and columns in the matrices. If we are to have only one row and column in the matrices corresponding to both (mn) and (nm), we must find a set of eigenfunctions ψ_{mn} of the form

$$\psi_{mn} = a_{mn}\psi_m(1)\,\psi_n(2) + b_{mn}\psi_m(2)\,\psi_n(1),$$

where the a_{mn}'s and b_{mn}'s are constants, which set must contain only one ψ_{mn} corresponding to both (mn) and (nm), and must be sufficient to enable one to

* The same time variable t must be taken in each owing to the fact that we write the Hamiltonian equation for the whole system : $\mathrm{H}(1) + \mathrm{H}(2) - \mathrm{W} = 0$, where H(1) and H(2) are the Hamiltonians for the two electrons separately, so that there is a common time t conjugate to minus the total energy W.

obtain the matrix representing any symmetrical function A of the two electrons. This means the ψ_{mn}'s must be chosen such that A times any chosen ψ_{mn} can be expanded in terms of the chosen ψ_{mn}'s in the form

$$A\psi_{mn} = \Sigma_{m'n'}\psi_{m'n'}A_{m'n',\, mn}, \tag{13}$$

where the $A_{m'n',\, mn}$'s are constants or functions of the time only.

There are two ways of choosing the set of ψ_{mn}'s to satisfy the conditions. We may either take $a_{mn} = b_{mn}$, which makes each ψ_{mn} a symmetrical function of the two electrons, so that the left-hand side of (13) is symmetrical and only symmetrical eigenfunctions will be required for its expansion, or we may take $a_{mn} = -b_{mn}$, which makes ψ_{mn} antisymmetrical, so that the left-hand side of (13) is antisymmetrical and only antisymmetrical eigenfunctions will be required for its expansion. Thus the symmetrical eigenfunctions alone or the antisymmetrical eigenfunctions alone give a complete solution of the problem. The theory at present is incapable of deciding which solution is the correct one. We are able to get complete solutions of the problem which make use of less than the total number of possible eigenfunctions at the expense of being able to represent only symmetrical functions of the two electrons by matrices.

These results may evidently be extended to any number of electrons. For r non-interacting electrons with co-ordinates $x_1, y_1, z_1 \ldots, x_r, y_r, z_r$, the symmetrical eigenfunctions are

$$\Sigma_{\alpha_1 \ldots \alpha_r}\psi_{n_1}(\alpha_1)\,\psi_{n_2}(\alpha_2)\ldots\psi_{n_r}(\alpha_r), \tag{14}$$

where $\alpha_1, \alpha_2 \ldots \alpha_r$ are any permutation of the integers $1, 2 \ldots r$, while the antisymmetrical ones may be written in the determinantal form

$$\begin{vmatrix} \psi_{n_1}(1), & \psi_{n_1}(2) & \ldots & \psi_{n_1}(r) \\ \psi_{n_2}(1), & \psi_{n_2}(2) & \ldots & \psi_{n_2}(r) \\ \cdot & \cdot & \cdot & \cdot \\ \psi_{n_r}(1), & \psi_{n_r}(2) & \ldots & \psi_{n_r}(r) \end{vmatrix}. \tag{15}$$

If there is interaction between the electrons, there will still be symmetrical and antisymmetrical eigenfunctions, although they can no longer be put in these simple forms. In any case the symmetrical ones alone or the antisymmetrical ones alone give a complete solution of the problem.

An antisymmetrical eigenfunction vanishes identically when two of the electrons are in the same orbit. This means that in the solution of the problem with antisymmetrical eigenfunctions there can be no stationary states with two or more electrons in the same orbit, which is just Pauli's exclusion principle.* The solution with symmetrical eigenfunctions, on the other hand, allows any number of electrons to be in the same orbit, so that this solution cannot be the correct one for the problem of electrons in an atom.†

§ 4. *Theory of the Ideal Gas.*

The results of the preceding section apply to any system containing several similar particles, in particular to an assembly of gas molecules. There will be two solutions of the problem, in one of which the eigenfunctions are symmetrical functions of the co-ordinates of all the molecules, and in the other antisymmetrical.

The wave equation for a single molecule of rest-mass m moving in free space is

$$\{p_x^2 + p_y^2 + p_z^2 - \mathrm{W}^2/c^2 + m^2c^2\}\,\psi = 0$$

$$\left\{\frac{\partial^2}{\partial x^2} + \frac{\partial^2}{\partial y^2} + \frac{\partial^2}{\partial z^2} - \frac{1}{c^2}\frac{\partial^2}{\partial t^2} - \frac{m^2c^2}{h^2}\right\}\psi = 0,$$

and its solution is of the form

$$\psi_{\alpha_1\alpha_2\alpha_3} = \exp.\, i\,(\alpha_1 x + \alpha_2 y + \alpha_3 z - \mathrm{E}t)/h, \tag{16}$$

where α_1, α_2, α_3 and E are constants satisfying

$$\alpha_1^2 + \alpha_2^2 + \alpha_3^2 - \mathrm{E}^2/c^2 + m^2c^2 = 0.$$

The eigenfunction (16) represents an atom having the momentum components α_1, α_2, α_3 and the energy E.

We must now obtain some restriction on the possible eigenfunctions due to the presence of boundary walls. It is usually assumed that the eigenfunction, or wave function associated with a molecule, vanishes at the boundary, but we should expect to be able to deduce this, if it is true, from the general theory. We assume, as a natural generalisation of the methods of the preceding section, that there must be only just sufficient eigenfunctions for one to be able to represent by a matrix any function of the co-ordinates that has a physical meaning. Suppose for definiteness that each molecule is confined between two boundaries at $x = 0$ and $x = 2\pi$. Then only those functions of x that are defined only for $0 < x < 2\pi$ have a physical meaning and must be capable of being represented by matrices. (This will require fewer eigenfunctions than if every

* Pauli, 'Zeits. f. Phys.,' vol. 31, p. 765 (1925).

† Prof. Born has informed me that Heisenberg has independently obtained results equivalent to these. (Added in proof)—see Heisenberg, 'Zeit. fur Phys.,' vol. 38, p. 411 (1926).

function of x had to be capable of being represented by a matrix.) These functions $f(x)$ can always be expanded as Fourier series of the form

$$f(x) = \Sigma_n a_n e^{inx}, \tag{17}$$

where the a_n's are constants and the n's integers. If we choose from the eigenfunctions (16) those for which α_1/h is an integer, then $f(x)$ times any chosen eigenfunction can be expanded as a series in the chosen eigenfunctions whose coefficients are functions of t only, and hence $f(x)$ can be represented by a matrix. Thus these chosen eigenfunctions are sufficient, and are easily seen to be only just sufficient, for the matrix representation of any function of x of the form (17). Instead of choosing those eigenfunctions with integral values for α_1/h, we could equally well take those with α_1/h equal to half an odd integer, or more generally with $\alpha_1/h = n + \varepsilon$, where n is an integer and ε is any real number. The theory is incapable of deciding which are the correct ones. For statistical problems, though, they all lead to the same results.

When y and z are also bounded by $0 < y < 2\pi$, $0 < z < 2\pi$, we find for the number of waves associated with molecules whose energies lie between E and $E + dE$ the value

$$\frac{4\pi}{c^3 h^3}(E^2 - m^2c^4)^{\frac{1}{2}}\, E\, dE.$$

This value is in agreement with the ordinary assumption that the wave function vanishes at the boundary. It reduces, when one neglects relativity mechanics, to the familiar expression

$$\frac{2\pi}{h^3}(2m)^{\frac{3}{2}}\, E_1^{\frac{1}{2}} dE_1, \tag{18}$$

where $E_1 = E - mc^2$ is the kinetic energy. For an arbitrary volume of gas V the expression must be multiplied by $V/(2\pi)^3$.

To pass to the eigenfunctions for the assembly of molecules, between which there is assumed to be no interaction, we multiply the eigenfunctions for the separate molecules, and then take either the symmetrical eigenfunctions, of the form (14), or the antisymmetrical ones, of the form (15). We must now make the new assumption that all stationary states of the assembly (each represented by one eigenfunction) have the same *a priori* probability. If now we adopt the solution of the problem that involves symmetrical eigenfunctions, we should find that all values for the number of molecules associated with any wave have the same *a priori* probability, which gives just the Einstein-Bose statistical mechanics.* On the other hand, we should obtain a different

* Bose, 'Zeits. f. Phys.,' vol. 26, p. 178 (1924); Einstein, 'Sitzungsb. d. Preuss. Ac.,' p. 261 (1924) and p. 3 (1925).

statistical mechanics if we adopted the solution with antisymmetrical eigenfunctions, as we should then have either 0 or 1 molecule associated with each wave. The solution with symmetrical eigenfunctions must be the correct one when applied to light quanta, since it is known that the Einstein-Bose statistical mechanics leads to Planck's law of black-body radiation. The solution with antisymmetrical eigenfunctions, though, is probably the correct one for gas molecules, since it is known to be the correct one for electrons in an atom, and one would expect molecules to resemble electrons more closely than light-quanta.

We shall now work out, according to well-known principles, the equation of state of the gas on the assumption that the solution with antisymmetrical eigenfunctions is the correct one, so that not more than one molecule can be associated with each wave. Divide the waves into a number of sets such that the waves in each set are associated with molecules of about the same energy. Let A_s be the number of waves in the sth set, and let E_s be the kinetic energy of a molecule associated with one of them. Then the probability of a distribution (or the number of antisymmetrical eigenfunctions corresponding to distributions) in which N_s molecules are associated with waves in the sth set is

$$W = \Pi_s \frac{A_s!}{N_s!\,(A_s - N_s)!},$$

giving for the entropy

$$S = k \log W = k\Sigma_s \{A_s(\log A_s - 1) - N_s(\log N_s - 1) - (A_s - N_s)[\log(A_s - N_s) - 1]\}.$$

This is to be a maximum, so that

$$0 = \delta S = k\Sigma_s \{-\log N_s + \log(A_s - N_s)\}\,\delta N_s = k\Sigma_s \log(A_s/N_s - 1)\,.\,\delta N_s,$$

for all variations δN_s that leave the total number of molecules $N = \Sigma_s N_s$ and the total energy $E = \Sigma_s E_s N_s$ unaltered, so that

$$\Sigma_s \delta N_s = 0, \qquad \Sigma_s E_s \delta N_s = 0.$$

We thus obtain

$$\log(A_s/N_s - 1) = \alpha + \beta E_s,$$

where α and β are constants, which gives

$$N_s = \frac{A_s}{e^{\alpha + \beta E_s} + 1}. \tag{19}$$

By making a variation in the total energy E and putting $\delta E/\delta S = T$, the temperature, we readily find that $\beta = 1/kT$, so that (19) becomes

$$N_s = \frac{A_s}{e^{\alpha + E_s/kT} + 1}.$$

This formula gives the distribution in energy of the molecules. On the Einstein-Bose theory the corresponding formula is

$$N_s = \frac{A_s}{e^{\alpha + E_s/kT} - 1}.$$

If the sth set of waves consists of those associated with molecules whose energies lie between E_s and $E_s + dE_s$, we have from (18) [where E_s now means the E_1 of equation (18)],

$$A_s = 2\pi V (2m)^{\frac{3}{2}} E_s^{\frac{1}{2}} dE_s/(2\pi h)^3,$$

where V is the volume of the gas. This gives

$$N = \Sigma N_s = \frac{2\pi V (2m)^{\frac{3}{2}}}{(2\pi h)^3} \int_0^\infty \frac{E_s^{\frac{1}{2}} dE_s}{e^{\alpha + E_s/kT} + 1}$$

and

$$E = \Sigma E_s N_s = \frac{2\pi V (2m)^{\frac{3}{2}}}{(2\pi h)^3} \int_0^\infty \frac{E_s^{\frac{3}{2}} dE_s}{e^{\alpha + E_s/kT} + 1}.$$

By eliminating α from these two equations and using the formula $PV = \frac{2}{3}E$, where P is the pressure, which holds for any statistical mechanics, the equation of state may be obtained.

The saturation phenomenon of the Einstein-Bose theory does not occur in the present theory. The specific heat can easily be shown to tend steadily to zero as $T \to 0$, instead of first increasing until the saturation point is reached and then decreasing, as in the Einstein-Bose theory.

§ 5. *Theory of Arbitrary Perturbations.*

In this section we shall consider the problem of an atomic system subjected to a perturbation from outside (*e.g.*, an incident electromagnetic field) which can vary with the time in an arbitrary manner. Let the wave equation for the undisturbed system be

$$(H - W)\psi = 0, \tag{20}$$

where H is a function of the p's and q's only. Its general solution is of the form

$$\psi = \Sigma_n c_n \psi_n, \tag{21}$$

where the c_n's are constants. We shall suppose the ψ_n's to be chosen so that one is associated with each stationary state of the atom, and to be multiplied by the proper constants to make the matrices that represent real quantities Hermitic.

Now suppose a perturbation to be applied, beginning at the time $t = 0$. The wave equation for the disturbed system will be of the form

$$(H - W + A)\psi = 0, \tag{22}$$

where A is a function of the p's, q's and t, and is real. It will be shown that we can obtain a solution of this equation of the form

$$\psi = \Sigma_n a_n \psi_n, \tag{23}$$

where the a_n's are functions of t only, which may have the arbitrary values c_n at the time $t = 0$. We shall consider the general solution (21) of equation (20) to represent an assembly of the undisturbed atoms in which $|c_n|^2$ is the number of atoms in the nth state, and shall assume that (23) represents in the same way an assembly of the disturbed atoms, $|a_n(t)|^2$ being the number in the nth state at any time t. We take $|a_n|^2$ instead of any other function of a_n because, as will be shown later, this makes the total number of atoms remain constant.

The condition that ψ defined by equation (23) shall satisfy equation (22) is

$$\begin{aligned} 0 &= \Sigma_n (H - W + A)\, a_n \psi_n \\ &= \Sigma_n a_n (H - W + A)\, \psi_n - ih\Sigma_n \dot{a}_n \psi_n, \end{aligned} \tag{24}$$

since H and A commute with a_n,† while $Wa_n - a_n W = ih\dot{a}_n$ identically. Suppose $A\psi_n$ to be expanded in the form

$$A\psi_n = \Sigma_m A_{mn} \psi_m,$$

where the coefficients A_{mn} are functions of t only, and satisfy $A_{mn}{}^* = A_{nm}$, where the * denotes the conjugate imaginary. Equation (24) now becomes, since $(H - W)\psi_n = 0$,

$$\Sigma_{mn} a_n A_{mn} \psi_m - ih\Sigma_m \dot{a}_m \psi_m = 0.$$

Taking out the coefficient of ψ_m, we find

$$ih\dot{a}_m = \Sigma_n a_n A_{mn}, \tag{25}$$

which is a simple differential equation showing how the a_m's vary with the time.

Taking conjugate imaginaries, we find

$$-ih\dot{a}_m{}^* = \Sigma_n a_n{}^* A_{mn}{}^* = \Sigma_n a_n{}^* A_{nm}.$$

Hence, if $N_m = a_m a_m{}^*$ is the number of atoms in the mth state, we have

$$\begin{aligned} ih\dot{N}_m &= ih(\dot{a}_m a_m{}^* + \dot{a}_m{}^* a_m) \\ &= \Sigma_n (a_n A_{mn} a_m{}^* - a_n{}^* A_{nm} a_m). \end{aligned}$$

† The statement a commutes with b means $ab = ba$ identically.

This gives

$$ih\Sigma_m \dot{\mathrm{N}}_m = \Sigma_{nm} (a_m{}^* \mathrm{A}_{mn} a_n - a_n{}^* \mathrm{A}_{nm} a_m) = 0,$$

as required.

If the perturbation consists of incident electromagnetic radiation moving in the direction of the x-axis and plane polarised with its electric vector in the direction of the y-axis, the perturbing term A in the Hamiltonian is, with neglect of relativity mechanics, $\kappa/c \,.\, \dot{\eta}$,† where η is the total polarisation in the direction of the y-axis and O, κ, O, O are the components of the potential of the incident radiation. We can expand $\eta\psi_n$ and $\dot{\eta}\psi_n$ in the form

$$\eta\psi_n = \Sigma_m \eta_{mn} e^{i(\mathrm{W}_m - \mathrm{W}_n)t/h}\, \psi_m,$$
$$\dot{\eta}\psi_n = \Sigma_m \dot{\eta}_{mn} e^{i(\mathrm{W}_m - \mathrm{W}_n)t/h}\, \psi_m,$$

where the η_{mn}'s and $\dot{\eta}_{mn}$'s are constants, and $\dot{\eta}_{mn} = i\,(\mathrm{W}_m - \mathrm{W}_n)/h \,.\, \eta_{mn}$. Our previous A_{mn} is now $\kappa/c \,.\, \dot{\eta}_{mn} e^{i(\mathrm{W}_m - \mathrm{W}_n)t/h}$, and equation (25) becomes

$$ihc\dot{a}_m = \Sigma_n a_n \kappa \dot{\eta}_{mn} e^{i(\mathrm{W}_m - \mathrm{W}_n)t/h}. \tag{26}$$

We can integrate this equation to the first order in κ by replacing the a_n's on the right-hand side by their values c_n at the time $t = 0$. This gives

$$a_m = c_m + 1/ihc \,.\, \Sigma_n c_n \dot{\eta}_{mn} \int_0^t \kappa\,(s)\, e^{i(\mathrm{W}_m - \mathrm{W}_n)s/h}\, ds. \tag{27}$$

To obtain a second approximation, we write for the a_n's on the right-hand side of (26) their values given by (27). We thus find for the value of a_m at the time T,

$$\begin{aligned} a_m &= c_m + 1/ihc \,.\, \Sigma_n c_n \dot{\eta}_{mn} \int_0^{\mathrm{T}} \kappa\,(t)\, e^{i(\mathrm{W}_m - \mathrm{W}_n)t/h}\, dt \\ &\quad - 1/h^2c^2 \,.\, \Sigma_{nk} c_k \dot{\eta}_{nk} \dot{\eta}_{mn} \int_0^{\mathrm{T}} \kappa\,(t)\, e^{i(\mathrm{W}_m - \mathrm{W}_n)t/h}\, dt \int_0^t \kappa\,(s)\, e^{i(\mathrm{W}_n - \mathrm{W}_k)s/h}\, ds \qquad (28) \\ &= c_m + c_m{}' + c_m{}'', \end{aligned}$$

say, where $c_m{}'$ and $c_m{}''$ denote the first- and second-order terms respectively.

This gives for the number of atoms in the state m at the time T

$$\mathrm{N}_m = a_m a_m{}^* = c_m c_m{}^* + c_m{}' c_m{}^* + c_m c_m{}'^* + c_m{}' c_m{}'^* + c_m{}'' c_m{}^* + c_m c_m{}''^*.$$

If we wish to obtain effects that are independent of the initial phases of the atoms, we must substitute $c_m \exp.\, i\gamma_m$ for c_m and average over all values of γ_m

† We have neglected a term involving κ^2. This approximation is legitimate, even though we later evaluate the number of transitions that occur in a time T to the order κ^2, provided T is large compared with the periods of the atom.

from 0 to 2π. This makes the first-order terms in N_m, namely, $c_m'c_m^*$ and $c_m c_m'^*$, vanish, while the second-order terms give

$$1/h^2c^2 \,.\, \Sigma_n c_n c_n^* \dot{\eta}_{mn} \dot{\eta}_{mn}^* \int_0^T \kappa(t)\, e^{i(W_m - W_n)t/h}\, dt \,.\, \int_0^T \kappa(t)\, e^{-i(W_m - W_n)t/h}\, dt$$

$$-1/h^2c^2 \,.\, \Sigma_n c_m c_m^* \dot{\eta}_{nm} \dot{\eta}_{mn} \int_0^T \kappa(t)\, e^{i(W_m - W_n)t/h}\, dt \int_0^t \kappa(s)\, e^{i(W_n - W_m)s/h}\, ds$$

$$-1/h^2c^2 \,.\, \Sigma_n c_m c_m^* \dot{\eta}_{nm}^* \dot{\eta}_{mn}^* \int_0^T \kappa(t)\, e^{-i(W_m - W_n)t/h}\, dt \int_0^t \kappa(s)\, e^{i(W_m - W_n)s/h}\, ds,$$

which reduces to

$$1/h^2c^2 \,.\, \Sigma_n\{|c_n|^2 - |c_m|^2\} |\dot{\eta}_{nm}|^2 \left| \int_0^T \kappa(t)\, e^{i(W_m - W_n)t/h}\, dt \right|^2. \qquad (29)$$

This gives ΔN_m, the increase in the number of atoms in the state m from the time $t = 0$ to the time $t = T$. The term in the summation that has the suffix n may be regarded as due to transitions between the state m and the state n.

If we resolve the radiation from the time $t = 0$ to the time $t = T$ into its harmonic components, we find for the intensity of frequency ν per unit frequency range the value

$$I_\nu = 2\pi\nu^2c^{-1} \left| \int_0^T \kappa(t)\, e^{2\pi i\nu t}\, dt \right|^2.$$

Hence the term in expression (29) for ΔN_m due to transitions between state m and state n may be written

$$1/2\pi h^2\nu^2 c \,.\, \{|c_n|^2 - |c_m|^2\} |\dot{\eta}_{nm}|^2 I_\nu,$$

where

$$2\pi\nu = (W_m - W_n)/h,$$

or

$$2\pi/h^2c \,.\, \{|c_n|^2 - |c_m|^2\} |\eta_{nm}|^2 I_\nu.$$

If one averages over all directions and states of polarisation of the incident radiation, this becomes

$$2\pi/3h^2c \,.\, \{|c_n|^2 - |c_m|^2\} |P_{nm}|^2 I_\nu,$$

where

$$|P_{nm}|^2 = |\xi_{nm}|^2 + |\eta_{nm}|^2 + |\zeta_{nm}|^2,$$

ξ, η and ζ being the three components of total polarisation. Thus one can say that the radiation has caused $2\pi/3h^2c \,.\, |c_n|^2 |P_{nm}|^2 I_\nu$ transitions from state n to state m, and $2\pi/3h^2c \,.\, |c_m|^2 |P_{nm}|^2 I_\nu$ transitions from state m to state n, the probability coefficient for either process being

$$B_{n \to m} = B_{m \to n} = 2\pi/3h^2c \,.\, |P_{nm}|^2,$$

in agreement with the ordinary Einstein theory.

The present theory thus accounts for the absorption and stimulated emission of radiation, and shows that the elements of the matrices representing the total polarisation determine the transition probabilities. One cannot take spontaneous emission into account without a more elaborate theory involving the positions of the various atoms and the interference of their individual emissions, as the effects will depend upon whether the atoms are distributed at random, or arranged in a crystal lattice, or all confined in a volume small compared with a wave-length. The last alternative mentioned, which is of no practical interest, appears to be the simplest theoretically.

It should be observed that we get the simple Einstein results only because we have averaged over all initial phases of the atoms. The following argument shows, however, that the initial phases are of real physical importance, and that in consequence the Einstein coefficients are inadequate to describe the phenomena except in special cases. If initially all the atoms are in the normal state, then it is easily seen that the expression (29) for ΔN_m holds without the averaging process, so that in this case the Einstein coefficients are adequate. If we now consider the case when some of the atoms are initially in an excited state, we may suppose that they were brought into this state by radiation incident on the atoms before the time $t = 0$. The effect of the subsequent incident radiation must then depend on its phase relationships with the earlier incident radiation, since a correct way of treating the problem would be to resolve both incident radiations into a single Fourier integral. If we do not wish the earlier radiation to appear explicitly in the calculation, we must suppose that it impresses certain phases on the atoms it excites, and that these phases are important for determining the effect of the subsequent radiation. It would thus not be permissible to average over these phases, but one would have to work directly from equation (28).

13

ON THE ELECTRON THEORY OF METALS

Arnold Sommerfeld

This article has been translated expressly for this Benchmark volume by R. Bruce Lindsay, Brown University, from Naturwissenschaften ***15:****825–832 (1927), by permission of the publisher, Springer-Verlag, Heidelberg. Figure 1 has been reproduced from the original article.*

I. THE ELECTRON GAS AND ITS STATISTICS

An interpretation of the electric current in conductors as due to the convection of electrified particles, in connection with the theories of electrolysis, was strongly suggested even before a genuine theory of electrons existed. This was presented by Giese, Schuster, and others, and then further developed by Riecke and Drude in particular. Its principal success was the theoretical derivation of the Wiedemann-Franz law by Drude and an at least qualitative explanation of thermoelectric phenomena. But the more precise statistical development that H. A. Lorentz[1] gave of Drude's theory led to a value of the numerical constant in the Wiedemann-Franz law too small by about 30%. It also appeared to demand, in the case of the Thomson effect, for example, rather arbitrary assumptions concerning the temperature dependence of the electron density in the metal. On the fundamental problem of contact potential and the Volta potential series, the theory proved helpless. There arose the additional difficulty that in the application of the classical equipartition of energy, the electrons should make an appreciable contribution to the specific heats of metals. This, however, is in contradiction to observation. Consequently, during the past twenty years the idea of the "electron gas" in a metal increasingly has fallen into discredit.

In the meantime, these difficulties are closely associated with the classical (Boltzmann) form of statistics. The latest development in quantum theory has led to countersuggestions against this form of statistics. This was done first by Bosé and Einstein,[2] who introduced a new method of counting the cases of equal probability to be employed in statistics. [*Editor's Note:* See Paper 9 in this volume.] While in the classical statistics the state to which the individual arbitrarily chosen corpuscle (molecule or electron) is to be assigned is determined probability-wise by dice theory or lot, in the new statistics the number of corpuscles to be assigned to the arbitrarily chosen state is to be determined by probability considerations. Thus in the classical statistics, the states are treated as equally probable, whereas in the new statistics, their occupation numbers are so treated—for example, n_0 for the state of energy zero, n_1 for the next higher energy state, n_k for the kth quantum state. Another, though perhaps less profound difference lies in the fact that in the classical statistics in the limit one can go to arbitrarily small state domains; in the new statistics the size of the "elementary cell" is given by quantum theory through Planck's constant *h*.

Einstein and later Schrödinger have pointed out that the new statistics corresponds to the standpoint of wave mechanics in accordance with which every quantum state is replaced by an eigenfunction of the space in question.

Statistical questions took a new turn when the young Italian physicist E. Fermi[3] combined them with the Pauli exclusion principle. [*Editor's Note:* See Paper 11 in this volume.] This principle (called jokingly the "housing office" of the electron) says (and first for the single atom) that every completely defined quantum state can be occupied by one electron at most. Pauli could show that the structure of complex spectra is governed by this principle, and more than that, the theory of the periodic system of the elements finds its justification in the principle: The period numbers 2, 8, 18, 32 are just those electron numbers that are permitted as maxima in the *K, L, M,* and *N* shells, without having two electrons having the same quantum number. [*Editor's Note:* See Paper 10 in this volume.] The circumstance that the structure of atomic spectra repeats itself in the band spectra of molecules (Mecke, Mulliken) indicates that the Pauli principle also holds for the totality of the electrons joined in a molecule. Fermi has now taken the bold step of extending the Pauli principle to the totality of the molecules of a gas. Just as the electrons in an atom can be told apart and distribute themselves among the available quantum states, so the Fermi statistics demands that all of the molecules of a gas adjust themselves among the possible quantum states of both the kinetic and potential energies. How that really happens remains a mystery. However, the consequences of this postulate must be examined with care.

Fermi at first showed that there follows from his statistics a specific law for the "degeneracy" of a monatomic ideal gas. This law, remarkably enough, works in the opposite direction from the degeneracy following from the Einstein-Bosé statistics: Whereas according to the Einstein result the pressure of the gas at low temperatures should be less than that given by the complete gas equation, according to Fermi law it should be higher. Moreover, the Fermi statistics involves a zero-point energy and a zero-point pressure. We can easily see the reason for this. Even at absolute zero, the Pauli exclusion principle holds, according to which not more than one atom can have the energy zero (ground state of the kinetic energy). Thus the vast majority of atoms are forced to reside in their higher states, accordingly in states corresponding not only to energy of motion but also to momentum. Hence the gas even at absolute zero and in the neighborhood of the same must possess energy E_0 and pressure p_0. Fig. 1 indicates by the dotted line the energy of complete gas as a function of the temperature. The solid line, on the other hand, gives the similar dependence of the degenerate gas according to Fermi. (According to Einstein, on the other hand, the relations in the neighborhood of absolute zero are complicated by the appearance of a two-phase equilibrium between the atoms with zero energy and those with energy greater than zero.)

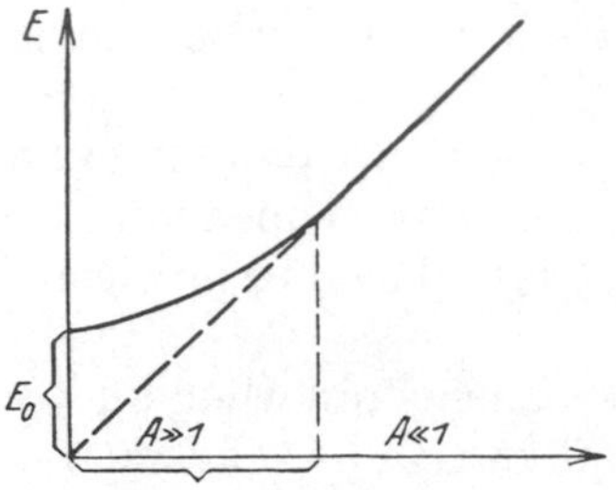

Figure 1. The energy of a monatomic gas according to Fermi. At high temperatures $A << 1$, linear dependence of E on T, as in classical statistics. At lower temperatures, $A >> 1$, characteristic deviation from classical result and zero point energy. For the meaning of A, see equations 3, 4, and 5 in the text.

The Fermi statistics has recently received strong support through an investigation by Pauli[4] concerning magnetism. The question here is about the weak residual and temperature-independent paramagnetism of metals, especially the alkalis. Curie's law, which has been found to hold for the gases oxygen and nitrous oxide, as well as for the paramagnetic salts, and has been beautifully established theoretically by Langevin, demands, as is well known, a systematic increase in magnetization with decreasing temperature. Atomic physics leads one to expect that the magnetic moment of atoms producing paramagnetism should not exceed a certain quantity, the so-called Bohr magneton: Because of this, the observed temperature-independent and abnormally weak paramagnetism of metals has been a complete puzzle up to now.

The solution of this problem starts with the old representation of an electron gas but handles the distribution over states not according to classical statistics but according to the Fermi statistics. Immediately it becomes clear that the electron gas, even at normal temperature, is completely degenerate. The reason for this is the extraordinarily small mass m of the electron. As part of the degeneracy criterion for an arbitrary gas of molecular weight m the characteristic quantity $2\pi mkT$ (k = Boltzmann's gas constant, and T is the absolute temperature) enters decisively. If accordingly helium gas ($m = 4$), as calculated by Fermi, shows a marked degeneracy at $T = 5°$, the electron gas ($m = 1/1800$) must already exhibit the same degree of degeneracy at $T = (1800)\,(4)\,(5)° = 36{,}000°$. At room temperature $T = 300°$, this electron gas is completely degenerate. Of course, the appearance of degeneracy depends not only on the temperature but also on the density in the sense that large density, as well as low temperature, favors degeneracy. We can estimate the density of the electron gas by assuming with Pauli that to a first approximation in every metal atom, especially every alkali atom, the valence electron is free. The number of electrons per unit volume will then be equal to the number of atoms per unit volume. In this way, Pauli calculates the paramagnetism of the alkalis in excellent agreement with experiment. And in fact, the magnetic moment at the basis of the observed paramagnetism arises from the electrons and not from the metal atoms. For the electron is itself a magnet (hypothesis of the spinning eletron devised by Uhlenbeck and Goudsmit, which is confirmed by the Pauli hypothesis). The temperature independence of the paramagnetism of metals is explained by the fact that, as is clear from Fig. 1, in the degeneracy region all properties are temperature independent to a first approximation. The relatively small magnitude of paramagnetism we can understand from the fact that simultaneously all the lower energy states are completely occupied, which results in a mutual compensation of positive and negative magnetism and ony a small residual total magnetism.

By the use of this Pauli explanation, we can at once do away with the above-mentioned difficulty concerning the specific heat of the free electrons. Since the electron gas is degenerate, its internal energy is to a first approximation independent of the temperature, and its specific heat is accordingly equal to zero. If we introduce a second approximation and thus consider the weak rise in the energy curve in the degeneracy region, we get a specific heat that gradually increases with the temperature but at room temperature and calculated per mole of the metal amounts to merely $R/100$, whereas on classical statistics, it would be effectively equal to R, where R is the universal gas constant. The value R is, of course, quite inconsistent with the observed specific heats, but the contribution $R/100$ is quite permissible from the standpoint of experiment.

II. ELECTRICAL AND THERMAL CONDUCTIVITY—THE WIEDEMANN-FRANZ LAW

Now that the above-mentioned difficulty has been removed, the question arises whether this conception of an electron gas can be applied successfully to electrical and thermal conductivity. In a lecture I gave during the current summer semester, I tried to tackle this problem and succeeded in achieving noteworthy results, which are briefly summarized in the following.

I introduce first a few words concerning the mathematical method, which is based on that of Lorentz. Let ξ, η, ζ be the rectangular components of the velocity v of an electron and $f(\xi, \eta, \zeta)$ the velocity distribution in the sense that $f(\xi, \eta, \zeta)\, d\omega$ is the number of electrons in unit volume having velocities in the velocity range $d\omega$ between ξ and $\xi + d\xi$, η and $\eta + d\eta$, and ζ and $\zeta + d\zeta$. Then under the influence of an electrical field F in the x direction (or under the influence of a temperature drop in the thermal case), a density change takes place in the x direction, as indicated here:

$$f(\xi, \eta, \zeta) = f_0 + \xi\chi; \tag{1}$$

Here f_0 denotes the normal velocity distribution without the presence of the field or the temperature drop. It is spherically symmetrical—that is, depends only on $v = \sqrt{\xi^2+\eta^2+\zeta^2}$. The same holds to a first approximation for the perturbation function χ . From a consideration of the collisions of the free electrons with the metal atoms (considered to be at rest), we get

$$\chi = -\frac{l}{v}\left(\frac{eF}{m}\frac{1}{v}\frac{\partial f_0}{\partial v} + \frac{\partial f_0}{\partial x}\right) \tag{2}$$

Here l is the free path of the electrons, accordingly a quantity inversely proportional to the effective cross-section of the metal atoms for collision. Lorentz worked with the Maxwell distribution

$$f_0 = A e^{-\frac{\varepsilon}{kT}}, \qquad \varepsilon = \frac{m}{2} v^2 \tag{3}$$

We, however, must use for f_0 the Fermi distribution

$$f_0 = \frac{1}{\frac{1}{A} e^{\frac{\varepsilon}{kT}} + 1}. \tag{4}$$

(The corresponding Einstein-Bosé law in which the +1 in the denominator in equation 4 is replaced by −1 has no application to the degenerate electron gas, since according to the Einstein-Bosé law, degeneracy is equivalent to the falling out of the electrons.)

For $A << 1$, equation 4 goes over into the form of equation 3, since in this case the 1 in the denominator can be neglected. We, however, have always to deal with the case $A >> 1$ (degeneracy). We shall pay attention to the other limiting case $A << 1$ only in order to have at hand the results of classical statistics for comparison purposes.

The thermodynamic significance of A is given by the formula

$$\log A = \frac{\mu}{kT} \tag{5}$$

as Pauli has recently shown and Heitler before him. Here μ in Gibbs' notation signifies the thermodynamic potential for unit mass of the electron.

The electric current J and the energy current W, corresponding to the electric charge and the kinetic energy of the electrons transported through unit area per unit time in the x direction, are then expressed as follows in terms of the distribution function f:

$$J = e\int \xi f \, d\Omega, \qquad W = \frac{m}{2}\int \xi v^2 f \, d\Omega. \tag{6}$$

We calculate the number of electrons per unit volume n by the formula

$$n = \int f \, d\Omega = \int f_0 \, d\Omega. \tag{7}$$

In equations 6 and 7 we have written $d\Omega$ in place of the former $d\omega$ in order to signify that $d\Omega$ is the phase element in so-called momentum space (instead of velocity space). Given in units of h^3 we have

$$d\Omega = 4\pi\left(\frac{m}{h}\right)^3 v^2 dv\,. \tag{8}$$

A. Electric Conduction

We obtain the case of pure electric conduction if we set

$$\frac{\partial f_0}{\partial x} = 0\,, \quad \text{thus} \quad \frac{\partial A}{\partial x} = 0\,, \quad \frac{\partial T}{\partial x} = 0$$

in equation 2 and also treat the field intensity F as independent of x. The electric conductivity σ—that is, the ratio J/F—is obtained from equations 6 and 7 for the two limiting cases $A << 1$ and $A >> 1$, as follows:

$$A \ll 1 \qquad\qquad\qquad A \gg 1$$

$$\sigma = \frac{4}{3}\frac{e^2 l\, n}{\sqrt{2\pi m k T}}\,, \qquad \sigma = \frac{4\pi}{3}\frac{e^2 l}{h}\left(\frac{3n}{4\pi}\right)^{\frac{2}{3}}$$

The first limiting case (no degeneracy, ideal gas) corresponds to the classical statistics. The corresponding value of σ agrees with that of Lorentz (and essentially with that of Drude). The second limiting case (complete degeneracy) yields a value of σ that essentially deviates from the former. The appearance of h in the latter emphasizes its quantum theory origin; the dependence on the electron density n is different; the explicit dependence on T has vanished. The last fact must at first surprise us, for experiment shows that $1/\sigma$ (the specific resistance) increases linearly with T, in fact for several metals is approximately proportional to T. But this proportionality is not an exact law. Also because we have neglected the temperature motion of the metal atoms, the free path l is certainly still temperature dependent. We accordingly assume (as W. Wien did in 1913) that the presence of l in our formula takes care of the temperature dependence of the resistance. In a well-known theoretical investigation by J. Frenkel,[5] this temperature dependence is explicitly expressed. However, we do not see any particular preference of this over an implicit temperature dependence expressed through l.

B. Thermal Conductivity

We get the case of pure heat conduction when we set $J = 0$ in equation 6 and then from that calculate the value of F produced by the differences in concentration of the electrons. (Because of equations 1 and 2, F occurs in J and W.) We then insert the result in equation 6 for W. The thermal conductivity κ is then determined according to its definition

$$W = -\kappa\frac{\partial T}{\partial x}$$

The results for the two limiting cases are as follows:

$A \ll 1$

$$\varkappa = \frac{8}{3} \frac{l n k^2 T}{\sqrt{2 \pi m k T}}$$

$A \gg 1$

$\varkappa = 0$ . . . . first approximation

$$\varkappa = \frac{4\pi^3}{9} \frac{l k^2 T}{h} \left(\frac{3n}{4\pi}\right)^{\frac{2}{3}}$$ . . . . . . second approximation.

The value of κ in the classical case $A << 1$ naturally agrees with that due to Lorentz. In the oppositve limit, the value $\kappa = 0$ appears at first to be absurd, but one can understand it at once on the basis of our Fig. 1. If the electron gas is completely degenerate, then to a first approximation the energy is independent of the temperature. Hence a fall in temperature is associated with no energy current to a first approximation. We must proceed to a second approximation in order to secure a heat flow and a coefficient of thermal conductivity. From the standpoint of calculation, this means that we should no longer neglect $(\log A)^{-2}$ with respect to unity. In this way we find the finite, nonvanishing value of κ given above.

C. Wiedemann-Franz Law

The ratio κ/σ (from sections A and B above) does not contain the unknown quantity l or n and $n^{2/3}$. We have

$A \ll 1$	$A \gg 1$
$\frac{\varkappa}{\sigma} = 2\left(\frac{k}{e}\right)^2 T$	$\frac{\varkappa}{\sigma} = \frac{\pi^2}{3}\left(\frac{k}{e}\right)^2 T.$
[From Drude: $3\left(\frac{k}{e}\right)^2 T$]	

In both limiting cases, we confirm the statement of the Wiedemann-Franz law: κ/σ is a universal quantity proportional to the absolute temperature. The difference lies only in the value of the numerical factor, which is 2 for the Lorentz theory and $\pi^2/3$ for the degenerate gas. As has already been mentioned in the introduction, the Lorentz factor 2 does not agree as well with the very precise measurements of Jäger and Diesselhorst as factor 3 from the primitive Drude theory. Our factor $\pi^2/3$, however, agrees better with observation than that of Drude. From the observations for the twelve metals—A1, Cu, Ag, Au, Ni, Zn, Cd, Pb, Sn, Pt, Pd, Fe—the mean value for $T = 293 + 18$ comes out to be

$$\frac{\varkappa}{\sigma} \cdot 10^{-10} = 7{,}11 .$$

The theoretical value calculated with the use of $k = 1.37 \times 10^{-16}$, $e = 1.59 \times 10^{-20}$, and for the same value of the temperature comes out as follows:

	Present theory	Drude	Lorentz
$\frac{\varkappa}{\sigma} \cdot 10^{-10} =$	7.1	6.3	4.2

It is a happy circumstance that the fundamental significance of the Wiedemann-Franz law is once more thoroughly established in the new statistics.

III. CONTACT POTENTIALS—THE VOLTAIC SERIES[6]

Here we deal with a simple electrostatic case. Two metals 1 and 2, equally annealed and carrying no current, are placed in contact. As was mentioned in the beginning of section IIB, from the condition $J = 0$, we can calculate the value of the electrical field intensity F, which must prevail at every place in such a combination of conductors, as soon as the difference in concentration of the electrons has achieved equilibrium. Naturally F is zero within 1 and 2 and discontinuously variable in the transition layer between 1 and 2. We call

$$\Phi_i = -\int_1^2 F\,dx \tag{9}$$

the inner contact potential or the inner voltaic difference. We also treat the electrons as free here, neglecting the action of undefined molecular forces (Helmholtz). The result is again quite different in our two limiting cases.

$A \ll 1$	$A \gg 1$
$\Phi_i = \frac{k}{e} T \log \frac{n_1}{n_2}$	$\Phi_i = \frac{h^2}{2\,me}\left\{\left(\frac{3\,n_1}{4\,\pi}\right)^{\frac{2}{3}} - \left(\frac{3\,n_2}{4\,\pi}\right)^{\frac{2}{3}}\right\}.$

While the classical formula ($A << 1$, Lorentz), on the basis of the contact difference in potential of several volts, leads to wholly absurd values for the ratio n_1/n_2 of electron densities, we calculate from the new formula the correct order of magnitude of Φ_i if we identify the electron number n with the number of metal atoms. We test this, for example, for the difference K→Ag, in which n is associated with K and n_2 with Ag.

In general we have

$$n = \frac{d}{M} L, \tag{10}$$

in which d is the density, M the atomic weight of the metal in question, and L is the Loschmidt number—that is, the number of atoms in M grams. Accordingly, M/L is equal to the mass of an atom. The right-hand side of the preceding equation is equal to the number of atoms per unit volume and hence, by our assumption, equal to the number of electrons per unit volume. For the numerical calculation of Φ_i we write (paying due attention to the negative sign of the electron charge e)

$$\Phi_i = \frac{h^2}{2\,m\,|\,e\,|}\left(\frac{3\,L}{4\,\pi}\right)^{\frac{2}{3}}\left\{\left(\frac{d}{M}\right)_{Ag}^{\frac{2}{3}} - \left(\frac{d}{M}\right)_{K}^{\frac{2}{3}}\right\} \tag{11}$$

and find

$$\Phi_i = 41\left\{\left(\frac{10.5}{107.9}\right)^{\frac{2}{3}} - \left(\frac{0,86}{39.1}\right)^{\frac{2}{3}}\right\}\ \text{Volt} = 5.7\ \text{Volt},$$

This agrees in order of magnitude with the observed value but not in sign.

We get corresponding results for the Voltaic series. As equation 11 immediately indicates, according to the theory presented here, the order of the metals in the series should agree with the order of the values of d/M. This is, in general, the case, as the following table indicates.

Voltaic Series	+Rb	K	Na	Al	Zn	Pb	Sn		
$\frac{d}{M} \times 10^3$	18	22	42	(100	108)	54	61		

Voltaic Series	Sb	Bi	Fe	Cu	Ag	Au	Dt	Pd	C–
$\frac{d}{M} \times 10^3$	(55	47)	(140	140)	97	98	110	112	190

The strong discrepancies in the numerical order are put in parentheses. However, it should be noted that according to our formula (11), the alkalis stand at the negative end of the series and carbon at the positive end.

Our formula for Φ_i, in contrast to the classical one, permits us to speak of the contact difference of potential with respect to vacuum by setting $n_2 = 0$ in equation 11. Hence the energy gain by emission of an electron with vacuum becomes

$$W_i = e\,\Phi_{i\,\mathrm{vac}} = \frac{h^2}{2\,m}\left(\frac{3\,n}{4\,\pi}\right)^{\frac{2}{3}} = \frac{h^2}{2\,m}\left(\frac{3\,L}{4\,\pi}\right)^{\frac{2}{3}}\left(\frac{M}{d}\right)^{\frac{2}{3}}. \tag{12}$$

This energy gain is overridden by an energy loss, which at ordinary temperatures forces the electrons to remain in the metal. We designate this part of the energy balance as W_a and call it the external work of emission (later called the external work function). Since our considerations are limited to what goes on in the interior of the metal, they do not permit us to calculate the work function W_a. Its origin must be sought in the forces of attraction exerted on the emerging electrons by the positively charged metal atoms left behind, the so-called image force. On the other hand, the internal work function W_i owes its origin to the kinetic pressure of the election gas (zero-point pressure), which is greater in proportion to the electron density n. This pressure assists the emission of the electron from the metal and provides the positive work (equation 12).

The total work needed to get the electron out of the metal is $W = W_a - W_i$ (total work function). It is noteworthy that this measured work as given by the Richardson (thermionic emission) effect is not provided solely by the attractive forces of the metal, corresponding to what is here given as W_a, but there has to be subtracted from this a substantial amount arising from the zero-point pressure, corresponding to W_i above. From photoelectric observations $W_a \sim 2W_i$.

The same considerations apply to the measured potential differences in the Volta effect. Here we must also consider W_a and W_i and have the resulting formula

$$e\,\Phi_{1\rightarrow 2} = (W_a - W_i)_2 - (W_a - W_i)_1\,.$$

IV. THOMSON AND PELTIER EFFECTS, THERMOELECTRICITY

The theoretical treatment of these phenomena shares with the Wiedemann-Franz law the special nature that it is independent of the free path.

A. Thomson Effect

We assume that en electric current J and a heat flow W [equation 6] pass simultaneously through a metal and calculate the energy remaining per unit volume per unit of time.

$$Q = JF - \frac{\partial W}{\partial x}, \tag{13}$$

Here the first part on the right represents the electric work performed on the electrons that are moved by the field F and the second part the energy conveyed by the heat flow. The energy Q can be divided into the three parts Q_1, Q_2, Q_3, which are respectively proportional to J^2, J^1, and J^0. We have

$$Q = Q_1 + Q_2 + Q_3$$

$$Q_1 = \frac{J^2}{\sigma}, \quad Q_2 = -\mu J \frac{\partial T}{\partial x}, \quad Q_3 = \frac{\partial}{\partial x}\left(\varkappa \frac{\partial T}{\partial x}\right).$$

Q_1 is the Joule heat (equal to the specific resistance times the square of the current). Q_3 is the heat conveyed into the region in question by heat conduction. We are interested in the reciprocal energy contribution Q_2, in particular the quantity μ, which is called the Thomson coefficient (and also the specific heat of electricity). By calculating equation 13 with the use of equations 1, 2, 4, and 6 and leaving out the portion Q_1 and Q_3, we have

$A \ll 1$

$$\mu = \frac{3}{2}\frac{k}{e}\left\{1 - \frac{2}{3} T \frac{d \log n}{dT}\right\};$$

$A \gg 1$

$\mu = 0$ . . . . to first approximation

$$\mu = \frac{2\pi^2}{3}\frac{mk^2}{eh^2}\left(\frac{4\pi}{3n}\right)^{\frac{2}{3}} T\left\{1 - \frac{2}{3} T \frac{d \log n}{dT}\right\}$$

. . . . to second approximation.

The value of μ in case $A<<1$ is, as must necessarily be the case, identical with that derived by Lorentz on the basis of classical statistics. It is well known that this value is *much too great* unless one reduces it by making an arbitrary assumption about the factor $\{\ \}$. Here the Fermi statistics is also of help. To a first approximation—that is, with the neglect of $(\log A)^{-2}$ compared with unity—it yields no Thomson heat. In the second approximation, it gives a value for μ much smaller than the classical one. Thus we have

$$\frac{\mu}{\mu_{Kl}} = \frac{4\pi^2}{9}\frac{mk}{h^2}\left(\frac{4\pi}{3n}\right)^{\frac{2}{3}} T \sim \frac{1}{200}.$$

In the calculation of the value 1/200, T has been taken as 300, and n is the number of copper atoms in a cubic centimeter. The proportionality with T

demanded by the theory in question has been confirmed by the measurements[7] of Borelius and Gunneson very beautifully in part, particularly for the noble metals and at higher temperatures. Whether the deviation of the observed values from strict proportionality, which occurs at lower temperatures, is connected with our $\{\ \}$ factor, we shall not attempt to decide. Possibly this factor might play a role in the generally negative sign of the Thomson effect in lead and tin. If we neglect this factor (or treat it as unity), the general increase with T is given by our formula by

$$\frac{2\pi^2}{3}\frac{mk^2}{eh^2}\left(\frac{4\pi}{3n}\right)^{\frac{2}{3}}.$$

With numbers given by the Volta effect for $n = d \cdot L/M$ in microvolts and degrees Celsius, we have

	Cu	Ag	Au
Calculated	2×10^{-3}	3×10^{-3}	3×10^{-3}
Observed	7.5×10^{-3}	5.5×10^{-3}	4.5×10^{-3} .

The observed rate of increase is taken from the upper part of the curves of Borelius and Gunneson. The calculated and observed absolute values of the Thomson coefficients agree to the same extent as the gradient with temperature.

We therefore reach the conclusion that the order of magnitude of the Thomson effect and its temperature dependence is reproduced accurately by the new statistics. The abnormally high oberved values for arsenic, antimony, and bismuth, which always lie apreciably below the values given by the classical theory, now provide a problem.

B. Peltier Effect

The Peltier effect is closely connected with the Thomson effect. This is the heat production π per unit time at the place where two different metals are soldered together and carry the same unit current with the same temperature on both sides of the junction. Proceeding from the same equation 13 as before, we have

$$A \ll 1$$

$$\Pi = \frac{k}{e}T\log\frac{n_1}{n_2}$$

$$A \gg 1$$

$$\Pi = 0 \quad \ldots\ldots\ldots\ldots\ldots \quad \text{first approximation}$$

$$\Pi = \frac{2\pi^2}{3}\frac{mk^2T^2}{eh^2}\left[\left(\frac{4\pi}{3n_2}\right)^{\frac{2}{3}} - \left(\frac{4\pi}{3n_1}\right)^{\frac{2}{3}}\right] \quad \text{second approximation.}$$

With respect to numerical values, we can say the same about the Peltier effect as about the Thomson effect: The classical theory yields far too high a value. It gives the same formula as for the Volta effect, of the order of magnitude of several volts, if we fit the ratio n_2/n_1 to the observed values of the Volta effect. On the other hand, the new formula yields the correct order of magnitude of a few hundred microvolts. For example, in the case of copper joined to silver, we get from our theory in microvolts

$\pi = 420$ compared with the observed value of 576.

The temperature dependence demanded by our theory (proportionality with T^2) corresponds in general with the observed measurements.

C. Thermoelectric Force

The Peltier and Thomson effects yield the energy that a heat current produces when flowing across two junctions at different temperatures. We shall not consider the problem of the closed, current-carrying thermoelectric chain but rather the open one and calculate the potential difference. Let T' and T'' be the temperatures of the junction 1→2 and 2→1, respectively. The beginning P and the end Q of the thermoelectric chain may have the same temperature T and be made of the same material (that denoted by 1). Since we have assumed that the chain carries no current, we must proceed as in section III with the condition $J = 0$ and from that derive the value of F, only now for the condition of changing the temperature and not, as in the former case, the condition of changing electron numbers. We then use F to get the potential difference

$$\Phi = -\int_P^Q F\,dx.$$

We then obtain

$A \ll 1$

$$\Phi = \frac{k}{e}\int_{T'}^{T''} \log\frac{n_2}{n_1}\,dT,$$

$A \gg 1$

$\Phi = 0$ . . . . first approximation

$$\Phi = \frac{2\pi^2}{3}\frac{mk^2}{eh^2}\int_{T'}^{T''}\left\{\left(\frac{4\pi}{3n_1}\right)^{\frac{2}{3}} - \left(\frac{4\pi}{3n_2}\right)^{\frac{2}{3}}\right\} T\,dT \text{ second approximation.}$$

Between the Peltier heat π and the Thomson effect coefficient μ on the one hand and the thermal potential difference Φ on the other hand, there exist well-known thermodynamic relations, which are naturally satisfied by our new numerical values.

For the case in which the electron number n can be considered as independent of the temperature, we get

$A \ll 1$

$$\Phi = \frac{k}{e}\log\frac{n_2}{n_1}(T'' - T')$$

$A \gg 1$

$$\Phi = \frac{2\pi^2}{3}\frac{mk^2}{eh^2}\left\{\left(\frac{4\pi}{3n_1}\right)^{\frac{2}{3}} - \left(\frac{4\pi}{3n_2}\right)^{\frac{2}{3}}\right\}\left(T''^2 - T'^2\right).$$

According to the classical theory, in this case there should be simple proportionality to the temperature difference

$$\vartheta = T'' - T'$$

This is, however, contrary to observation. The new theory yields, rather, a quadratic term in addition to the linear one. Thus

$$T''^2 - T'^2 = (T'' - T')(T'' + T') = \vartheta(2T' + \vartheta),$$

so that the ratio of the quadratic to the linear term is equal to

$$\frac{1}{2T'} \sim \frac{1}{550} \text{ for } T' = 0°\text{C}.$$

This agrees very well with the average of observations.[8] Moreover, there is also good agreement with the ratio of the thermal potential difference Φ to the Volta potential difference, which we shall call Φ_V. If we calculate the thermal potential difference for $\Phi = 1°$ temperatue difference and for $T' = 273°$ and, as seems appropriate from an order of magnitude point of view, set Φ_V equal to our "internal Volta potential difference" Φ_i [equation 11], we get, if we express n as in section III,

$$\frac{\Phi}{\Phi_V} = \frac{16\pi^3}{9}\left(\frac{4\pi}{3L}\right)^{\frac{1}{3}} \frac{m^2k^2}{h^4L} 2T' \left(\frac{M_1}{d_1}\right)^{\frac{2}{3}} \left(\frac{M_2}{d_2}\right)^{\frac{2}{3}}$$

$$= 29 \cdot 10^{-9} \left(\frac{M_1 M_2}{d_1 d_2}\right)^{\frac{2}{3}}.$$

If, for example, we take M_1, M_2, d_1, d_2 for silver, we have

$$\frac{\Phi}{\Phi_V} = 6 \cdot 10^{-7}.$$

For two metals for which the Volta potential difference is about 2 volts, the thermal potential difference becomes about 1 microvolt—about the right order of magnitude.

On the other hand, from the corresponding classical formula we get

$$\frac{\Phi}{\Phi_V} = \frac{T'' - T'}{T} = \frac{1}{273},$$

for 1° temperature difference and at 0°C. This ratio is decidely too great to fit the observed value.

Moreover, we note that according to the new formula, the thermoelectric force for 1° temperature difference vanishes at absolute zero as the Nernst heat theorem demands, contrary to the result from classical statistics.

V. CONCLUSION

We have here considered only a small part of the domain of metallic conduction. But the general conclusion is doubtless that by means of the new statistics, the contradictions of the old theory are removed and the experimental results are correctly reproduced, in part quantitatively and in part qualitatively. Further questions concern the emission of electrons from hot metals (Richardson effect), the thermomagnetic and galvanomagnetic phenomena, especially the Hall effect, and possibly also ferromagnetism, which occurs only in conductors and hence must be affected by the statistics of free electrons. The author hopes to take up these matters shortly.

The foregoing considerations are characterized by the absence of any special hypothesis concerning the interaction between the electrons and the metal atoms. We have taken over the most primitive representations of the old theory and merely reworked them using the Fermi statistics in place of the classical statistics. Through the agreement with the vast amount of observational material, the Fermi statistics has obtained a satisfactory empirical basis.

The weak point of our considerations is the introduction of the free path as a geometrical quantity, given by the configuration of the metal lattice. In reality we know that the free path depends on the velocity of the electrons. Therefore it is not permissible to treat l as a constant as in equation 6. Rather a mean value must be formed over the velocity domain in question; this domain is indeed more extensive in the Fermi statistics than in the classical statistics. It is not self-evident that the various mean values of l arising in this way will all be the same and hence will cancel out in the Wiedemann-Franz law.

It is necessary in the refinement of the theory to introduce the free path in a more physical way, as in the sense of wave mechanics, by following the scattering of the de Broglie waves in the lattice of metal atoms and hence paying attention to the thermal agitation of this lattice.

The partial agreement of our results with experience leaves very little doubt that such refinements will confirm in all essential points our method of calculation based on the classical electron theory.

NOTES

1. Amsterdam Academy, January and March, 1905; see also *Theory of Electrons,* Teubner, 1909, pages 63-67 and 266-273.
2. Bose, Zs. F. Phys. *26,* 178, 1924; A. Einstein, Sitzungsberichte d. preuss. Akad. d. Wiss., 1924, No. 22 and 1925, Nos. 1 and 3.
3. "On the Quantization of the Ideal Monatomic Gas," Zs. f. Phys. *36,* 902, 1926; see also P. A. M. Dirac, Proc. Roy. Soc., *112,* 661, 1926.
4. "On Gas Degeneracy and Paramagnetism" Zs. F. Physik, *41,* 81, 1927.
5. Zs. f. Physik, *29,* 214, 1924.
6. In this part, I have profited greatly from the critical advice of my friend K. F. Herzfeld, for which I am very grateful.
7. *Annalen du Physik, 65,* 520, 1921.
8. Bädecker, Die elektrischen Erscheinangen in metallischen Lectern, Sammlung Wissenschaft, Braunschweig, 1911.

14

Reprinted from *Roy. Soc. (London) Proc.*, ser. A, **114**:243–265 (1927)

The Quantum Theory of the Emission and Absorption of Radiation.

By P. A. M. Dirac, St. John's College, Cambridge, and Institute for Theoretical Physics, Copenhagen.

(Communicated by N. Bohr, For. Mem. R.S.—Received February 2, 1927.)

§ 1. *Introduction and Summary.*

The new quantum theory, based on the assumption that the dynamical variables do not obey the commutative law of multiplication, has by now been developed sufficiently to form a fairly complete theory of dynamics. One can treat mathematically the problem of any dynamical system composed of a number of particles with instantaneous forces acting between them, provided it is describable by a Hamiltonian function, and one can interpret the mathematics physically by a quite definite general method. On the other hand, hardly anything has been done up to the present on quantum electrodynamics. The questions of the correct treatment of a system in which the forces are propagated with the velocity of light instead of instantaneously, of the production of an electromagnetic field by a moving electron, and of the reaction of this field on the electron have not yet been touched. In addition, there is a serious difficulty in making the theory satisfy all the requirements of the restricted principle of relativity, since a Hamiltonian function can no longer be used. This relativity question is, of course, connected with the previous ones, and it will be impossible to answer any one question completely without at the same time answering them all. However, it appears to be possible to build up a fairly satisfactory theory of the emission of radiation and of the reaction of the radiation field on the emitting system on the basis of a kinematics and dynamics which are not strictly relativistic. This is the main object of the present paper. The theory is non-relativistic only on account of the time being counted throughout as a c-number, instead of being treated symmetrically with the space co-ordinates. The relativity variation of mass with velocity is taken into account without difficulty.

The underlying ideas of the theory are very simple. Consider an atom interacting with a field of radiation, which we may suppose for definiteness to be confined in an enclosure so as to have only a discrete set of degrees of freedom. Resolving the radiation into its Fourier components, we can consider the energy and phase of each of the components to be dynamical variables describing the radiation field. Thus if E_r is the energy of a component labelled r and θ_r is the corresponding phase (defined as the time since the wave was in a standard phase), we can suppose each E_r and θ_r to form a pair of canonically conjugate variables. In the absence of any interaction between the field and the atom, the whole system of field plus atom will be describable by the Hamiltonian

$$H = \Sigma_r E_r + H_0 \qquad (1)$$

equal to the total energy, H_0 being the Hamiltonian for the atom alone, since the variables E_r, θ_r obviously satisfy their canonical equations of motion

$$\dot{E}_r = -\frac{\partial H}{\partial \theta_r} = 0, \quad \dot{\theta}_r = \frac{\partial H}{\partial E_r} = 1.$$

When there is interaction between the field and the atom, it could be taken into account on the classical theory by the addition of an interaction term to the Hamiltonian (1), which would be a function of the variables of the atom and of the variables E_r, θ_r that describe the field. This interaction term would give the effect of the radiation on the atom, and also the reaction of the atom on the radiation field.

In order that an analogous method may be used on the quantum theory, it is necessary to assume that the variables E_r, θ_r are q-numbers satisfying the standard quantum conditions $\theta_r E_r - E_r \theta_r = ih$, etc., where h is $(2\pi)^{-1}$ times the usual Planck's constant, like the other dynamical variables of the problem. This assumption immediately gives light-quantum properties to the radiation.* For if ν_r is the frequency of the component r, $2\pi\nu_r\theta_r$ is an angle variable, so that its canonical conjugate $E_r/2\pi\nu_r$ can only assume a discrete set of values differing by multiples of h, which means that E_r can change only by integral multiples of the quantum $(2\pi h)\,\nu_r$. If we now add an interaction term (taken over from the clasical theory) to the Hamiltonian (1), the problem can be solved according to the rules of quantum mechanics, and we would expect to obtain the correct results for the action of the radiation and the atom on one another. It will be shown that we actually get the correct laws for the emission and absorption of radiation, and the correct values for Einstein's A's and B's. In the author's previous theory,† where the energies and phases of the components of radiation were c-numbers, only the B's could be obtained, and the reaction of the atom on the radiation could not be taken into account.

It will also be shown that the Hamiltonian which describes the interaction of the atom and the electromagnetic waves can be made identical with the Hamiltonian for the problem of the interaction of the atom with an assembly of particles moving with the velocity of light and satisfying the Einstein-Bose statistics, by a suitable choice of the interaction energy for the particles. The number of particles having any specified direction of motion and energy, which can be used as a dynamical variable in the Hamiltonian for the particles, is equal to the number of quanta of energy in the corresponding wave in the Hamiltonian for the waves. There is thus a complete harmony between the wave and light-quantum descriptions of the interaction. We shall actually build up the theory from the light-quantum point of view, and show that the Hamiltonian transforms naturally into a form which resembles that for the waves.

The mathematical development of the theory has been made possible by the author's general transformation theory of the quantum matrices.‡ Owing to the fact that we count the time as a c-number, we are allowed to use the notion of the value of any dynamical variable at any instant of time. This value is

* Similar assumptions have been used by Born and Jordan [' Z. f. Physik,' vol. 34, p. 886 (1925)] for the purpose of taking over the classical formula for the emission of radiation by a dipole into the quantum theory, and by Born, Heisenberg and Jordan [' Z. f. Physik,' vol. 35, p. 606 (1925)] for calculating the energy fluctuations in a field of black-body radiation.

† ' Roy. Soc. Proc.,' A, vol. 112, p. 661, § 5 (1926). This is quoted later by, *loc. cit.*, I.

‡ ' Roy. Soc. Proc.,' A, vol. 113, p. 621 (1927). This is quoted later by *loc. cit.*, II. An essentially equivalent theory has been obtained independently by Jordan [' Z. f. Physik,' vol. 40, p. 809 (1927)]. See also, F. London, ' Z. f. Physik,' vol. 40, p. 193 (1926).

a q-number, capable of being represented by a generalised " matrix " according to many different matrix schemes, some of which may have continuous ranges of rows and columns, and may require the matrix elements to involve certain kinds of infinities (of the type given by the δ functions*). A matrix scheme can be found in which any desired set of constants of integration of the dynamical system that commute are represented by diagonal matrices, or in which a set of variables that commute are represented by matrices that are diagonal at a specified time.† The values of the diagonal elements of a diagonal matrix representing any q-number are the characteristic values of that q-number. A Cartesian co-ordinate or momentum will in general have all characteristic values from $-\infty$ to $+\infty$, while an action variable has only a discrete set of characteristic values. (We shall make it a rule to use unprimed letters to denote the dynamical variables or q-numbers, and the same letters primed or multiply primed to denote their characteristic values. Transformation functions or eigenfunctions are functions of the characteristic values and not of the q-numbers themselves, so they should always be written in terms of primed variables.)

If $f(\xi, \eta)$ is any function of the canonical variables ξ_k, η_k, the matrix representing f at any time t in the matrix scheme in which the ξ_k at time t are diagonal matrices may be written down without any trouble, since the matrices representing the ξ_k and η_k themselves at time t are known, namely,

$$\left.\begin{aligned} \xi_k(\xi'\xi'') &= \xi_k'\,\delta(\xi'\xi''), \\ \eta_k(\xi'\xi'') &= -i\hbar\,\delta(\xi_1'-\xi_1'')\ldots\delta(\xi_{k-1}'-\xi_{k-1}'')\,\delta'(\xi_k'-\xi_k'')\,\delta(\xi_{k+1}'-\xi_{k+1}'')\ldots \end{aligned}\right\}. \quad (2)$$

Thus if the Hamiltonian H is given as a function of the ξ_k and η_k, we can at once write down the matrix $H(\xi'\,\xi'')$. We can then obtain the transformation function, (ξ'/α') say, which transforms to a matrix scheme (α) in which the Hamiltonian is a diagonal matrix, as (ξ'/α') must satisfy the integral equation

$$\int H(\xi'\xi'')\,d\xi''\,(\xi''/\alpha') = W(\alpha')\,.\,(\xi'/\alpha'), \quad (3)$$

of which the characteristic values $W(\alpha')$ are the energy levels. This equation is just Schrödinger's wave equation for the eigenfunctions (ξ'/α'), which becomes an ordinary differential equation when H is a simple algebraic function of the

* *Loc. cit.* II, § 2.

† One can have a matrix scheme in which a set of variables that commute are at all times represented by diagonal matrices if one will sacrifice the condition that the matrices must satisfy the equations of motion. The transformation function from such a scheme to one in which the equations of motion are satisfied will involve the time explicitly. See p. 628 in *loc. cit.*, II.

ξ_k and η_k on account of the special equations (2) for the matrices representing ξ_k and η_k. Equation (3) may be written in the more general form

$$\int H(\xi'\xi'')\,d\xi''\,(\xi''/\alpha') = ih\,\partial(\xi'/\alpha')/\partial t, \tag{3'}$$

in which it can be applied to systems for which the Hamiltonian involves the time explicitly.

One may have a dynamical system specified by a Hamiltonian H which cannot be expressed as an algebraic function of any set of canonical variables, but which can all the same be represented by a matrix $H(\xi'\xi'')$. Such a problem can still be solved by the present method, since one can still use equation (3) to obtain the energy levels and eigenfunctions. We shall find that the Hamiltonian which describes the interaction of a light-quantum and an atomic system is of this more general type, so that the interaction can be treated mathematically, although one cannot talk about an interaction potential energy in the usual sense.

It should be observed that there is a difference between a light-wave and the de Broglie or Schrödinger wave associated with the light-quanta. Firstly, the light-wave is always real, while the de Broglie wave associated with a light-quantum moving in a definite direction must be taken to involve an imaginary exponential. A more important difference is that their intensities are to be interpreted in different ways. The number of light-quanta per unit volume associated with a monochromatic light-wave equals the energy per unit volume of the wave divided by the energy $(2\pi h)\nu$ of a single light-quantum. On the other hand a monochromatic de Broglie wave of amplitude a (multiplied into the imaginary exponential factor) must be interpreted as representing a^2 light-quanta per unit volume for all frequencies. This is a special case of the general rule for interpreting the matrix analysis,* according to which, if (ξ'/α') or $\psi_{\alpha'}(\xi_k')$ is the eigenfunction in the variables ξ_k of the state α' of an atomic system (or simple particle), $|\psi_{\alpha'}(\xi_k')|^2$ is the probability of each ξ_k having the value ξ_k', [or $|\psi_{\alpha'}(\xi_k')|^2\,d\xi_1'\,d\xi_2'\ldots$ is the probability of each ξ_k lying between the values ξ_k' and $\xi_k' + d\xi_k'$, when the ξ_k have continuous ranges of characteristic values] on the assumption that all phases of the system are equally probable. The wave whose intensity is to be interpreted in the first of these two ways appears in the theory only when one is dealing with an assembly of the associated particles satisfying the Einstein-Bose statistics. There is thus no such wave associated with electrons.

* *Loc. cit.*, II, §§ 6, 7.

§ 2. *The Perturbation of an Assembly of Independent Systems.*

We shall now consider the transitions produced in an atomic system by an arbitrary perturbation. The method we shall adopt will be that previously given by the author,† which leads in a simple way to equations which determine the probability of the system being in any stationary state of the unperturbed system at any time.‡ This, of course, gives immediately the probable number of systems in that state at that time for an assembly of the systems that are independent of one another and are all perturbed in the same way. The object of the present section is to show that the equations for the rates of change of these probable numbers can be put in the Hamiltonian form in a simple manner, which will enable further developments in the theory to be made.

Let H_0 be the Hamiltonian for the unperturbed system and V the perturbing energy, which can be an arbitrary function of the dynamical variables and may or may not involve the time explicitly, so that the Hamiltonian for the perturbed system is $H = H_0 + V$. The eigenfunctions for the perturbed system must satisfy the wave equation

$$ih\, \partial\psi/\partial t = (H_0 + V)\, \psi,$$

where $(H_0 + V)$ is an operator. If $\psi = \Sigma_r a_r \psi_r$ is the solution of this equation that satisfies the proper initial conditions, where the ψ_r's are the eigenfunctions for the unperturbed system, each associated with one stationary state labelled by the suffix r, and the a_r's are functions of the time only, then $|a_r|^2$ is the probability of the system being in the state r at any time. The a_r's must be normalised initially, and will then always remain normalised. The theory will apply directly to an assembly of N similar independent systems if we multiply each of these a_r's by $N^{\frac{1}{2}}$ so as to make $\Sigma_r |a_r|^2 = N$. We shall now have that $|a_r|^2$ is the probable number of systems in the state r.

The equation that determines the rate of change of the a_r's is§

$$ih\dot{a}_r = \Sigma_s V_{rs} a_s, \tag{4}$$

where the V_{rs}'s are the elements of the matrix representing V. The conjugate imaginary equation is

$$-ih\dot{a}_r^* = \Sigma_s V_{rs}^* a_s^* = \Sigma_s a_s^* V_{sr}. \tag{4'}$$

† *Loc. cit.* I.

‡ The theory has recently been extended by Born ['Z. f. Physik,' vol. 40, p. 167 (1926)] so as to take into account the adiabatic changes in the stationary states that may be produced by the perturbation as well as the transitions. This extension is not used in the present paper.

§ *Loc. cit.*, I, equation (25).

If we regard a_r and $ih\, a_r^*$ as canonical conjugates, equations (4) and (4′) take the Hamiltonian form with the Hamiltonian function $F_1 = \Sigma_{rs} a_r^* V_{rs} a_s$, namely,

$$\frac{da_r}{dt} = \frac{1}{ih}\frac{\partial F_1}{\partial a_r^*}, \quad ih\frac{da_r^*}{dt} = -\frac{\partial F_1}{\partial a_r}.$$

We can transform to the canonical variables N_r, ϕ_r by the contact transformation

$$a_r = N_r^{\frac{1}{2}} e^{-i\phi_r/h}, \quad a_r^* = N_r^{\frac{1}{2}} e^{i\phi_r/h}.$$

This transformation makes the new variables N_r and ϕ_r real, N_r being equal to $a_r a_r^* = |a_r|^2$, the probable number of systems in the state r, and ϕ_r/h being the phase of the eigenfunction that represents them. The Hamiltonian F_1 now becomes

$$F_1 = \Sigma_{rs} V_{rs} N_r^{\frac{1}{2}} N_s^{\frac{1}{2}} e^{i(\phi_r - \phi_s)/h},$$

and the equations that determine the rate at which transitions occur have the canonical form

$$\dot{N}_r = -\frac{\partial F_1}{\partial \phi_r}, \quad \dot{\phi}_r = \frac{\partial F_1}{\partial N_r}.$$

A more convenient way of putting the transition equations in the Hamiltonian form may be obtained with the help of the quantities

$$b_r = a_r\, e^{-iW_r t/h}, \quad b_r^* = a_r^*\, e^{iW_r t/h},$$

W_r being the energy of the state r. We have $|b_r|^2$ equal to $|a_r|^2$, the probable number of systems in the state r. For $\dot{b}_r$ we find

$$\begin{aligned} ih\,\dot{b}_r &= W_r b_r + ih\dot{a}_r e^{-iW_r t/h} \\ &= W_r b_r + \Sigma_s V_{rs} b_s e^{i(W_s - W_r)t/h} \end{aligned}$$

with the help of (4). If we put $V_{rs} = v_{rs} e^{i(W_r - W_s)t/h}$, so that v_{rs} is a constant when V does not involve the time explicitly, this reduces to

$$\begin{aligned} ih\,\dot{b}_r &= W_r b_r + \Sigma_s v_{rs} b_s \\ &= \Sigma_s H_{rs} b_s, \end{aligned} \tag{5}$$

where $H_{rs} = W_r\,\delta_{rs} + v_{rs}$, which is a matrix element of the total Hamiltonian $H = H_0 + V$ with the time factor $e^{i(W_r - W_s)t/h}$ removed, so that H_{rs} is a constant when H does not involve the time explicitly. Equation (5) is of the same form as equation (4), and may be put in the Hamiltonian form in the same way.

It should be noticed that equation (5) is obtained directly if one writes down the Schrödinger equation in a set of variables that specify the stationary states of the unperturbed system. If these variables are ξ_h, and if $H(\xi'\xi'')$ denotes a matrix element of the total Hamiltonian H in the (ξ) scheme, this Schrödinger equation would be

$$ih\, \partial\psi(\xi')/\partial t = \Sigma_{\xi''} H(\xi'\xi'')\, \psi(\xi''), \tag{6}$$

like equation (3'). This differs from the previous equation (5) only in the notation, a single suffix r being there used to denote a stationary state instead of a set of numerical values ξ_k' for the variables ξ_k, and b_r being used instead of $\psi(\xi')$. Equation (6), and therefore also equation (5), can still be used when the Hamiltonian is of the more general type which cannot be expressed as an algebraic function of a set of canonial variables, but can still be represented by a matrix $H(\xi'\xi'')$ or H_{rs}.

We now take b_r and $ih\, b_r^*$ to be canonically conjugate variables instead of a_r and $ih\, a_r^*$. The equation (5) and its conjugate imaginary equation will now take the Hamiltonian form with the Hamiltonian function

$$F = \Sigma_{rs} b_r^* H_{rs} b_s. \tag{7}$$

Proceeding as before, we make the contact transformation

$$b_r = N_r^{\frac{1}{2}} e^{-i\theta_r/h}, \qquad b_r^* = N_r^{\frac{1}{2}} e^{i\theta_r/h}, \tag{8}$$

to the new canonical variables N_r, θ_r, where N_r is, as before, the probable number of systems in the state r, and θ_r is a new phase. The Hamiltonian F will now become

$$F = \Sigma_{rs} H_{rs} N_r^{\frac{1}{2}} N_s^{\frac{1}{2}} e^{i(\theta_r - \theta_s)/h},$$

and the equations for the rates of change of N_r and θ_r will take the canonical form

$$\dot{N}_r = -\frac{\partial F}{\partial \theta_r}, \qquad \dot{\theta}_r = \frac{\partial F}{\partial N_r}.$$

The Hamiltonian may be written

$$F = \Sigma_r W_r N_r + \Sigma_{rs} v_{rs} N_r^{\frac{1}{2}} N_s^{\frac{1}{2}} e^{i(\theta_r - \theta_s)/h}. \tag{9}$$

The first term $\Sigma_r W_r N_r$ is the total proper energy of the assembly, and the second may be regarded as the additional energy due to the perturbation. If the perturbation is zero, the phases θ_r would increase linearly with the time, while the previous phases ϕ_r would in this case be constants.

§ 3. *The Perturbation of an Assembly satisfying the Einstein-Bose Statistics.*

According to the preceding section we can describe the effect of a perturbation on an assembly of independent systems by means of canonical variables and Hamiltonian equations of motion. The development of the theory which naturally suggests itself is to make these canonical variables q-numbers satisfying the usual quantum conditions instead of c-numbers, so that their Hamiltonian equations of motion become true quantum equations. The Hamiltonian function will now provide a Schrödinger wave equation, which must be solved and interpreted in the usual manner. The interpretation will give not merely the probable number of systems in any state, but the probability of any given distribution of the systems among the various states, this probability being, in fact, equal to the square of the modulus of the normalised solution of the wave equation that satisfies the appropriate initial conditions. We could, of course, calculate directly from elementary considerations the probability of any given distribution when the systems are independent, as we know the probability of each system being in any particular state. We shall find that the probability calculated directly in this way does not agree with that obtained from the wave equation except in the special case when there is only one system in the assembly. In the general case it will be shown that the wave equation leads to the correct value for the probability of any given distribution when the systems obey the Einstein-Bose statistics instead of being independent.

We assume the variables b_r, ihb_r^* of § 2 to be canonical q-numbers satisfying the quantum conditions

$$b_r \,.\, ih\, b_r^* - ih\, b_r^* \,.\, b_r = ih$$

or

$$b_r b_r^* - b_r^* b_r = 1,$$

and

$$b_r b_s - b_s b_r = 0, \qquad b_r^* b_s^* - b_s^* b_r^{\ *} = 0,$$

$$b_r b_s^* - b_s^* b_r = 0 \qquad (s \neq r).$$

The transformation equations (8) must now be written in the quantum form

$$\left.\begin{aligned} b_r &= (\mathrm{N}_r + 1)^{\frac{1}{2}}\, e^{-i\theta_r/h} = e^{-i\theta_r/h}\mathrm{N}_r^{\frac{1}{2}} \\ b_r^* &= \mathrm{N}_r^{\frac{1}{2}} e^{i\theta_r/h} = e^{i\theta_r/h}(\mathrm{N}_r + 1)^{\frac{1}{2}}, \end{aligned}\right\} \tag{10}$$

in order that the N_r, θ_r may also be canonical variables. These equations show that the N_r can have only integral characteristic values not less than zero,† which provides us with a justification for the assumption that the variables are q-numbers in the way we have chosen. The numbers of systems in the different states are now ordinary quantum numbers.

† See § 8 of the author's paper 'Roy. Soc. Proc.,' A, vol. 111, p. 281 (1926). What are there called the c-number values that a q-number can take are here given the more precise name of the characteristic values of that q-number.

The Hamiltonian (7) now becomes

$$F = \Sigma_{rs} b_r^* H_{rs} b_s = \Sigma_{rs} N_r^{\frac{1}{2}} e^{i\theta_r/h} H_{rs} (N_s + 1)^{\frac{1}{2}} e^{-i\theta_s/h}$$

$$= \Sigma_{rs} H_{rs} N_r^{\frac{1}{2}} (N_s + 1 - \delta_{rs})^{\frac{1}{2}} e^{i(\theta_r - \theta_s)/h} \qquad (11)$$

in which the H_{rs} are still c-numbers. We may write this F in the form corresponding to (9)

$$F = \Sigma_r W_r N_r + \Sigma_{rs} v_{rs} N_r^{\frac{1}{2}} (N_s + 1 - \delta_{rs})^{\frac{1}{2}} e^{i(\theta_r - \theta_s)/h} \qquad (11')$$

in which it is again composed of a proper energy term $\Sigma_r W_r N_r$ and an interaction energy term.

The wave equation written in terms of the variables N_r is†

$$ih \frac{\partial}{\partial t} \psi (N_1', N_2', N_3' \ldots) = F\psi (N_1', N_2', N_3' \ldots), \qquad (12)$$

where F is an operator, each θ_r occurring in F being interpreted to mean $ih\, \partial/\partial N_r'$. If we apply the operator $e^{\pm i\theta_r/h}$ to any function $f(N_1', N_2', \ldots N_r', \ldots)$ of the variables $N_1', N_2', \ldots$ the result is

$$e^{\pm i\theta_r/h} f(N_1', N_2', \ldots N_r', \ldots) = e^{\mp \partial/\partial N_r'} f(N_1', N_2', \ldots N_r' \ldots)$$

$$= f(N_1', N_2', \ldots N_r' \mp 1, \ldots).$$

If we use this rule in equation (12) and use the expression (11) for F we obtain‡

$$ih \frac{\partial}{\partial t} \psi (N_1', N_2', N_3' \ldots)$$

$$= \Sigma_{rs} H_{rs} N_r'^{\frac{1}{2}} (N_s' + 1 - \delta_{rs})^{\frac{1}{2}} \psi (N_1', N_2' \ldots N_r' - 1, \ldots N_s' + 1, \ldots). \qquad (13)$$

We see from the right-hand side of this equation that in the matrix representing F, the term in F involving $e^{i(\theta_r - \theta_s)/h}$ will contribute only to those matrix elements that refer to transitions in which N_r decreases by unity and N_s increases by unity, *i.e.*, to matrix elements of the type $F(N_1', N_2' \ldots N_r' \ldots N_s'; N_1', N_2' \ldots N_r' - 1 \ldots N_s' + 1 \ldots)$. If we find a solution $\psi (N_1', N_2' \ldots)$ of equation (13) that is normalised [*i.e.*, one for which $\Sigma_{N_1', N_2' \ldots} | \psi (N_1', N_2' \ldots)|^2 = 1$] and that satisfies the proper initial conditions, then $| \psi (N_1', N_2' \ldots)|^2$ will be the probability of that distribution in which N_1' systems are in state 1, N_2' in state 2, ... at any time.

Consider first the case when there is only one system in the assembly. The probability of its being in the state q is determined by the eigenfunction

† We are supposing for definiteness that the label r of the stationary states takes the values 1, 2, 3, ...

‡ When $s = r$, $\psi (N_1', N_2' \ldots N_r' - 1 \ldots N_s' + 1)$ is to be taken to mean $\psi (N_1' N_2' \ldots N_r' \ldots)$.

$\psi(N_1', N_2', \ldots)$ in which all the N′'s are put equal to zero except N_q', which is put equal to unity. This eigenfunction we shall denote by $\psi\{q\}$. When it is substituted in the left-hand side of (13), all the terms in the summation on the right-hand side vanish except those for which $r = q$, and we are left with

$$ih \frac{\partial}{\partial t}\psi\{q\} = \Sigma_r H_{qs}\psi\{s\},$$

which is the same equation as (5) with $\psi\{q\}$ playing the part of b_q. This establishes the fact that the present theory is equivalent to that of the preceding section when there is only one system in the assembly.

Now take the general case of an arbitrary number of systems in the assembly, and assume that they obey the Einstein-Bose statistical mechanics. This requires that, in the ordinary treatment of the problem, only those eigenfunctions that are symmetrical between all the systems must be taken into account, these eigenfunctions being by themselves sufficient to give a complete quantum solution of the problem.† We shall now obtain the equation for the rate of change of one of these symmetrical eigenfunctions, and show that it is identical with equation (13).

If we label each system with a number n, then the Hamiltonian for the assembly will be $H_A = \Sigma_n H(n)$, where $H(n)$ is the H of § 2 (equal to $H_0 + V$) expressed in terms of the variables of the nth system. A stationary state of the assembly is defined by the numbers $r_1, r_2 \ldots r_n \ldots$ which are the labels of the stationary states in which the separate systems lie. The Schrödinger equation for the assembly in a set of variables that specify the stationary states will be of the form (6) [with H_A instead of H], and we can write it in the notation of equation (5) thus:—

$$ih\dot{b}(r_1 r_2 \ldots) = \Sigma_{s_1, s_2 \ldots} H_A(r_1 r_2 \ldots ; s_1 s_2 \ldots)\, b(s_1 s_2 \ldots), \tag{14}$$

where $H_A(r_1 r_2 \ldots ; s_1 s_2 \ldots)$ is the general matrix element of H_A [with the time factor removed]. This matrix element vanishes when more than one s_n differs from the corresponding r_n; equals $H_{r_m s_m}$ when s_m differs from r_m and every other s_n equals r_n; and equals $\Sigma_n H_{r_n r_n}$ when every s_n equals r_n. Substituting these values in (14), we obtain

$$ih\dot{b}(r_1 r_2 \ldots) = \Sigma_m \Sigma_{s_m \neq r_m} H_{r_m s_m} b(r_1 r_2 \ldots r_{m-1} s_m r_{m+1} \ldots) + \Sigma_n H_{r_n r_n} b(r_1 r_2 \ldots). \tag{15}$$

We must now restrict $b(r_1 r_2 \ldots)$ to be a symmetrical function of the variables $r_1, r_2 \ldots$ in order to obtain the Einstein-Bose statistics. This is permissible since if $b(r_1 r_2 \ldots)$ is symmetrical at any time, then equation (15) shows that

† *Loc. cit.*, I, § 3.

$\dot{b}(r_1 r_2 \ldots)$ is also symmetrical at that time, so that $b\ (r_1 r_2 \ldots)$ will remain symmetrical.

Let N_r denote the number of systems in the state r. Then a stationary state of the assembly describable by a symmetrical eigenfunction may be specified by the numbers $N_1, N_2 \ldots N_r \ldots$ just as well as by the numbers $r_1, r_2 \ldots r_n \ldots$, and we shall be able to transform equation (15) to the variables $N_1, N_2 \ldots$. We cannot actually take the new eigenfunction $b\ (N_1, N_2 \ldots)$ equal to the previous one $b\ (r_1 r_2 \ldots)$, but must take one to be a numerical multiple of the other in order that each may be correctly normalised with respect to its respective variables. We must have, in fact,

$$\Sigma_{r_1, r_2 \ldots} |b(r_1 r_2 \ldots)|^2 = 1 = \Sigma_{N_1, N_2 \ldots} |b(N_1, N_2 \ldots)|^2,$$

and hence we must take $|b(N_1, N_2 \ldots)|^2$ equal to the sum of $|b(r_1 r_2 \ldots)|^2$ for all values of the numbers $r_1, r_2 \ldots$ such that there are N_1 of them equal to 1, N_2 equal to 2, etc. There are $N!/N_1!\, N_2! \ldots$ terms in this sum, where $N = \Sigma_r N_r$ is the total number of systems, and they are all equal, since $b(r_1 r_2 \ldots)$ is a symmetrical function of its variables $r_1, r_2 \ldots$. Hence we must have

$$b(N_1, N_2 \ldots) = (N!/N_1!\, N_2! \ldots)^{\frac{1}{2}}\, b(r_1 r_2 \ldots).$$

If we make this substitution in equation (15), the left-hand side will become $ih\, (N_1!\, N_2! \ldots /N!)^{\frac{1}{2}}\, \dot{b}\ (N_1, N_2 \ldots)$. The term $H_{r_m s_m} b\ (r_1 r_2 \ldots r_{m-1} s_m r_{m+1} \ldots)$ in the first summation on the right-hand side will become

$$[N_1!\, N_2! \ldots (N_r - 1)! \ldots (N_s + 1)! \ldots /N!]^{\frac{1}{2}}\, H_{rs} b\, (N_1, N_2 \ldots N_r - 1 \ldots N_s + 1 \ldots), \quad (16)$$

where we have written r for r_m and s for s_m. This term must be summed for all values of s except r, and must then be summed for r taking each of the values $r_1, r_2 \ldots$. Thus each term (16) gets repeated by the summation process until it occurs a total of N_r times, so that it contributes

$$\begin{aligned} N_r [N_1!\, N_2! \ldots (N_r - 1)! \ldots (N_s + 1)! \ldots /N!]^{\frac{1}{2}}\, H_{rs} b\, (N_1, N_2 \ldots N_r - 1 \ldots N_s + 1 \ldots) \\ = N_r^{\frac{1}{2}} (N_s + 1)^{\frac{1}{2}} (N_1!\, N_2! \ldots /N!)^{\frac{1}{2}}\, H_{rs} b\, (N_1, N_2 \ldots N_r - 1 \ldots N_s + 1 \ldots) \end{aligned}$$

to the right-hand side of (15). Finally, the term $\Sigma_n H_{r_n r_n} b\, (r_1, r_2 \ldots)$ becomes

$$\Sigma_r N_r H_{rr} \,.\, b\, (r_1 r_2 \ldots) = \Sigma_r N_r H_{rr} \,.\, (N_1!\, N_2! \ldots /N!)^{\frac{1}{2}}\, b\, (N_1, N_2 \ldots).$$

Hence equation (15) becomes, with the removal of the factor $(N_1!\, N_2! \ldots /N!)^{\frac{1}{2}}$,

$$\begin{aligned} ih\dot{b}\ (N_1, N_2 \ldots) = \Sigma_r \Sigma_{s \neq r} N_r^{\frac{1}{2}} (N_s + 1)^{\frac{1}{2}} H_{rs} b\, (N_1, N_2 \ldots N_r - 1 \ldots N_s + 1 \ldots) \\ + \Sigma_r N_r H_{rr} b\, (N_1, N_2 \ldots), \quad (17) \end{aligned}$$

which is identical with (13) [except for the fact that in (17) the primes have been omitted from the N's, which is permissible when we do not require to refer to the N's as q-numbers]. We have thus established that the Hamiltonian (11) describes the effect of a perturbation on an assembly satisfying the Einstein-Bose statistics.

§ 4. *The Reaction of the Assembly on the Perturbing System.*

Up to the present we have considered only perturbations that can be represented by a perturbing energy V added to the Hamiltonian of the perturbed system, V being a function only of the dynamical variables of that system and perhaps of the time. The theory may readily be extended to the case when the perturbation consists of interaction with a perturbing dynamical system, the reaction of the perturbed system on the perturbing system being taken into account. (The distinction between the perturbing system and the perturbed system is, of course, not real, but it will be kept up for convenience.)

We now consider a perturbing system, described, say, by the canonical variables J_k, ω_k, the J's being its first integrals when it is alone, interacting with an assembly of perturbed systems with no mutual interaction, that satisfy the Einstein-Bose statistics. The total Hamiltonian will be of the form

$$H_T = H_P(J) + \Sigma_n H(n),$$

where H_P is the Hamiltonian of the perturbing system (a function of the J's only) and $H(n)$ is equal to the proper energy $H_0(n)$ plus the perturbation energy $V(n)$ of the nth system of the assembly. $H(n)$ is a function only of the variables of the nth system of the assembly and of the J's and w's, and does not involve the time explicitly.

The Schrödinger equation corresponding to equation (14) is now

$$i\hbar\dot{b}(J', r_1 r_2 \ldots) = \Sigma_{J''}\Sigma_{s_1, s_2}\ldots H_T(J', r_1 r_2 \ldots;\ J'', s_1 s_2 \ldots)\, b(J'', s_1 s_2 \ldots),$$

in which the eigenfunction b involves the additional variables J_k'. The matrix element $H_T(J', r_1 r_2 \ldots;\ J'', s_1 s_2 \ldots)$ is now always a constant. As before, it vanishes when more than one s_n differs from the corresponding r_n. When s_m differs from r_m and every other s_n equals r_n, it reduces to $H(J'r_m;\ J''s_m)$, which is the $(J'r_m;\ J''s_m)$ matrix element (with the time factor removed) of $H = H_0 + V$, the proper energy plus the perturbation energy of a single system of the assembly; while when every s_n equals r_n, it has the value $H_P(J')\,\delta_{J'J''} + \Sigma_n H(J'r_n;\ J''r_n)$. If, as before, we restrict the eigenfunctions to be symmetrical in the variables r_1, r_2 ..., we can again transform to the variables N_1, N_2 ..., which will lead, as before, to the result

$$i\hbar\dot{b}(J', N_1', N_2' \ldots) = H_P(J')\, b(J', N'_1, N_2' \ldots)$$
$$+\Sigma_{J''}\Sigma_{r,s} N_r'^{\frac{1}{2}}(N_s'+1-\delta_{rs})^{\frac{1}{2}} H(J'r; J''s)\, b(J'', N_1', N_2' \ldots N_r'-1 \ldots N_s'+1 \ldots) \quad (18)$$

This is the Schrödinger equation corresponding to the Hamiltonian function

$$F = H_P(J) + \Sigma_{r,s} H_{rs} N_r^{\frac{1}{2}}(N_s+1-\delta_{rs})^{\frac{1}{2}} e^{i(\theta_r-\theta_s)/h}, \quad (19)$$

in which H_{rs} is now a function of the J's and w's, being such that when represented by a matrix in the (J) scheme its (J′ J″) element is $H(J'r; J''s)$. (It should be noticed that H_{rs} still commutes with the N's and θ's.)

Thus the interaction of a perturbing system and an assembly satisfying the Einstein-Bose statistics can be described by a Hamiltonian of the form (19). We can put it in the form corresponding to (11′) by observing that the matrix element $H(J'r; J''s)$ is composed of the sum of two parts, a part that comes from the proper energy H_0, which equals W_r when $J_k'' = J_k'$ and $s = r$ and vanishes otherwise, and a part that comes from the interaction energy V, which may be denoted by $v(J'r; J''s)$. Thus we shall have

$$H_{rs} = W_r\,\delta_{rs} + v_{rs},$$

where v_{rs} is that function of the J's and w's which is represented by the matrix whose (J′ J″) element is $v(J'r; J''s)$, and so (19) becomes

$$F = H_P(J) + \Sigma_r W_r N_r + \Sigma_{r,s} v_{rs} N_r^{\frac{1}{2}}(N_s + 1 - \delta_{rs})^{\frac{1}{2}} e^{i(\theta_r-\theta_s)/h}. \quad (20)$$

The Hamiltonian is thus the sum of the proper energy of the perturbing system $H_P(J)$, the proper energy of the perturbed systems $\Sigma_r W_r N_r$ and the perturbation energy $\Sigma_{r,s} v_{rs} N_r^{\frac{1}{2}}(N_s + 1 - \delta_{rs})^{\frac{1}{2}} e^{i(\theta_r-\theta_s)/h}$.

§ 5. *Theory of Transitions in a System from One State to Others of the Same Energy.*

Before applying the results of the preceding sections to light-quanta, we shall consider the solution of the problem presented by a Hamiltonian of the type (19). The essential feature of the problem is that it refers to a dynamical system which can, under the influence of a perturbation energy which does not involve the time explicitly, make transitions from one state to others of the same energy. The problem of collisions between an atomic system and an electron, which has been treated by Born,* is a special case of this type. Born's method is to find a *periodic* solution of the wave equation which consists, in so far as it involves the co-ordinates of the colliding electron, of plane waves,

* Born, 'Z. f. Physik,' vol. 38, p. 803 (1926).

representing the incident electron, approaching the atomic system, which are scattered or diffracted in all directions. The square of the amplitude of the waves scattered in any direction with any frequency is then assumed by Born to be the probability of the electron being scattered in that direction with the corresponding energy.

This method does not appear to be capable of extension in any simple manner to the general problem of systems that make transitions from one state to others of the same energy. Also there is at present no very direct and certain way of interpreting a periodic solution of a wave equation to apply to a non-periodic physical phenomenon such as a collision. (The more definite method that will now be given shows that Born's assumption is not quite right, it being necessary to multiply the square of the amplitude by a certain factor.)

An alternative method of solving a collision problem is to find a *non-periodic* solution of the wave equation which consists initially simply of plane waves moving over the whole of space in the necessary direction with the necessary frequency to represent the incident electron. In course of time waves moving in other directions must appear in order that the wave equation may remain satisfied. The probability of the electron being scattered in any direction with any energy will then be determined by the rate of growth of the corresponding harmonic component of these waves. The way the mathematics is to be interpreted is by this method quite definite, being the same as that of the beginning of § 2.

We shall apply this method to the general problem of a system which makes transitions from one state to others of the same energy under the action of a perturbation. Let H_0 be the Hamiltonian of the unperturbed system and V the perturbing energy, which must not involve the time explicitly. If we take the case of a continuous range of stationary states, specified by the first integrals, α_k say, of the unperturbed motion, then, following the method of § 2, we obtain

$$ih\,\dot{a}(\alpha') = \int V(\alpha'\alpha'')\,d\alpha'' \,.\, a(\alpha''), \tag{21}$$

corresponding to equation (4). The probability of the system being in a state for which each α_k lies between α_k' and $\alpha_k' + d\alpha_k'$ at any time is $|a(\alpha')|^2 d\alpha_1' \,.\, d\alpha_2' \ldots$ when $a(\alpha')$ is properly normalised and satisfies the proper initial conditions. If initially the system is in the state α^0, we must take the initial value of $a(\alpha')$ to be of the form $a^0 \,.\, \delta(\alpha' - \alpha^0)$. We shall keep a^0 arbitrary, as it would be inconvenient to normalise $a(\alpha')$ in the present case. For a first approximation we may substitute for $a(\alpha'')$ in the right-hand side of (21) its initial value. This gives

$$i h\, \dot{a}(\alpha') = a^0 V(\alpha'\alpha^0) = \alpha^0 v(\alpha'\alpha^0)\, e^{i[W(\alpha') - W(\alpha^0)]t/h},$$

where $v(\alpha'\alpha^0)$ is a constant and $W(\alpha')$ is the energy of the state α'. Hence

$$i h\, a(\alpha') = a^0 \delta(\alpha' - \alpha^0) + a^0 v(\alpha'\alpha^0) \frac{e^{i[W(\alpha') - W(\alpha^0)]t/h} - 1}{i[W(\alpha') - W(\alpha^0)]/h}. \qquad (22)$$

For values of the α_k' such that $W(\alpha')$ differs appreciably from $W(\alpha^0)$, $a(\alpha')$ is a periodic function of the time whose amplitude is small when the perturbing energy V is small, so that the eigenfunctions corresponding to these stationary states are not excited to any appreciable extent. On the other hand, for values of the α_k' such that $W(\alpha') = W(\alpha^0)$ and $\alpha_k' \neq \alpha_k^0$ for some k, $a(\alpha')$ increases uniformly with respect to the time, so that the probability of the system being in the state α' at any time increases proportionally with the square of the time. Physically, the probability of the system being in a state with exactly the same proper energy as the initial proper energy $W(\alpha^0)$ is of no importance, being infinitesimal. We are interested only in the integral of the probability through a small range of proper energy values about the initial proper energy, which, as we shall find, increases linearly with the time, in agreement with the ordinary probability laws.

We transform from the variables $\alpha_1, \alpha_2 \ldots \alpha_u$ to a set of variables that are arbitrary independent functions of the α's such that one of them is the proper energy W, say, the variables $W, \gamma_1, \gamma_2, \ldots \gamma_{u-1}$. The probability at any time of the system lying in a stationary state for which each γ_k lies between γ_k' and $\gamma_k' + d\gamma_k'$ is now (apart from the normalising factor) equal to

$$d\gamma_1' \cdot d\gamma_2' \ldots d\gamma_{u-1}' \int |a(\alpha')|^2 \frac{\partial(\alpha_1', \alpha_2' \ldots \alpha_u')}{\partial(W', \gamma_1' \ldots \gamma_{u-1}')} dW'. \qquad (23)$$

For a time that is large compared with the periods of the system we shall find that practically the whole of the integral in (23) is contributed by values of W' very close to $W^0 = W(\alpha^0)$. Put

$$a(\alpha') = a(W', \gamma') \quad \text{and} \quad \partial(\alpha_1', \alpha_2' \ldots \alpha_u')/\partial(W', \gamma_1' \ldots \gamma_{u-1}') = J(W', \gamma').$$

Then for the integral in (23) we find, with the help of (22) (provided $\gamma_k' \neq \gamma_k^0$ for some k)

$$\int |a(W', \gamma')|^2 J(W', \gamma')\, dW'$$

$$= |a^0|^2 \int |v(W', \gamma'; W^0, \gamma^0)|^2 J(W', \gamma') \frac{[e^{i(W' - W^0)t/h} - 1][e^{-i(W'-W^0)t/h} - 1]}{(W' - W^0)^2} dW'$$

$$= 2|a^0|^2 \int |v(W', \gamma'; W^0, \gamma^0)|^2 J(W', \gamma')[1 - \cos(W' - W^0)t/h]/(W' - W^0)^2 . dW'$$

$$= 2|a^0|^2 t/h \,. \int |v(W^0 + hx/t, \gamma'; W^0, \gamma^0)|^2 J(W^0 + hx/t, \gamma')(1 - \cos x)/x^2 . dx,$$

if one makes the substitution $(W'-W^0)t/h = x$. For large values of t this reduces to

$$2\,|a^0|^2 t/h\,.\,|v(W^0,\gamma';\,W^0,\gamma^0)|^2\,J(W^0,\gamma')\int_{-\infty}^{\infty}(1-\cos x)/x^2\,.\,dx$$

$$= 2\pi\,|a^0|^2 t/h\,.\,|v(W^0,\gamma';\,W^0,\gamma^0)|^2\,J(W^0,\gamma').$$

The probability per unit time of a transition to a state for which each γ_k lies between γ_k' and $\gamma_k' + d\gamma_k'$ is thus (apart from the normalising factor)

$$2\pi\,|a^0|^2/h\,.\,|v(W^0,\gamma';\,W^0,\gamma^0)|^2\,J(W^0,\gamma')\,d\gamma_1'\,.\,d\gamma_2'\,\ldots\,d\gamma_{u-1}', \qquad (24)$$

which is proportional to the square of the matrix element associated with that transition of the perturbing energy.

To apply this result to a simple collision problem, we take the α's to be the components of momentum p_x, p_y, p_z of the colliding electron and the γ's to be θ and ϕ, the angles which determine its direction of motion. If, taking the relativity change of mass with velocity into account, we let P denote the resultant momentum, equal to $(p_x^2 + p_y^2 + p_z^2)^{\frac{1}{2}}$, and E the energy, equal to $(m^2c^4 + P^2c^2)^{\frac{1}{2}}$, of the electron, m being its rest-mass, we find for the Jacobian

$$J = \frac{\partial(p_x, p_y, p_z)}{\partial(E, \theta, \phi)} = \frac{EP}{c^2}\sin\theta.$$

Thus the $J(W^0, \gamma')$ of the expression (24) has the value

$$J(W^0,\gamma') = E'P'\sin\theta'/c^2, \qquad (25)$$

where E′ and P′ refer to that value for the energy of the scattered electron which makes the total energy equal the initial energy W^0 (*i.e.*, to that value required by the conservation of energy).

We must now interpret the initial value of $a(\alpha')$, namely, $a^0\,\delta(\alpha'-\alpha^0)$, which we did not normalise. According to § 2 the wave function in terms of the variables α_k is $b(\alpha') = a(\alpha')\,e^{-iW't/h}$, so that its initial value is

$$a^0\,\delta(\alpha'-\alpha^0)\,e^{-iW't/h} = a^0\,\delta(p_x'-p_x^0)\,\delta(p_y'-p_y^0)\,\delta(p_z'-p_z^0)\,e^{-iW't/h}.$$

If we use the transformation function*

$$(x'/p') = (2\pi h)^{-3/2} e^{i\Sigma_{xyz}p_x'x'/h},$$

and the transformation rule

$$\psi(x') = \int (x'/p')\,\psi(p')\,dp_x'\,dp_y'\,dp_z',$$

we obtain for the initial wave function in the co-ordinates x, y, z the value

$$a^0\,(2\pi h)^{-3/2}\,e^{i\Sigma_{xyz}p_x^0 x'/h}\,e^{-iW't/h}.$$

* The symbol x is used for brevity to denote x, y, z.

This corresponds to an initial distribution of $|a^0|^2 (2\pi h)^{-3}$ electrons per unit volume. Since their velocity is $P^0 c^2/E^0$, the number per unit time striking a unit surface at right-angles to their direction of motion is $|a^0|^2 P^0 c^2/(2\pi h)^3 E^0$. Dividing this into the expression (24) we obtain, with the help of (25),

$$4\pi^2 (2\pi h)^2 \frac{E' E^0}{c^4} |v(p'; p^0)|^2 \frac{P'}{P^0} \sin\theta' \, d\theta' \, d\phi'. \tag{26}$$

This is the effective area that must be hit by an electron in order that it shall be scattered in the solid angle $\sin\theta' \, d\theta' \, d\phi'$ with the energy E′. This result differs by the factor $(2\pi h)^2/2mE' \, . \, P'/P^0$ from Born's.* The necessity for the factor P'/P^0 in (26) could have been predicted from the principle of detailed balancing, as the factor $|v(p'; p^0)|^2$ is symmetrical between the direct and reverse processes.†

§6. *Application to Light-Quanta.*

We shall now apply the theory of §4 to the case when the systems of the assembly are light-quanta, the theory being applicable to this case since light-quanta obey the Einstein-Bose statistics and have no mutual interaction. A light-quantum is in a stationary state when it is moving with constant momentum in a straight line. Thus a stationary state r is fixed by the three components of momentum of the light-quantum and a variable that specifies its state of polarisation. We shall work on the assumption that there are a finite number of these stationary states, lying very close to one another, as it would be inconvenient to use continuous ranges. The interaction of the light-quanta with an atomic system will be described by a Hamiltonian of the form (20), in which $H_P(J)$ is the Hamiltonian for the atomic system alone, and the coefficients v_{rs} are for the present unknown. We shall show that this form for the Hamiltonian, with the v_{rs} arbitrary, leads to Einstein's laws for the emission and absorption of radiation.

The light-quantum has the peculiarity that it apparently ceases to exist when it is in one of its stationary states, namely, the zero state, in which its momentum, and therefore also its energy, are zero. When a light-quantum is absorbed it can be considered to jump into this zero state, and when one is emitted it can be considered to jump from the zero state to one in which it is

* In a more recent paper ('Nachr. Gesell. d. Wiss.,' Gottingen, p. 146 (1926)) Born has obtained a result in agreement with that of the present paper for non-relativity mechanics, by using an interpretation of the analysis based on the conservation theorems. I am indebted to Prof. N. Bohr for seeing an advance copy of this work.

† See Klein and Rosseland, 'Z. f. Physik,' vol. 4, p. 46, equation (4) (1921).

physically in evidence, so that it appears to have been created. Since there is no limit to the number of light-quanta that may be created in this way, we must suppose that there are an infinite number of light-quanta in the zero state, so that the N_0 of the Hamiltonian (20) is infinite. We must now have θ_0, the variable canonically conjugate to N_0, a constant, since

$$\dot{\theta}_0 = \partial F/\partial N_0 = W_0 + \text{terms involving } N_0^{-\frac{1}{2}} \text{ or } (N_0 + 1)^{-\frac{1}{2}}$$

and W_0 is zero. In order that the Hamiltonian (20) may remain finite it is necessary for the coefficients v_{r0}, v_{0r} to be infinitely small. We shall suppose that they are infinitely small in such a way as to make $v_{r0}N_0^{\frac{1}{2}}$ and $v_{0r}N_0^{\frac{1}{2}}$ finite, in order that the transition probability coefficients may be finite. Thus we put

$$v_{r0}(N_0 + 1)^{\frac{1}{2}} e^{-i\theta_0/h} = v_r, \quad v_{0r}N_0^{\frac{1}{2}}e^{i\theta_0/h} = v_r^*,$$

where v_r and v_r^* are finite and conjugate imaginaries. We may consider the v_r and v_r^* to be functions only of the J's and w's of the atomic system, since their factors $(N_0 + 1)^{\frac{1}{2}} e^{-i\theta_0/h}$ and $N_0^{\frac{1}{2}}e^{i\theta_0/h}$ are practically constants, the rate of change of N_0 being very small compared with N_0. The Hamiltonian (20) now becomes

$$F = H_P(J) + \Sigma_r W_r N_r + \Sigma_{r \neq 0}[v_r N_r^{\frac{1}{2}} e^{i\theta_r/h} + v_r^*(N_r + 1)^{\frac{1}{2}} e^{-i\theta_r/h}]$$
$$+ \Sigma_{r\neq 0}\Sigma_{s\neq 0} v_{rs} N_r^{\frac{1}{2}}(N_s + 1 - \delta_{rs})^{\frac{1}{2}} e^{i(\theta_r - \theta_s)/h}. \qquad (27)$$

The probability of a transition in which a light-quantum in the state r is absorbed is proportional to the square of the modulus of that matrix element of the Hamiltonian which refers to this transition. This matrix element must come from the term $v_r N_r^{\frac{1}{2}} e^{i\theta_r/h}$ in the Hamiltonian, and must therefore be proportional to $N_r'^{\frac{1}{2}}$ where N_r' is the number of light-quanta in state r before the process. The probability of the absorption process is thus proportional to N_r'. In the same way the probability of a light-quantum in state r being emitted is proportional to $(N_r' + 1)$, and the probability of a light-quantum in state r being scattered into state s is proportional to $N_r'(N_s' + 1)$. Radiative processes of the more general type considered by Einstein and Ehrenfest,† in which more than one light-quantum take part simultaneously, are not allowed on the present theory.

To establish a connection between the number of light-quanta per stationary state and the intensity of the radiation, we consider an enclosure of finite volume, A say, containing the radiation. The number of stationary states for light-quanta of a given type of polarisation whose frequency lies in the

† 'Z. f. Physik,' vol. 19, p. 301 (1923).

range ν_r to $\nu_r + d\nu_r$ and whose direction of motion lies in the solid angle $d\omega_r$ about the direction of motion for state r will now be $\mathrm{A}\nu_r^2 d\nu_r d\omega_r/c^3$. The energy of the light-quanta in these stationary states is thus $\mathrm{N}_r' \,.\, 2\pi h\nu_r \,.\, \mathrm{A}\nu_r^2 d\nu_r d\omega_r/c^3$. This must equal $\mathrm{A}c^{-1}\mathrm{I}_r d\nu_r d\omega_r$, where I_r is the intensity per unit frequency range of the radiation about the state r. Hence

$$\mathrm{I}_r = \mathrm{N}_r' (2\pi h) \nu_r^3/c^2, \qquad (28)$$

so that N_r' is proportional to I_r and $(\mathrm{N}_r' + 1)$ is proportional to $\mathrm{I}_r + (2\pi h)\nu_r^3/c^2$. We thus obtain that the probability of an absorption process is proportional to I_r, the incident intensity per unit frequency range, and that of an emission process is proportional to $\mathrm{I}_r + (2\pi h)\nu_r^3/c^2$, which are just Einstein's laws.* In the same way the probability of a process in which a light-quantum is scattered from a state r to a state s is proportional to $\mathrm{I}_r [\mathrm{I}_s + (2\pi h)\nu_r^3/c^2]$, which is Pauli's law for the scattering of radiation by an electron.†

§7. *The Probability Coefficients for Emission and Absorption.*

We shall now consider the interaction of an atom and radiation from the wave point of view. We resolve the radiation into its Fourier components, and suppose that their number is very large but finite. Let each component be labelled by a suffix r, and suppose there are σ_r components associated with the radiation of a definite type of polarisation per unit solid angle per unit frequency range about the component r. Each component r can be described by a vector potential κ_r chosen so as to make the scalar potential zero. The perturbation term to be added to the Hamiltonian will now be, according to the classical theory with neglect of relativity mechanics, $c^{-1}\Sigma_r \kappa_r \dot{\mathrm{X}}_r$, where X_r is the component of the total polarisation of the atom in the direction of κ_r, which is the direction of the electric vector of the component r.

We can, as explained in § 1, suppose the field to be described by the canonical variables N_r, θ_r, of which N_r is the number of quanta of energy of the component r, and θ_r is its canonically conjugate phase, equal to $2\pi h\nu_r$ times the θ_r of § 1. We shall now have $\kappa_r = a_r \cos \theta_r/h$, where a_r is the amplitude of κ_r, which can be connected with N_r as follows:—The flow of energy per unit area per unit time for the component r is $\frac{1}{2}\pi c^{-1} a_r^2 \nu_r^2$. Hence the intensity

* The ratio of stimulated to spontaneous emission in the present theory is just twice its value in Einstein's. This is because in the present theory either polarised component of the incident radiation can stimulate only radiation polarised in the same way, while in Einstein's the two polarised components are treated together. This remark applies also to the scattering process.

† Pauli, 'Z. f. Physik,' vol. 18, p. 272 (1923).

per unit frequency range of the radiation in the neighbourhood of the component r is $I_r = \frac{1}{2}\pi c^{-1} a_r^2 \nu_r^2 \sigma_r$. Comparing this with equation (28), we obtain $a_r = 2(h\nu_r/c\sigma_r)^{\frac{1}{2}} N_r^{\frac{1}{2}}$, and hence

$$\kappa_r = 2(h\nu_r/c\sigma_r)^{\frac{1}{2}} N_r^{\frac{1}{2}} \cos \theta_r/h.$$

The Hamiltonian for the whole system of atom plus radiation would now be, according to the classical theory,

$$F = H_P(J) + \Sigma_r (2\pi h\nu_r) N_r + 2c^{-1}\Sigma_r (h\nu_r/c\sigma_r)^{\frac{1}{2}} \dot{X}_r N_r^{\frac{1}{2}} \cos \theta_r/h, \qquad (29)$$

where $H_P(J)$ is the Hamiltonian for the atom alone. On the quantum theory we must make the variables N_r and θ_r canonical q-numbers like the variables J_k, w_k that describe the atom. We must now replace the $N_r^{\frac{1}{2}} \cos \theta_r/h$ in (29) by the real q-number

$$\tfrac{1}{2}\{N_r^{\frac{1}{2}} e^{i\theta r/h} + e^{-i\theta r/h} N_r^{\frac{1}{2}}\} = \tfrac{1}{2}\{N_r^{\frac{1}{2}} e^{i\theta r/h} + (N_r + 1)^{\frac{1}{2}} e^{-i\theta r/h}\}$$

so that the Hamiltonian (29) becomes

$$F = H_P(J) + \Sigma_r (2\pi h\nu_r) N_r + h^{\frac{1}{2}} c^{-\frac{3}{2}} \Sigma_r (\nu_r/\sigma_r)^{\frac{1}{2}} \dot{X}_r \{N_r^{\frac{1}{2}} e^{i\theta r/h} + (N_r + 1)^{\frac{1}{2}} e^{-i\theta r/h}\}. \qquad (30)$$

This is of the form (27), with

$$v_r = v_r^* = h^{\frac{1}{2}} c^{-\frac{3}{2}} (\nu_r/\sigma_r)^{\frac{1}{2}} \dot{X}_r \qquad (31)$$

and

$$v_{rs} = 0 \qquad (r, s \neq 0).$$

The wave point of view is thus consistent with the light-quantum point of view and gives values for the unknown interaction coefficient v_{rs} in the light-quantum theory. These values are not such as would enable one to express the interaction energy as an algebraic function of canonical variables. Since the wave theory gives $v_{rs} = 0$ for $r, s \neq 0$, it would seem to show that there are no direct scattering processes, but this may be due to an incompleteness in the present wave theory.

We shall now show that the Hamiltonian (30) leads to the correct expressions for Einstein's A's and B's. We must first modify slightly the analysis of § 5 so as to apply to the case when the system has a large number of discrete stationary states instead of a continuous range. Instead of equation (21) we shall now have

$$ih\,\dot{a}(\alpha') = \Sigma_{\alpha''} V(\alpha'\alpha'')\, a(\alpha'').$$

If the system is initially in the state α^0, we must take the initial value of $a(\alpha')$ to be $\delta_{\alpha'\alpha^0}$, which is now correctly normalised. This gives for a first approximation

$$ih\,\dot{a}(\alpha') = V(\alpha'\alpha^0) = v(\alpha'\alpha^0)\, e^{i[W(\alpha') - W(\alpha^0)]t/h},$$

which leads to

$$ih\, a(\alpha') = \delta_{\alpha'\alpha^0} + v(\alpha'\alpha^0) \frac{e^{i[W(\alpha') - W(\alpha^0)]t/h} - 1}{i[W(\alpha') - W(\alpha^0)]/h},$$

corresponding to (22). If, as before, we transform to the variables W, γ_1, $\gamma_2 \dots \gamma_{u-1}$, we obtain (when $\gamma' \neq \gamma^0$)

$$a(W'\gamma') = v(W', \gamma'; W^0, \gamma^0)\,[1 - e^{i(W'-W^0)t/h}]/(W'-W^0).$$

The probability of the system being in a state for which each γ_k equals γ_k' is $\Sigma_{W'} \, | a(W' \gamma')|^2$. If the stationary states lie close together and if the time t is not too great, we can replace this sum by the integral $(\Delta W)^{-1} \int |a(W'\gamma')|^2 dW'$, where ΔW is the separation between the energy levels. Evaluating this integral as before, we obtain for the probability per unit time of a transition to a state for which each $\gamma_k = \gamma_k'$

$$2\pi/h\Delta W \,.\, |v(W^0, \gamma'; W^0, \gamma^0)|^2. \tag{32}$$

In applying this result we can take the γ's to be any set of variables that are independent of the total proper energy W and that together with W define a stationary state.

We now return to the problem defined by the Hamiltonian (30) and consider an absorption process in which the atom jumps from the state J^0 to the state J' with the absorption of a light-quantum from state r. We take the variables γ' to be the variables J' of the atom together with variables that define the direction of motion and state of polarisation of the absorbed quantum, but not its energy. The matrix element $v(W^0, \gamma'; W^0, \gamma^0)$ is now

$$h^{1/2} c^{-3/2} (\nu_r/\sigma_r)^{1/2} \dot{X}_r (J^0 J') N_r^0,$$

where $\dot{X}_r(J^0J')$ is the ordinary (J^0J') matrix element of $\dot{X}_r$. Hence from (32) the probability per unit time of the absorption process is

$$\frac{2\pi}{h\Delta W} \cdot \frac{h\nu_r}{c^3\sigma_r} |\dot{X}_r(J^0J')|^2 N_r^0.$$

To obtain the probability for the process when the light-quantum comes from any direction in a solid angle $d\omega$, we must multiply this expression by the number of possible directions for the light-quantum in the solid angle $d\omega$, which is $d\omega\, \sigma_r \Delta W/2\pi h$. This gives

$$d\omega \frac{\nu_r}{hc^3} |\dot{X}_r(J_0J')|^2 N_r^0 = d\omega \frac{1}{2\pi h^2 c \nu_r^2} |\dot{X}_r(J^0J')|^2 I_r$$

with the help of (28). Hence the probability coefficient for the absorption process is $1/2\pi h^2 c \nu_r^2 \,.\, |\dot{X}_r(J^0J')|^2$, in agreement with the usual value for Einstein's absorption coefficient in the matrix mechanics. The agreement for the emission coefficients may be verified in the same manner.

The present theory, since it gives a proper account of spontaneous emission, must presumably give the effect of radiation reaction on the emitting system, and enable one to calculate the natural breadths of spectral lines, if one can overcome the mathematical difficulties involved in the general solution of the wave problem corresponding to the Hamiltonian (30). Also the theory enables one to understand how it comes about that there is no violation of the law of the conservation of energy when, say, a photo-electron is emitted from an atom under the action of extremely weak incident radiation. The energy of interaction of the atom and the radiation is a q-number that does not commute with the first integrals of the motion of the atom alone or with the intensity of the radiation. Thus one cannot specify this energy by a c-number at the same time that one specifies the stationary state of the atom and the intensity of the radiation by c-numbers. In particular, one cannot say that the interaction energy tends to zero as the intensity of the incident radiation tends to zero. There is thus always an unspecifiable amount of interaction energy which can supply the energy for the photo-electron.

I would like to express my thanks to Prof. Niels Bohr for his interest in this work and for much friendly discussion about it.

Summary.

The problem is treated of an assembly of similar systems satisfying the Einstein-Bose statistical mechanics, which interact with another different system, a Hamiltonian function being obtained to describe the motion. The theory is applied to the interaction of an assembly of light-quanta with an ordinary atom, and it is shown that it gives Einstein's laws for the emission and absorption of radiation.

The interaction of an atom with electromagnetic waves is then considered, and it is shown that if one takes the energies and phases of the waves to be q-numbers satisfying the proper quantum conditions instead of c-numbers, the Hamiltonian function takes the same form as in the light-quantum treatment. The theory leads to the correct expressions for Einstein's A's and B's.

Part IV

ENERGY OF ELEMENTARY PARTICLES

Editor's Comments on Papers 15 Through 20

The recent development of atomic physics has shown great concern for the so-called elementary particles, which form the constituent parts of atoms. When the nuclear atom model was first proposed in 1911 by Ernest Rutherford (1871–1937), the two fundamental particles were the negatively charged electron and the positively charged nucleus of the hydrogen atom. Realizing the probable special significance of the latter particle, Rutherford in 1920 decided to call it the *proton*. For a time, it was believed that the two elementary atomic particles were the electron and the proton. The most energetic electrons produced in J. J. Thomson's cathode-ray experiments[1] had energies of the order of 4.5×10^{-8} erg. The energy unit normally used in atomic physics is the electron (ev) or 1.6×10^{-12} erg. The electron energy just mentioned then becomes approximately 3×10^{4} ev. A still more common unit in modern research with atomic accelerating devices is the Mev, which is 10^{6} ev. The early cathode-ray experiments of Thomson therefore had energies of the order of 0.03 Mev. The electrons in the form of beta rays emitted naturally from radioactive substances have, of course, a wide range of energies, from almost zero to as much as 5 Mev. Protons are not emitted directly by radioactive substances, though Rutherford in 1920 produced them by bombarding nitrogen with alpha particles (helium nuclei). J. D. Cockcroft (1897–1967) and E. T. S. Walton (1903–) in their early 1932 experiments on high-voltage atomic accelerators produced protons with energies of about 0.5 Mev. This figure has been enormously increased in more recent times (see Papers 31 through 35).

For a time it was believed that the proton provided the mass of the nucleus and that the negative electrons would serve to reduce the net positive charge on the nucleus to the appropriate value for the corresponding atom in its normal state. The atomic number was defined (and still is) as the number of protons in the nucleus. This is equal to the number of electrons revolving around the nucleus in the normal un-ionized state of the atom. Further research in nuclear physics showed that there can be no free electrons in the nucleus. The fact that electrons are emitted from the nuclei of radioactive atoms in the form of beta rays then had to be explained in terms of the disintegration of an uncharged particle in the nucleus. The existence of such a particle with mass approximately equal to that of the proton was suspected as early as 1920. It was actually discovered and identified in 1932 by James Chadwick (1891–1974) and called the "neutron." Chadwick's first brief article on the subject is reproduced here as Paper 17. The references at the end of this section indicate the slightly later article giving the full details of the discovery.[2] With the discovery of the neutron, the nucleus was finally considered to be made up of protons and neutrons, the number of the latter being sufficient to provide (with the protons) for the mass of the nucleus without adding to its positive charge. The emission of a beta ray was then considered due to the disintegration of a neutron into an electron and a proton.

Interest in the neutron was further excited by the isolation in 1931 by Harold C. Urey (1893–1981), Ferdinand G. Brickwedde (1903–), and George M. Murphy (1903–) of the hydrogen isotope of mass 2, which was named *deuterium*. The corresponding nucleus, composed of a proton and a neutron, was named the *deuteron*. Heavy hydrogen in the form of its constituency of heavy water has proved of great theoretical and practical importance in modern physics and chemistry. The fundamental article announcing the existence of the deuteron is listed in the bibliography at the end of this section.

In the meantime, Dirac had concerned himself with the development of a relativistic theory of the electron. In a remarkable paper published in 1928, he was able not only to deduce electron spin and magnetic moment but also to derive the experimentally verified relativistic energy levels for the electron in the hydrogen atom. At the same time, his theory predicted the existence of a sea of negative kinetic energy states for the electron, which at first sight appear to have no physical meaning. The puzzle was ultimately resolved by assuming that normally all the negative energy states are occupied and can then show no physical effect, but that it is possible by the use of a sufficiently energetic photon to remove an electron from such a negative energy state, leaving a hole behind. Such a hole should behave as a positively charged particle of the same mass as the electron. This was the prediction of the positron, which is produced along with an electron by the disappearance of a photon with energy mc^2 (where m is the mass of the electron and c is the velocity of light in free space, that is, 3×10^{10}cm/sec) or 0.5 Mev. Such an electron-positron pair can then be annihilated with the emission of a photon of this energy.

The existence of positrons was demonstrated in 1933 by Carl David Anderson (1905-   ) in a cloud chamber experiment. Pair production of electrons and positrons has now become a familiar phenomenon with the use of high-energy particle accelerators.

Dirac's fundamental 1928 article is reproduced here in full as Paper 15. Anderson's announcement of the discovery of the positron is reproduced as Paper 16.

It had early been discovered that the electrons in the form of beta rays emitted by radioactive substances have a continuous range of energy, whereas the nuclei from which they come have discrete energy states. This was an apparent violation of the principle of conservation of energy, which for the most part has been successfully assumed to hold in atomic processes. It is true that in 1924 Bohr, Kramers, and Slater in a study of radiation from atoms thought it might be wise to assume that conservation of energy in such processes is of purely statistical character.[3] This point of view was later invalidated by a Compton scattering experiment performed in 1924 by Walther Bothe (1891–1957) and Hans Geiger (1882–1945). To preserve conservation of energy in the emissions of beta rays, Wolfgang Pauli proposed the existence of particles of zero charge and practically zero mass, which are emitted along with the beta-ray electrons. Enrico Fermi suggested that the new particle should be called a neutrino. Its emission would also guarantee conservation of momentum and angular momentum (spin), as well as of energy. We reproduce here as Paper 18 Pauli's proposal at the Solvay Congress at Brussels in 1933. Neutrinos were first detected experimentally in 1953. They occur in two forms: (1) those associated with beta-ray emission and (2) those connected with the decay of μ-mesons (muons) into pimesons (pions). (See Paper 19 for meson.) Moreover, for each neutrino there is an antineutrino, the distinction being that for the neutrino, the spin is right-handed with respect to the direction of motion, and for the antineutrino it is left-handed. These elementary particles play a decisive role in all so-called weak interactions in nuclear physics.

We go on to discuss the emergence of still another elementary particle, the meson. The prediction of the existence of such a particle was made by the Japanese physicist Hideki Yukawa (1907–1981) in 1935. He suggested that the strong interaction between neutrons and protons, responsible for the stable existence of a nucleus requires the mediation of a particle capable of carrying both negative and positive charges between the nucleons. His calculations also carried the prediction that such particles would have mass considerably in excess of the mass of the electron and would obey the Bose-Einstein statistics. Such particles were found in 1947 by C.F. Powell (1903–) in cosmic rays, and probably even earlier (see below), with mass equal to 270 times the mass of the electron. They received the name *meson,* or more particularly π-mesons, now known as pions. Another type of meson (known now as the muon, or μ-meson), was identified by Carl Anderson in cosmic rays in 1937. It has a mass 207 times that of the electron. Mesons turn up frequently in cosmic radiation and are also produced copiously in nuclear collisions resulting from the use of high-energy accelerating devices (See Part VI). Later studies of cosmic rays turned up still more massive mesons called K mesons, with masses of the order of a thousand times the electron mass. These are the playthings of the modern nuclear physicist, who may also be called the high-energy physicist. K mesons correspond to an intrinsic energy of nearly 500 Mev. The basic 1935 article of Yukawa is reproduced here as Paper 19.

Mention of cosmic rays in connection with the meson suggets that we should include something about the history of this form of radiation, which was inadvertently omitted in the previous Benchmark volume *Early Concepts of Energy in Atomic Physics.*[5] Early workers in radioactivity realized around 1900 that their ionization chambers were affected by external radiation independent of that due to radioactive sources. It was soon decided that this radiation was of extraterrestrial origin and was very penetrating. Some of the early precise work on the penetrating radiation was done by Victor F. Hess (1883–1964) in Austria; he first used balloon ascensions in its study. Hess's work provided convincing evidence of the source of this radiation outside the earth.[6]

The penetrating radiation began to be studied in earnest in the

1920s and was renamed *cosmic radiation* in allusion to its source. The subject has become an important part of the physics of high-energy elementary particles, since practically all such particles have been found in cosmic radiation, and before the days of high-energy atomic accelerators, cosmic rays provided the principal source of elementary particles. The whole subject has become so elaborate that only this brief historical reference is possible here. We do, however, include here an important 1937 article on the nature of cosmic-ray particles by Seth H. Neddermeyer (1907-   ) and Carl D. Anderson. They detected particles corresponding to the mesons predicted by Yukawa. We reproduce their article here as Paper 20.

Since 1950 concern with the elementary particle problem has focused on the various types of interaction between particles—for example, strong interaction forces between the neutron and the proton, the weak interactions involving the emission of beta rays, the classical electromagnetic interaction and gravitational force. At the same time, the high-energy-producing machines (see Part VI) turned up a relatively enormous number of particles masquerading as elementary. The present volume will not go into these details. We may merely comment that there has been a recent attempt to consider all elementary particles as composed of a subclass called quarks, which may have charges that are fractional parts of the charge in the electron. The theory seems to work fairly well in accounting for the behavior of certain of the older elementary particles, but so far no free quarks have been observed. Modern high-energy physics is, however, so elaborate that one hesitates to make any categorical statements about its success in explaining experience. Our main concern in the present volume has been to stress the role of energy in the atomic domain leading up to the present-day situation. (See reference to article on quarks.)

REFERENCES

1. J. J. Thomson, "Cathode Rays," *Philosophical Magazine,* ser. 5, **44** (1897):293-316. Reprinted as Paper 20 in R. Lindsay, ed., 1979, *Early Concepts of Energy in Atomic Physics,* Benchmark Papers on Energy, vol. 7, Stroudsburg, Pa.: Dowden, Hutchinson & Ross.
2. J. Chadwick, "The Existence of a Neutron," *Proceedings of the Royal Society (London)* **A136** (1932):692-708.
3. N. Bohr, H. A. Kramers, and J. C. Slater, "The Quantum Theory of Radiation," *Philosophical Magazine* **47** (1924):785-802.
4. W. W. G. Bothe and H. Geiger, "Ein weg zur experimentellen nachprüfung der theorie von Bohr, Kramers und Slater," *Zeitschrift für Physik* **26** (1924):44.
5. R. Lindsay, ed., 1979, *Early Concepts of Energy in Atomic Physics,*

Benchmark Papers on Energy, vol. 7, Stroudsburg, Pa.: Dowden, Hutchinson & Ross.

6. V. F. Hess, "Measurements of the Penetrating Radiation in Two Balloon Ascensions," *Sitzungsberichte der Kaiserlichen Akademie der Wissenschaften in Wien,* **120** (1911):1575–1585.
7. "Quarks," 1982, *Encyclopedia of Science and Technology,* 5th edition, vol. 11, New York: McGraw-Hill, pp. 188–191.

BIBLIOGRAPHY

Dirac, P. A. M., 1977, Recollections of an Exciting Era, History of Twentieth Century Physics, Proceedings of the International School of Physics "Enrico Fermi," Course 57, Academic, New York, pp. 141-146.

Feather, N., 1936, *An Introduction to Nuclear Physics,* Cambridge University Press, New York.

Kallen, G., 1964, Elementary Particle Physics, Addison-Wesley, Reading, Mass.

Rutherford, E., J. Chadwick, and C. D. Ellis, 1930, *Radiation from Radioactive Substances,* Cambridge University Press, New York, 1930.

Urey, H. C., F. G. Brickwedde, and G. M. Murphy, 1932, A Hydrogen Isotope of Mars 2 and Its Concentration, *Physical Review* **40:**1–15.

15

Reprinted from *Roy. Soc. (London) Proc.*, ser. A, **117**:610–624 (1927)

The Quantum Theory of the Electron.

By P. A. M. Dirac, St. John's College, Cambridge.

(Communicated by R. H. Fowler, F.R.S.—Received January 2, 1928.)

The new quantum mechanics, when applied to the problem of the structure of the atom with point-charge electrons, does not give results in agreement with experiment. The discrepancies consist of " duplexity " phenomena, the observed number of stationary states for an electron in an atom being twice the number given by the theory. To meet the difficulty, Goudsmit and Uhlenbeck have introduced the idea of an electron with a spin angular momentum of half a quantum and a magnetic moment of one Bohr magneton. This model for the electron has been fitted into the new mechanics by Pauli,* and Darwin,† working with an equivalent theory, has shown that it gives results in agreement with experiment for hydrogen-like spectra to the first order of accuracy.

The question remains as to why Nature should have chosen this particular model for the electron instead of being satisfied with the point-charge. One would like to find some incompleteness in the previous methods of applying quantum mechanics to the point-charge electron such that, when removed, the whole of the duplexity phenomena follow without arbitrary assumptions. In the present paper it is shown that this is the case, the incompleteness of the previous theories lying in their disagreement with relativity, or, alternatetively, with the general transformation theory of quantum mechanics. It appears that the simplest Hamiltonian for a point-charge electron satisfying the requirements of both relativity and the general transformation theory leads to an explanation of all duplexity phenomena without further assumption. All the same there is a great deal of truth in the spinning electron model, at least as a first approximation. The most important failure of the model seems to be that the magnitude of the resultant orbital angular momentum of an electron moving in an orbit in a central field of force is not a constant, as the model leads one to expect.

* Pauli, 'Z. f. Physik,' vol. 43, p. 601 (1927).
† Darwin, 'Roy. Soc. Proc.,' A, vol. 116, p. 227 (1927).

§ 1. *Previous Relativity Treatments.*

The relativity Hamiltonian according to the classical theory for a point electron moving in an arbitrary electro-magnetic field with scalar potential A_0 and vector potential A is

$$F \equiv \left(\frac{W}{c} + \frac{e}{c}A_0\right)^2 + \left(\mathbf{p} + \frac{e}{c}\mathbf{A}\right)^2 + m^2c^2,$$

where $\mathbf{p}$ is the momentum vector. It has been suggested by Gordon* that the operator of the wave equation of the quantum theory should be obtained from this F by the same procedure as in non-relativity theory, namely, by putting

$$W = ih\frac{\partial}{\partial t},$$

$$p_r = -ih\frac{\partial}{\partial x_r}, \qquad r = 1, 2, 3,$$

in it. This gives the wave equation

$$F\psi \equiv \left[\left(ih\frac{\partial}{c\,\partial t} + \frac{e}{c}A_0\right)^2 + \Sigma_r\left(-ih\frac{\partial}{\partial x_r} + \frac{e}{c}A_r\right)^2 + m^2c^2\right]\psi = 0, \quad (1)$$

the wave function ψ being a function of x_1, x_2, x_3, t. This gives rise to two difficulties.

The first is in connection with the physical interpretation of ψ. Gordon, and also independently Klein,† from considerations of the conservation theorems, make the assumption that if ψ_m, ψ_n are two solutions

$$\rho_{mn} = -\frac{e}{2mc^2}\left\{ih\left(\psi_m\frac{\partial\bar{\psi}_n}{\partial t} - \bar{\psi}_n\frac{\partial\psi_m}{\partial t}\right) + 2eA_0\psi_m\bar{\psi}_n\right\}$$

and

$$\mathbf{I}_{mn} = -\frac{e}{2m}\left\{-ih\left(\psi_m \operatorname{grad}\bar{\psi}_n - \bar{\psi}_n \operatorname{grad}\psi_m\right) + 2\frac{e}{c}\mathbf{A}_m\psi_m\bar{\psi}_n\right\}$$

are to be interpreted as the charge and current associated with the transition $m \rightarrow n$. This appears to be satisfactory so far as emission and absorption of radiation are concerned, but is not so general as the interpretation of the non-relativity quantum mechanics, which has been developed‡ sufficiently to enable one to answer the question: What is the probability of any dynamical variable

* Gordon, 'Z. f. Physik,' vol. 40, p. 117 (1926).

† Klein, 'Z. f. Physik,' vol. 41, p. 407 (1927).

‡ Jordan, 'Z. f. Physik,' vol. 40, p. 809 (1927); Dirac, 'Roy. Soc. Proc.,' A, vol. 113, p. 621 (1927).

at any specified time having a value lying between any specified limits, when the system is represented by a given wave function ψ_n? The Gordon-Klein interpretation can answer such questions if they refer to the position of the electron (by the use of ρ_{nn}), but not if they refer to its momentum, or angular momentum or any other dynamical variable. We should expect the interpretation of the relativity theory to be just as general as that of the non-relativity theory.

The general interpretation of non-relativity quantum mechanics is based on the transformation theory, and is made possible by the wave equation being of the form

$$(H - W)\,\psi = 0, \tag{2}$$

i.e., being linear in W or $\partial/\partial t$, so that the wave function at any time determines the wave function at any later time. The wave equation of the relativity theory must also be linear in W if the general interpretation is to be possible.

The second difficulty in Gordon's interpretation arises from the fact that if one takes the conjugate imaginary of equation (1), one gets

$$\left[\left(-\frac{W}{c} + \frac{e}{c}A_0\right)^2 + \left(-\mathbf{p} + \frac{e}{c}\mathbf{A}\right)^2 + m^2c^2\right]\psi = 0,$$

which is the same as one would get if one put $-e$ for e. The wave equation (1) thus refers equally well to an electron with charge e as to one with charge $-e$. If one considers for definiteness the limiting case of large quantum numbers one would find that some of the solutions of the wave equation are wave packets moving in the way a particle of charge $-e$ would move on the classical theory, while others are wave packets moving in the way a particle of charge e would move classically. For this second class of solutions W has a negative value. One gets over the difficulty on the classical theory by arbitrarily excluding those solutions that have a negative W. One cannot do this on the quantum theory, since in general a perturbation will cause transitions from states with W positive to states with W negative. Such a transition would appear experimentally as the electron suddenly changing its charge from $-e$ to e, a phenomenon which has not been observed. The true relativity wave equation should thus be such that its solutions split up into two non-combining sets, referring respectively to the charge $-e$ and the charge e.

In the present paper we shall be concerned only with the removal of the first of these two difficulties. The resulting theory is therefore still only an approximation, but it appears to be good enough to account for all the duplexity phenomena without arbitrary assumptions.

§ 2. *The Hamiltonian for No Field.*

Our problem is to obtain a wave equation of the form (2) which shall be invariant under a Lorentz transformation and shall be equivalent to (1) in the limit of large quantum numbers. We shall consider first the case of no field, when equation (1) reduces to

$$(-p_0^2 + \mathbf{p}^2 + m^2c^2)\,\psi = 0 \tag{3}$$

if one puts

$$p_0 = \frac{W}{c} = ih\frac{\partial}{c\,\partial t}.$$

The symmetry between p_0 and p_1, p_2, p_3 required by relativity shows that, since the Hamiltonian we want is linear in p_0, it must also be linear in p_1, p_2 and p_3. Our wave equation is therefore of the form

$$(p_0 + \alpha_1 p_1 + \alpha_2 p_2 + \alpha_3 p_3 + \beta)\,\psi = 0, \tag{4}$$

where for the present all that is known about the dynamical variables or operators α_1, α_2, α_3, β is that they are independent of p_0, p_1, p_2, p_3, *i.e.*, that they commute with t, x_1, x_2, x_3. Since we are considering the case of a particle moving in empty space, so that all points in space are equivalent, we should expect the Hamiltonian not to involve t, x_1, x_2, x_3. This means that α_1, α_2, α_3, β are independent of t, x_1, x_2, x_3, *i.e.*, that they commute with p_0, p_1, p_2, p_3. We are therefore obliged to have other dynamical variables besides the co-ordinates and momenta of the electron, in order that α_1, α_2, α_3, β may be functions of them. The wave function ψ must then involve more variables than merely x_1, x_2, x_3, t.

Equation (4) leads to

$$\begin{aligned} 0 &= (-p_0 + \alpha_1 p_1 + \alpha_2 p_2 + \alpha_3 p_3 + \beta)(p_0 + \alpha_1 p_1 + \alpha_2 p_2 + \alpha_3 p_3 + \beta)\,\psi \\ &= [-p_0^2 + \Sigma\,\alpha_1^2 p_1^2 + \Sigma\,(\alpha_1\alpha_2 + \alpha_2\alpha_1)\,p_1 p_2 + \beta^2 + \Sigma\,(\alpha_1\beta + \beta\alpha_1)\,p_1]\,\psi, \end{aligned} \tag{5}$$

where the Σ refers to cyclic permutation of the suffixes 1, 2, 3. This agrees with (3) if

$$\left.\begin{aligned} \alpha_r^2 &= 1, & \alpha_r\alpha_s + \alpha_s\alpha_r &= 0 \quad (r \neq s) \\ \beta^2 &= m^2c^2, & \alpha_r\beta + \beta\alpha_r &= 0 \end{aligned}\right\} \quad r, s = 1, 2, 3.$$

If we put $\beta = \alpha_4 mc$, these conditions become

$$\alpha_\mu^2 = 1 \qquad \alpha_\mu\alpha_\nu + \alpha_\nu\alpha_\mu = 0\ (\mu \neq \nu) \qquad \mu, \nu = 1, 2, 3, 4. \tag{6}$$

We can suppose the α_μ's to be expressed as matrices in some matrix scheme, the matrix elements of α_μ being, say, $\alpha_\mu(\zeta'\,\zeta'')$. The wave function ψ must now be a function of ζ as well as x_1, x_2, x_3, t. The result of α_μ multiplied into ψ will be a function $(\alpha_\mu \psi)$ of x_1, x_2, x_3, t, ζ defined by

$$(\alpha_\mu \psi)(x, t, \zeta) = \Sigma_{\zeta'} \alpha_\mu (\zeta \zeta') \psi (x, t, \zeta').$$

We must now find four matrices α_μ to satisfy the conditions (6). We make use of the matrices

$$\sigma_1 = \begin{pmatrix} 0 & 1 \\ 1 & 0 \end{pmatrix} \qquad \sigma_2 = \begin{pmatrix} 0 & -i \\ i & 0 \end{pmatrix} \qquad \sigma_3 = \begin{pmatrix} 1 & 0 \\ 0 & -1 \end{pmatrix}$$

which Pauli introduced* to describe the three components of spin angular momentum. These matrices have just the properties

$$\sigma_r^2 = 1 \qquad \sigma_r \sigma_s + \sigma_s \sigma_r = 0, \qquad (r \neq s), \tag{7}$$

that we require for our α's. We cannot, however, just take the σ's to be three of our α's, because then it would not be possible to find the fourth. We must extend the σ's in a diagonal manner to bring in two more rows and columns, so that we can introduce three more matrices ρ_1, ρ_2, ρ_3 of the same form as σ_1, σ_2, σ_3, but referring to different rows and columns, thus :—

$$\sigma_1 = \begin{Bmatrix} 0 & 1 & 0 & 0 \\ 1 & 0 & 0 & 0 \\ 0 & 0 & 0 & 1 \\ 0 & 0 & 1 & 0 \end{Bmatrix} \quad \sigma_2 = \begin{Bmatrix} 0 & -i & 0 & 0 \\ i & 0 & 0 & 0 \\ 0 & 0 & 0 & -i \\ 0 & 0 & i & 0 \end{Bmatrix} \quad \sigma_3 = \begin{Bmatrix} 1 & 0 & 0 & 0 \\ 0 & -1 & 0 & 0 \\ 0 & 0 & 1 & 0 \\ 0 & 0 & 0 & -1 \end{Bmatrix},$$

$$\rho_1 = \begin{Bmatrix} 0 & 0 & 1 & 0 \\ 0 & 0 & 0 & 1 \\ 1 & 0 & 0 & 0 \\ 0 & 1 & 0 & 0 \end{Bmatrix} \quad \rho_2 = \begin{Bmatrix} 0 & 0 & -i & 0 \\ 0 & 0 & 0 & -i \\ i & 0 & 0 & 0 \\ 0 & i & 0 & 0 \end{Bmatrix} \quad \rho_3 = \begin{Bmatrix} 1 & 0 & 0 & 0 \\ 0 & 1 & 0 & 0 \\ 0 & 0 & -1 & 0 \\ 0 & 0 & 0 & -1 \end{Bmatrix}.$$

The ρ's are obtained from the σ's by interchanging the second and third rows, and the second and third columns. We now have, in addition to equations (7)

$$\text{and also} \qquad \left.\begin{array}{lll} \rho_r^2 = 1 & \rho_r \rho_s + \rho_s \rho_r = 0 & (r \neq s), \\ & \rho_r \sigma_t = \sigma_t \rho_r. & \end{array}\right\}. \tag{7'}$$

* Pauli, *loc. cit.*

If we now take

$$\alpha_1 = \rho_1\sigma_1, \qquad \alpha_2 = \rho_1\sigma_2, \qquad \alpha_3 = \rho_1\sigma_3, \qquad \alpha_4 = \rho_3,$$

all the conditions (6) are satisfied, *e.g.*,

$$\alpha_1^2 = \rho_1\sigma_1\rho_1\sigma_1 = \rho_1^2\sigma_1^2 = 1$$

$$\alpha_1\alpha_2 = \rho_1\sigma_1\rho_1\sigma_2 = \rho_1^2\sigma_1\sigma_2 = -\rho_1^2\sigma_2\sigma_1 = -\alpha_2\alpha_1.$$

The following equations are to be noted for later reference

$$\left.\begin{aligned} \rho_1\rho_2 = i\rho_3 = -\rho_2\rho_1 \\ \sigma_1\sigma_2 = i\sigma_3 = -\sigma_2\sigma_1 \end{aligned}\right\}, \tag{8}$$

together with the equations obtained by cyclic permutation of the suffixes.

The wave equation (4) now takes the form

$$[p_0 + \rho_1(\boldsymbol{\sigma}, \mathbf{p}) + \rho_3 mc]\,\psi = 0, \tag{9}$$

where $\boldsymbol{\sigma}$ denotes the vector $(\sigma_1, \sigma_2, \sigma_3)$.

§ 3. *Proof of Invariance under a Lorentz Transformation.*

Multiply equation (9) by ρ_3 on the left-hand side. It becomes, with the help of (8),

$$[\rho_3 p_0 + i\rho_2(\sigma_1 p_1 + \sigma_2 p_2 + \sigma_3 p_3) + mc]\,\psi = 0.$$

Putting

$$p_0 = ip_4,$$

$$\rho_3 = \gamma_4, \qquad \rho_2\sigma_r = \gamma_r, \qquad r = 1, 2, 3, \tag{10}$$

we have

$$[i\Sigma\gamma_\mu p_\mu + mc]\,\psi = 0, \qquad \mu = 1, 2, 3, 4. \tag{11}$$

The p_μ transform under a Lorentz transformation according to the law

$$p_\mu' = \Sigma_\nu a_{\mu\nu} p_\nu,$$

where the coefficients $a_{\mu\nu}$ are c-numbers satisfying

$$\Sigma_\mu a_{\mu\nu} a_{\mu\tau} = \delta_{\nu\tau}, \qquad \Sigma_\tau a_{\mu\tau} a_{\nu\tau} = \delta_{\mu\nu}.$$

The wave equation therefore transforms into

$$[i\Sigma\gamma_\mu' p_\mu' + mc]\,\psi = 0, \tag{12}$$

where

$$\gamma_\mu' = \Sigma_\nu a_{\mu\nu}\gamma_\nu.$$

Now the γ_μ, like the α_μ, satisfy

$$\gamma_\mu^2 = 1, \qquad \gamma_\mu\gamma_\nu + \gamma_\nu\gamma_\mu = 0, \qquad (\mu \neq \nu).$$

These relations can be summed up in the single equation

$$\gamma_\mu \gamma_\nu + \gamma_\nu \gamma_\mu = 2\delta_{\mu\nu}.$$

We have

$$\begin{aligned}\gamma_\mu' \gamma_\nu' + \gamma_\nu' \gamma_\mu' &= \Sigma_{\tau\lambda}\, a_{\mu\tau}\, a_{\nu\lambda}\, (\gamma_\tau \gamma_\lambda + \gamma_\lambda \gamma_\tau) \\ &= 2\Sigma_{\tau\lambda}\, a_{\mu\tau}\, a_{\nu\lambda}\, \delta_{\tau\lambda} \\ &= 2\Sigma_\tau\, a_{\mu\tau}\, a_{\nu\tau} = 2\delta_{\mu\nu}.\end{aligned}$$

Thus the γ_μ' satisfy the same relations as the γ_μ. Thus we can put, analogously to (10)

$$\gamma_4' = \rho_3' \qquad \gamma_r' = \rho_2' \sigma_r'$$

where the ρ''s and σ''s are easily verified to satisfy the relations corresponding to (7), (7') and (8), if ρ_2' and ρ_1' are defined by $\rho_2' = -i\gamma_1'\gamma_2'\gamma_3'$, $\rho_1' = -i\rho_2'\rho_3'$.

We shall now show that, by a canonical transformation, the ρ''s and σ''s may be brought into the form of the ρ's and σ's. From the equation $\rho_3'^2 = 1$, it follows that the only possible characteristic values for ρ_3' are ± 1. If one applies to ρ_3' a canonical transformation with the transformation function ρ_1', the result is

$$\rho_1' \rho_3' (\rho_1')^{-1} = -\rho_3' \rho_1' (\rho_1')^{-1} = -\rho_3'.$$

Since characteristic values are not changed by a canonical transformation, ρ_3' must have the same characteristic values as $-\rho_3'$. Hence the characteristic values of ρ_3' are $+1$ twice and -1 twice. The same argument applies to each of the other ρ''s, and to each of the σ''s.

Since ρ_3' and σ_3' commute, they can be brought simultaneously to the diagonal form by a canonical transformation. They will then have for their diagonal elements each $+1$ twice and -1 twice. Thus, by suitably rearranging the rows and columns, they can be brought into the form ρ_3 and σ_3 respectively. (The possibility $\rho_3' = \pm \sigma_3'$ is excluded by the existence of matrices that commute with one but not with the other.)

Any matrix containing four rows and columns can be expressed as

$$c + \Sigma_r c_r \sigma_r + \Sigma_r c_r' \rho_r + \Sigma_{rs} c_{rs} \rho_r \sigma_s \qquad (13)$$

where the sixteen coefficients c, c_r, c_r', c_{rs} are c-numbers. By expressing σ_1' in this way, we see, from the fact that it commutes with $\rho_3' = \rho_3$ and anticommutes* with $\sigma_3' = \sigma_3$, that it must be of the form

$$\sigma_1' = c_1\sigma_1 + c_2\sigma_2 + c_{31}\rho_3\sigma_1 + c_{32}\rho_3\sigma_2,$$

* We say that a anticommutes with b when $ab = -ba$.

i.e., of the form

$$\sigma_1' = \begin{Bmatrix} 0 & a_{12} & 0 & 0 \\ a_{21} & 0 & 0 & 0 \\ 0 & 0 & 0 & a_{34} \\ 0 & 0 & a_{43} & 0 \end{Bmatrix}$$

The condition $\sigma_1'^2 = 1$ shows that $a_{12}a_{21} = 1$, $a_{34}a_{43} = 1$. If we now apply the canonical transformation : first row to be multiplied by $(a_{21}/a_{12})^{\frac{1}{2}}$ and third row to be multiplied by $(a_{43}/a_{34})^{\frac{1}{2}}$, and first and third columns to be divided by the same expressions, σ_1' will be brought into the form of σ_1, and the diagonal matrices σ_3' and ρ_3' will not be changed.

If we now express ρ_1' in the form (13) and use the conditions that it commutes with $\sigma_1' = \sigma_1$ and $\sigma_3' = \sigma_3$ and anticommutes with $\rho_3' = \rho_3$, we see that it must be of the form

$$\rho_1' = c_1'\rho_1 + c_2'\rho_2.$$

The condition $\rho_1'^2 = 1$ shows that $c_1'^2 + c_2'^2 = 1$, or $c_1' = \cos\theta$, $c_2' = \sin\theta$. Hence ρ_1' is of the form

$$\rho_1' = \begin{Bmatrix} 0 & 0 & e^{-i\theta} & 0 \\ 0 & 0 & 0 & e^{-i\theta} \\ e^{i\theta} & 0 & 0 & 0 \\ 0 & e^{i\theta} & 0 & 0 \end{Bmatrix}$$

If we now apply the canonical transformation : first and second rows to be multiplied by $e^{i\theta}$ and first and second columns to be divided by the same expression, ρ_1' will be brought into the form ρ_1, and σ_1, σ_3, ρ_3 will not be altered. ρ_2' and σ_2' must now be of the form ρ_2 and σ_2, on account of the relations $i\rho_2' = \rho_3'\rho_1'$, $i\sigma_2' = \sigma_3'\sigma_1'$.

Thus by a succession of canonical transformations, which can be combined to form a single canonical transformation, the ρ''s and σ''s can be brought into the form of the ρ's and σ's. The new wave equation (12) can in this way be brought back into the form of the original wave equation (11) or (9), so that the results that follow from this original wave equation must be independent of the frame of reference used.

§ 4. *The Hamiltonian for an Arbitrary Field.*

To obtain the Hamiltonian for an electron in an electromagnetic field with scalar potential A_0 and vector potential $\mathbf{A}$, we adopt the usual procedure of substituting $p_0 + e/c \,.\, A_0$ for p_0 and $\mathbf{p} + e/c \,.\, \mathbf{A}$ for $\mathbf{p}$ in the Hamiltonian for no field. From equation (9) we thus obtain

$$\left[p_0 + \frac{e}{c}A_0 + \rho_1\left(\boldsymbol{\sigma}, \mathbf{p} + \frac{e}{c}\mathbf{A}\right) + \rho_3 mc\right]\psi = 0. \tag{14}$$

This wave equation appears to be sufficient to account for all the duplexity phenomena. On account of the matrices ρ and σ containing four rows and columns, it will have four times as many solutions as the non-relativity wave equation, and twice as many as the previous relativity wave equation (1). Since half the solutions must be rejected as referring to the charge $+ e$ on the electron, the correct number will be left to account for duplexity phenomena. The proof given in the preceding section of invariance under a Lorentz transformation applies equally well to the more general wave equation (14).

We can obtain a rough idea of how (14) differs from the previous relativity wave equation (1) by multiplying it up analogously to (5). This gives, if we write e' for e/c

$$\begin{aligned} 0 &= [-(p_0 + c'A_0) + \rho_1(\boldsymbol{\sigma}, \mathbf{p} + e'\mathbf{A}) + \rho_3 mc] \\ &\qquad \times [(p_0 + e'A_0) + \rho_1(\boldsymbol{\sigma}, \mathbf{p} + e'\mathbf{A}) + \rho_3 mc]\,\psi \\ &= [-(p_0 + e'A_0)^2 + (\boldsymbol{\sigma}, \mathbf{p} + e'\mathbf{A})^2 + m^2c^2 \\ &\qquad + \rho_1\{(\boldsymbol{\sigma}, \mathbf{p} + e'\mathbf{A})(p_0 + e'A_0) - (p_0 + e'A_0)(\boldsymbol{\sigma}, \mathbf{p} + e'\mathbf{A})\}]\,\psi. \end{aligned} \tag{15}$$

We now use the general formula, that if $\mathbf{B}$ and $\mathbf{C}$ are any two vectors that commute with $\boldsymbol{\sigma}$

$$\begin{aligned} (\boldsymbol{\sigma}, \mathbf{B})(\boldsymbol{\sigma}, \mathbf{C}) &= \Sigma\, \sigma_1^2 B_1C_1 + \Sigma\,(\sigma_1\sigma_2 B_1C_2 + \sigma_2\sigma_1 B_2C_1) \\ &= (\mathbf{B}, \mathbf{C}) + i\,\Sigma\, \sigma_3 (B_1C_2 - B_2C_1) \\ &= (\mathbf{B}, \mathbf{C}) + i\,(\boldsymbol{\sigma}, \mathbf{B} \times \mathbf{C}). \end{aligned} \tag{16}$$

Taking $\mathbf{B} = \mathbf{C} = \mathbf{p} + e'\mathbf{A}$, we find

$$\begin{aligned} (\boldsymbol{\sigma}, \mathbf{p} + e'\mathbf{A})^2 &= (\mathbf{p} + e'\mathbf{A})^2 + i\,\Sigma\, \sigma_3 \\ &\qquad [(p_1 + e'A_1)(p_2 + e'A_2) - (p_2 + c'A_2)(p_1 + e'A_1)] \\ &= (\mathbf{p} + e'\mathbf{A})^2 + he'(\boldsymbol{\sigma}, \operatorname{curl} \mathbf{A}). \end{aligned}$$

Thus (15) becomes

$$0 = \Big[-(p_0 + e'A_0)^2 + (\mathbf{p} + e'\mathbf{A})^2 + m^2c^2 + e'h(\boldsymbol{\sigma}, \text{curl } \mathbf{A}) - ie'h\rho_1\left(\boldsymbol{\sigma}, \text{grad } A_0 + \frac{1}{c}\frac{\partial \mathbf{A}}{\partial t}\right)\Big]\psi$$

$$= [-(p_0 + e'A_0)^2 + (\mathbf{p} + e'\mathbf{A})^2 + m^2c^2 + e'h(\boldsymbol{\sigma}, \mathbf{H}) + ie'h\rho_1(\boldsymbol{\sigma}, \mathbf{E})]\,\psi,$$

where **E** and **H** are the electric and magnetic vectors of the field.

This differs from (1) by the two extra terms

$$\frac{eh}{c}(\boldsymbol{\sigma}, \mathbf{H}) + \frac{ieh}{c}\rho_1(\boldsymbol{\sigma}, \mathbf{E})$$

in F. These two terms, when divided by the factor $2m$, can be regarded as the additional potential energy of the electron due to its new degree of freedom. The electron will therefore behave as though it has a magnetic moment $eh/2mc\,.\,\boldsymbol{\sigma}$ and an electric moment $ieh/2mc\,.\,\rho_1\boldsymbol{\sigma}$. This magnetic moment is just that assumed in the spinning electron model. The electric moment, being a pure imaginary, we should not expect to appear in the model. It is doubtful whether the electric moment has any physical meaning, since the Hamiltonian in (14) that we started from is real, and the imaginary part only appeared when we multiplied it up in an artificial way in order to make it resemble the Hamiltonian of previous theories.

§ 5. *The Angular Momentum Integrals for Motion in a Central Field.*

We shall consider in greater detail the motion of an electron in a central field of force. We put $\mathbf{A} = 0$ and $e'A_0 = V(r)$, an arbitrary function of the radius r, so that the Hamiltonian in (14) becomes

$$F \equiv p_0 + V + \rho_1(\boldsymbol{\sigma}, \mathbf{p}) + \rho_3 mc.$$

We shall determine the periodic solutions of the wave equation $F\psi = 0$, which means that p_0 is to be counted as a parameter instead of an operator; it is, in fact, just $1/c$ times the energy level.

We shall first find the angular momentum integrals of the motion. The orbital angular momentum **m** is defined by

$$\mathbf{m} = \mathbf{x} \times \mathbf{p},$$

and satisfies the following " Vertauschungs " relations

$$\left.\begin{aligned} m_1x_1 - x_1m_1 &= 0, & m_1x_2 - x_2m_1 &= ihx_3 \\ m_1p_1 - p_1m_1 &= 0, & m_1p_2 - p_2m_1 &= ihp_3 \\ \mathbf{m} \times \mathbf{m} &= ih\mathbf{m}, & \mathbf{m}^2m_1 - m_1\mathbf{m}^2 &= 0, \end{aligned}\right\}, \qquad (17)$$

together with similar relations obtained by permuting the suffixes. Also $\mathbf{m}$ commutes with r, and with p_r, the momentum canonically conjugate to r.

We have

$$\begin{aligned} m_1\mathrm{F} - \mathrm{F}m_1 &= \rho_1 \{m_1 (\boldsymbol{\sigma}, \mathbf{p}) - (\boldsymbol{\sigma}, \mathbf{p}) m_1\} \\ &= \rho_1 (\boldsymbol{\sigma}, m_1\mathbf{p} - \mathbf{p}m_1) \\ &= ih\rho_1 (\sigma_2 p_3 - \sigma_3 p_2), \end{aligned}$$

and so

$$\mathbf{m}\mathrm{F} - \mathrm{F}\mathbf{m} = ih\rho_1 \, \boldsymbol{\sigma} \times \mathbf{p}. \tag{18}$$

Thus $\mathbf{m}$ is not a constant of the motion. We have further

$$\begin{aligned} \sigma_1\mathrm{F} - \mathrm{F}\sigma_1 &= \rho_1 \{\sigma_1 (\boldsymbol{\sigma}, \mathbf{p}) - (\boldsymbol{\sigma}, \mathbf{p}) \sigma_1\} \\ &= \rho_1 (\sigma_1 \boldsymbol{\sigma} - \boldsymbol{\sigma} \sigma_1, \mathbf{p}) \\ &= 2i\rho_1 (\sigma_3 p_2 - \sigma_2 p_3), \end{aligned}$$

with the help of (8), and so

$$\boldsymbol{\sigma}\mathrm{F} - \mathrm{F}\boldsymbol{\sigma} = -2i\rho_1 \, \boldsymbol{\sigma} \times \mathbf{p}.$$

Hence

$$(\mathbf{m} + \tfrac{1}{2}h\boldsymbol{\sigma})\,\mathrm{F} - \mathrm{F}\,(\mathbf{m} + \tfrac{1}{2}h\boldsymbol{\sigma}) = 0.$$

Thus $\mathbf{m} + \frac{1}{2}h\boldsymbol{\sigma}$ (= $\mathbf{M}$ say) is a constant of the motion. We can interpret this result by saying that the electron has a spin angular momentum of $\frac{1}{2}h\boldsymbol{\sigma}$, which, added to the orbital angular momentum $\mathbf{m}$, gives the total angular momentum $\mathbf{M}$, which is a constant of the motion.

The Vertauschungs relations (17) all hold when M's are written for the m's. In particular

$$\mathbf{M} \times \mathbf{M} = ih\mathbf{M} \quad \text{and} \quad \mathbf{M}^2\mathrm{M}_3 = \mathrm{M}_3\mathbf{M}^2.$$

M_3 will be an action variable of the system. Since the characteristic values of m_3 must be integral multiples of h in order that the wave function may be single-valued, the characteristic values of M_3 must be half odd integral multiples of h. If we put

$$\mathbf{M}^2 = (j^2 - \tfrac{1}{4})\,h^2, \tag{19}$$

j will be another quantum number, and the characteristic values of M_3 will extend from $(j - \frac{1}{2})\,h$ to $(-j + \frac{1}{2})\,h$.* Thus j takes integral values.

One easily verifies from (18) that $\mathbf{m}^2$ does not commute with F, and is thus not a constant of the motion. This makes a difference between the present theory and the previous spinning electron theory, in which $\mathbf{m}^2$ is constant, and defines the azimuthal quantum number k by a relation similar to (19). We shall find that our j plays the same part as the k of the previous theory.

* See 'Roy. Soc. Proc.,' A, vol. 111, p. 281 (1926).

§ 6. *The Energy Levels for Motion in a Central Field.*

We shall now obtain the wave equation as a differential equation in r, with the variables that specify the orientation of the whole system removed. We can do this by the use only of elementary non-commutative algebra in the following way.

In formula (16) take $\mathbf{B} = \mathbf{C} = \mathbf{m}$. This gives

$$\begin{aligned}(\boldsymbol{\sigma}, \mathbf{m})^2 &= \mathbf{m}^2 + i\,(\boldsymbol{\sigma}, \mathbf{m} \times \mathbf{m}) \qquad (20)\\ &= (\mathbf{m} + \tfrac{1}{2}h\,\boldsymbol{\sigma})^2 - h\,(\boldsymbol{\sigma}, \mathbf{m}) - \tfrac{1}{4}h^2\,\boldsymbol{\sigma}^2 - h\,(\boldsymbol{\sigma}, \mathbf{m})\\ &= \mathbf{M}^2 - 2h\,(\boldsymbol{\sigma}, \mathbf{m}) - \tfrac{3}{4}h^2.\end{aligned}$$

Hence

$$\{(\boldsymbol{\sigma}, \mathbf{m}) + h\}^2 = \mathbf{M}^2 + \tfrac{1}{4}h^2 = j^2h^2.$$

Up to the present we have defined j only through j^2, so that we could now, if we liked, take jh equal to $(\boldsymbol{\sigma}, \mathbf{m}) + h$. This would not be convenient since we want j to be a constant of the motion while $(\boldsymbol{\sigma}, \mathbf{m}) + h$ is not, although its square is. We have, in fact, by another application of (16),

$$(\boldsymbol{\sigma}, \mathbf{m})\,(\boldsymbol{\sigma}, \mathbf{p}) = i\,(\boldsymbol{\sigma}, \mathbf{m} \times \mathbf{p})$$

since $(\mathbf{m}, \mathbf{p}) = 0$, and similarly

$$(\boldsymbol{\sigma}, \mathbf{p})\,(\boldsymbol{\sigma}, \mathbf{m}) = i\,(\boldsymbol{\sigma}, \mathbf{p} \times \mathbf{m}),$$

so that

$$\begin{aligned}(\boldsymbol{\sigma}, \mathbf{m})\,(\boldsymbol{\sigma}, \mathbf{p}) + (\boldsymbol{\sigma}, \mathbf{p})\,(\boldsymbol{\sigma}, \mathbf{m}) &= i\Sigma\sigma_1\,(m_2p_3 - m_3p_2 + p_2m_3 - p_3m_2)\\ &= i\Sigma\sigma_1 \,.\, 2ihp_1 = -2h\,(\boldsymbol{\sigma}, \mathbf{p}),\end{aligned}$$

or

$$\{(\boldsymbol{\sigma}, \mathbf{m}) + h\}\,(\boldsymbol{\sigma}, \mathbf{p}) + (\boldsymbol{\sigma}, \mathbf{p})\,\{(\boldsymbol{\sigma}, \mathbf{m}) + h\} = 0.$$

Thus $(\boldsymbol{\sigma}, \mathbf{m}) + h$ anticommutes with one of the terms in F, namely, $\rho_1\,(\boldsymbol{\sigma}, \mathbf{p})$, and commutes with the other three. Hence $\rho_3\,\{(\boldsymbol{\sigma}, \mathbf{m}) + h\}$ commutes with all four, and is therefore a constant of the motion. But the square of $\rho_3\{(\boldsymbol{\sigma}, \mathbf{m}) + h\}$ must also equal j^2h^2. We therefore take

$$jh = \rho_3\,\{(\boldsymbol{\sigma}, \mathbf{m}) + h\}. \qquad (21)$$

We have, by a further application of (16)

$$(\boldsymbol{\sigma}, \mathbf{x})\,(\boldsymbol{\sigma}, \mathbf{p}) = (\mathbf{x}, \mathbf{p}) + i\,(\boldsymbol{\sigma}, \mathbf{m}).$$

Now a permissible definition of p_r is

$$(\mathbf{x}, \mathbf{p}) = rp_r + ih,$$

and from (21)

$$(\boldsymbol{\sigma}, \mathbf{m}) = \rho_3 jh - h.$$

Hence

$$(\boldsymbol{\sigma}, \mathbf{x})\,(\boldsymbol{\sigma}, \mathbf{p}) = rp_r + i\rho_3 jh. \qquad (22)$$

Introduce the quantity ε defined by

$$r\varepsilon = \rho_1 (\boldsymbol{\sigma}, \mathbf{x}). \tag{23}$$

Since r commutes with ρ_1 and with $(\boldsymbol{\sigma}, \mathbf{x})$, it must commute with ε. We thus have

$$r^2\varepsilon^2 = [\rho_1 (\boldsymbol{\sigma}, \mathbf{x})]^2 = (\boldsymbol{\sigma}, \mathbf{x})^2 = \mathbf{x}^2 = r^2$$

or

$$\varepsilon^2 = 1.$$

Since there is symmetry between $\mathbf{x}$ and $\mathbf{p}$ so far as angular momentum is concerned, $\rho_1 (\boldsymbol{\sigma}, \mathbf{x})$, like $\rho_1 (\boldsymbol{\sigma}, \mathbf{p})$, must commute with $\mathbf{M}$ and j. Hence ε commutes with $\mathbf{M}$ and j. Further, ε must commute with p_r, since we have

$$(\boldsymbol{\sigma}, \mathbf{x}) (\mathbf{x}, \mathbf{p}) - (\mathbf{x}, \mathbf{p}) (\boldsymbol{\sigma}, \mathbf{x}) = ih (\boldsymbol{\sigma}, \mathbf{x}),$$

which gives

$$r\varepsilon (rp_r + ih) - (rp_r + ih) r\varepsilon = ihr\varepsilon,$$

which reduces to

$$\varepsilon p_r - p_r\varepsilon = 0.$$

From (22) and (23) we now have

$$r\varepsilon\rho_1 (\boldsymbol{\sigma}, \mathbf{p}) = rp_r + i\rho_3 jh$$

or

$$\rho_1 (\boldsymbol{\sigma}, \mathbf{p}) = \varepsilon p_r + i\varepsilon\rho_3 jh/r.$$

Thus

$$\mathrm{F} = p_0 + \mathrm{V} + \varepsilon p_r + i\varepsilon\rho_3 jh/r + \rho_3 mc. \tag{24}$$

Equation (23) shows that ε anticommutes with ρ_3. We can therefore by a canonical transformation (involving perhaps the x's and p's as well as the σ's and ρ's) bring ε into the form of the ρ_2 of § 2 without changing ρ_3, and without changing any of the other variables occurring on the right-hand side of (24), since these other variables all commute with ε. $i\varepsilon\rho_3$ will now be of the form $i\rho_2\rho_3 = -\rho_1$, so that the wave equation takes the form

$$\mathrm{F}\psi \equiv [p_0 + \mathrm{V} + \rho_2 p_r - \rho_1 jh/r + \rho_3 mc]\, \psi = 0.$$

If we write this equation out in full, calling the components of ψ referring to the first and third rows (or columns) of the matrices ψ_α and ψ_β respectively, we get

$$(\mathrm{F}\psi)_\alpha \equiv (p_0 + \mathrm{V})\, \psi_\alpha - h\frac{\partial}{\partial r}\psi_\beta - \frac{jh}{r}\psi_\beta + mc\psi_\alpha = 0,$$

$$(\mathrm{F}\psi)_\beta \equiv (p_0 + \mathrm{V})\, \psi_\beta + h\frac{\partial}{\partial r}\psi_\alpha - \frac{jh}{r}\psi_\alpha - mc\psi_\beta = 0.$$

The second and fourth components give just a repetition of these two equations. We shall now eliminate ψ_α. If we write $h\mathrm{B}$ for $p_0 + \mathrm{V} + mc$, the first equation becomes

$$\left(\frac{\partial}{\partial r} + \frac{j}{r}\right)\psi_\beta = \mathrm{B}\psi_\alpha,$$

which gives on differentiating

$$\frac{\partial^2}{\partial r^2}\psi_\beta + \frac{j}{r}\frac{\partial}{\partial r}\psi_\beta - \frac{j}{r^2}\psi_\beta = \mathrm{B}\frac{\partial}{\partial r}\psi_\alpha + \frac{\partial \mathrm{B}}{\partial r}\psi_\alpha$$

$$= \frac{\mathrm{B}}{h}\left[-(p_0 + \mathrm{V} - mc)\psi_\beta + \frac{jh}{r}\psi_\alpha\right] + \frac{1}{h}\frac{\partial \mathrm{V}}{\partial r}\psi_\alpha$$

$$= -\frac{(p_0 + \mathrm{V})^2 - m^2c^2}{h^2}\psi_\beta + \left(\frac{j}{r} + \frac{1}{\mathrm{B}h}\frac{\partial \mathrm{V}}{\partial r}\right)\left(\frac{\partial}{\partial r} + \frac{j}{r}\right)\psi_\beta.$$

This reduces to

$$\frac{\partial^2}{\partial r^2}\psi_\beta + \left[\frac{(p_0 + \mathrm{V})^2 - m^2c^2}{h^2} - \frac{j(j+1)}{r^2}\right]\psi_\beta - \frac{1}{\mathrm{B}h}\frac{\partial \mathrm{V}}{\partial r}\left(\frac{\partial}{\partial r} + \frac{j}{r}\right)\psi_\beta = 0. \quad (25)$$

The values of the parameter p_0 for which this equation has a solution finite at $r = 0$ and $r = \infty$ are $1/c$ times the energy levels of the system. To compare this equation with those of previous theories, we put $\psi_\beta = r\chi$, so that

$$\frac{\partial^2}{\partial r^2}\chi + \frac{2}{r}\frac{\partial}{\partial r}\chi + \left[\frac{(p_0 + \mathrm{V})^2 - m^2c^2}{h^2} - \frac{j(j+1)}{r^2}\right]\chi - \frac{1}{\mathrm{B}h}\frac{\partial \mathrm{V}}{\partial r}\left(\frac{\partial}{\partial r} + \frac{j+1}{r}\right)\chi = 0. \quad (26)$$

If one neglects the last term, which is small on account of B being large, this equation becomes the same as the ordinary Schroedinger equation for the system, with relativity correction included. Since j has, from its definition, both positive and negative integral characteristic values, our equation will give twice as many energy levels when the last term is not neglected.

We shall now compare the last term of (26), which is of the same order of magnitude as the relativity correction, with the spin correction given by Darwin and Pauli. To do this we must eliminate the $\partial\chi/\partial r$ term by a further transformation of the wave function. We put

$$\chi = \mathrm{B}^{-\frac{1}{2}}\chi_1,$$

which gives

$$\frac{\partial^2}{\partial r^2}\chi_1 + \frac{2}{r}\frac{\partial}{\partial r}\chi_1 + \left[\frac{(p_0 + \mathrm{V})^2 - m^2c^2}{h^2} - \frac{j(j+1)}{r^2}\right]\chi_1$$

$$+ \left[\frac{1}{\mathrm{B}h}\frac{j}{r}\frac{\partial \mathrm{V}}{\partial r} - \frac{1}{2}\frac{1}{\mathrm{B}h}\frac{\partial^2 \mathrm{V}}{\partial r^2} + \frac{1}{4}\frac{1}{\mathrm{B}^2h^2}\left(\frac{\partial \mathrm{V}}{\partial r}\right)^2\right]\chi_1 = 0. \quad (27)$$

The correction is now, to the first order of accuracy

$$\frac{1}{Bh}\left(\frac{j}{r}\frac{\partial V}{\partial r} - \frac{1}{2}\frac{\partial^2 V}{\partial r^2}\right),$$

where $Bh = 2mc$ (provided p_0 is positive). For the hydrogen atom we must put $V = e^2/cr$. The first order correction now becomes

$$-\frac{e^2}{2mc^2r^3}(j+1). \tag{28}$$

If we write $-j$ for $j+1$ in (27), we do not alter the terms representing the unperturbed system, so

$$\frac{e^2}{2mc^2r^3}\,j \tag{28'}$$

will give a second possible correction for the same unperturbed term.

In the theory of Pauli and Darwin, the corresponding correcting term is

$$\frac{e^2}{2mhc^2r^3}(\boldsymbol{\sigma}, \mathbf{m})$$

when the Thomas factor $\frac{1}{2}$ is included. We must remember that in the Pauli-Darwin theory, the resultant orbital angular momentum k plays the part of our j. We must define k by

$$\mathbf{m}^2 = k\,(k+1)\,h^2$$

instead of by the exact analogue of (19), in order that it may have integral characteristic values, like j. We have from (20)

$$(\boldsymbol{\sigma}, \mathbf{m})^2 = k\,(k+1)\,h^2 - h\,(\boldsymbol{\sigma}, \mathbf{m})$$

or

$$\{(\boldsymbol{\sigma}, \mathbf{m}) + \tfrac{1}{2}h\}^2 = (k + \tfrac{1}{2})^2 h^2,$$

hence

$$(\boldsymbol{\sigma}, \mathbf{m}) = kh \text{ or } -(k+1)\,h.$$

The correction thus becomes

$$\frac{e^2}{2mc^2r^3}\,k \quad \text{or} \quad -\frac{e^2}{2mc^2r^3}(k+1),$$

which agrees with (28) and (28'). The present theory will thus, in the first approximation, lead to the same energy levels as those obtained by Darwin, which are in agreement with experiment.

16

Reprinted from *Phys. Rev.* **43:**491–494 (1933)

The Positive Electron

Carl D. Anderson, *California Institute of Technology, Pasadena, California*
(Received February 28, 1933)

Out of a group of 1300 photographs of cosmic-ray tracks in a vertical Wilson chamber 15 tracks were of positive particles which could not have a mass as great as that of the proton. From an examination of the energy-loss and ionization produced it is concluded that the charge is less than twice, and is probably exactly equal to, that of the proton. If these particles carry unit positive charge the curvatures and ionizations produced require the mass to be less than twenty times the electron mass. These particles will be called positrons. Because they occur in groups associated with other tracks it is concluded that they must be secondary particles ejected from atomic nuclei.

Editor

On August 2, 1932, during the course of photographing cosmic-ray tracks produced in a vertical Wilson chamber (magnetic field of 15,000 gauss) designed in the summer of 1930 by Professor R. A. Millikan and the writer, the tracks shown in Fig. 1 were obtained, which seemed to be interpretable only on the basis of the existence in this case of a particle carrying a positive charge but having a mass of the same order of magnitude as that normally possessed by a free negative electron. Later study of the photograph by a whole group of men of the Norman Bridge Laboratory only tended to strengthen this view. The reason that this interpretation seemed so inevitable is that the track appearing on the upper half of the figure cannot possibly have a mass as large as that of a proton for as soon as the mass is fixed the energy is at once fixed by the curvature. The energy of a proton of that curvature comes out 300,000 volts, but a proton of that energy according to well established and universally accepted determinations[1] has a total range of about 5 mm in air while that portion of the range actually visible in this case exceeds 5 cm without a noticeable change in curvature. The only escape from this conclusion would be to assume that at exactly the same instant (and the sharpness of the tracks determines that instant to within about a fiftieth of a second) two independent electrons happened to produce two tracks so placed as to give the impression of a single particle shooting through the lead plate. This assumption was dismissed on a probability basis, since a sharp track of this order of curvature under the experimental conditions prevailing occurred in the chamber only once in some 500 exposures, and since there was practically no chance at all that two such tracks should line up in this way. We also discarded as completely untenable the assumption of an electron of 20 million volts entering the lead on one side and coming out with an energy of 60 million volts on the other side. A fourth possibility is that a photon, entering the lead from above, knocked out of the nucleus of a lead atom two particles, one of which shot upward and the other downward. But in this case the upward moving one would be a positive of small mass so that either of the two possibilities leads to the existence of the positive electron.

In the course of the next few weeks other photographs were obtained which could be interpreted logically only on the positive-electron basis, and a brief report was then published[2] with due reserve in interpretation in view of the importance and striking nature of the announcement.

Magnitude of Charge and Mass

It is possible with the present experimental data only to assign rather wide limits to the

[1] Rutherford, Chadwick and Ellis, *Radiations from Radioactive Substances*, p. 294. Assuming $R \propto v^3$ and using data there given the range of a 300,000 volt proton in air S.T.P. is about 5 mm.

[2] C. D. Anderson, Science **76**, 238 (1932).

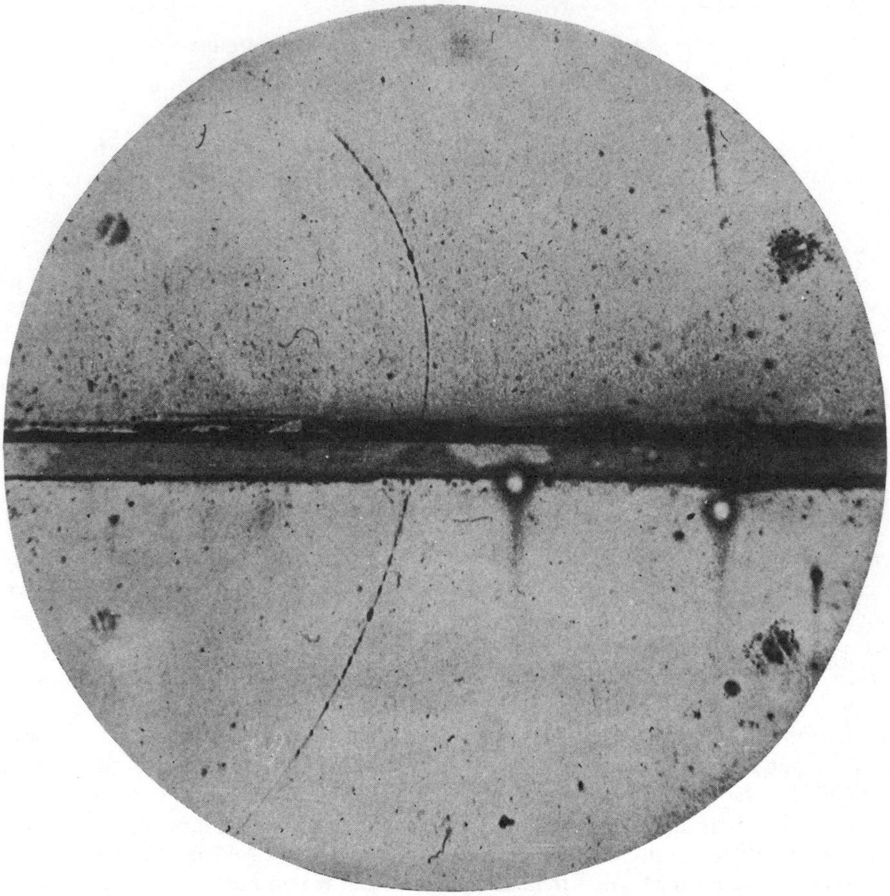

FIG. 1. A 63 million volt positron ($H\rho=2.1\times10^5$ gauss-cm) passing through a 6 mm lead plate and emerging as a 23 million volt positron ($H\rho=7.5\times10^4$ gauss-cm). The length of this latter path is at least ten times greater than the possible length of a proton path of this curvature.

magnitude of the charge and mass of the particle. The specific ionization was not in these cases measured, but it appears very probable, from a knowledge of the experimental conditions and by comparison with many other photographs of high- and low-speed electrons taken under the same conditions, that the charge cannot differ in magnitude from that of an electron by an amount as great as a factor of two. Furthermore, if the photograph is taken to represent a positive particle penetrating the 6 mm lead plate, then the energy lost, calculated for unit charge, is approximately 38 million electron-volts, this value being practically independent of the proper mass of the particle as long as it is not too many times larger than that of a free negative electron. This value of 63 million volts per cm energy-loss for the positive particle it was considered legitimate to compare with the measured mean of approximately 35 million volts[3] for negative electrons of 200–300 million volts energy since the rate of energy-loss for particles of small mass is expected to change only very slowly over an energy range extending from several million to several hundred million volts. Allowance being made for experimental uncertainties, an upper limit to the rate of loss of energy for the positive particle can then be set at less than four times that for an electron, thus fixing, by the usual relation between rate of ionization and

[3] C. D. Anderson, Phys. Rev. **43**, 381A (1933).

charge, an upper limit to the charge less than twice that of the negative electron. It is concluded, therefore, that the magnitude of the charge of the positive electron which we shall henceforth contract to positron is very probably equal to that of a free negative electron which from symmetry considerations would naturally then be called a negatron.

FIG. 2. A positron of 20 million volts energy ($H\rho = 7.1 \times 10^4$ gauss-cm) and a negatron of 30 million volts energy ($H\rho = 10.2 \times 10^4$ gauss-cm) projected from a plate of lead. The range of the positive particle precludes the possibility of ascribing it to a proton of the observed curvature.

It is pointed out that the effective depth of the chamber in the line of sight which is the same as the direction of the magnetic lines of force was 1 cm and its effective diameter at right angles to that line 14 cm, thus insuring that the particle crossed the chamber practically normal to the lines of force. The change in direction due to scattering in the lead,[3] in this case about 8° measured in the plane of the chamber, is a probable value for a particle of this energy though less than the most probable value.

The magnitude of the proper mass cannot as yet be given further than to fix an upper limit to it about twenty times that of the electron mass. If Fig. 1 represents a particle of unit charge passing through the lead plate then the curvatures, on the basis of the information at hand on ionization, give too low a value for the energy-loss unless the mass is taken less than twenty times that of the negative electron mass. Further determinations of $H\rho$ for relatively low energy particles before and after they cross a known amount of matter, together with a study of ballistic effects such as close encounters with electrons, involving large energy transfers, will enable closer limits to be assigned to the mass.

To date, out of a group of 1300 photographs of cosmic-ray tracks 15 of these show positive particles penetrating the lead, none of which can be ascribed to particles with a mass as large as that of a proton, thus establishing the existence of positive particles of unit charge and of mass small compared to that of a proton. In many other cases due either to the short section of track available for measurement or to the high energy of the particle it is not possible to differentiate with certainty between protons and positrons. A comparison of the six or seven hundred positive-ray tracks which we have taken is, however, still consistent with the view that the positive particle which is knocked out of the nucleus by the incoming primary cosmic ray is in many cases a proton.

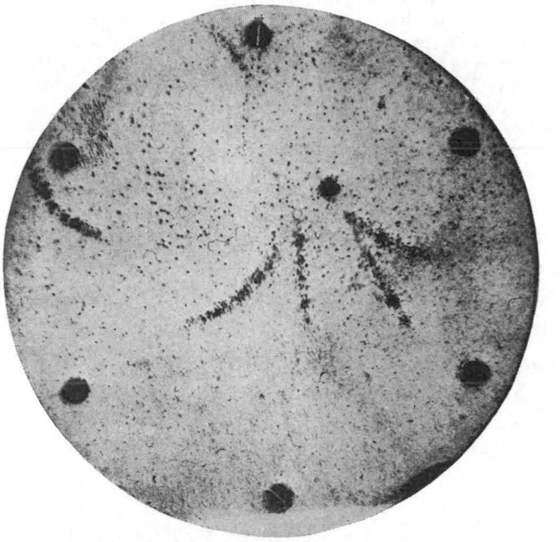

FIG. 3. A group of six particles projected from a region in the wall of the chamber. The track at the left of the central group of four tracks is a negatron of about 18 million volts energy ($H\rho = 6.2 \times 10^4$ gauss-cm) and that at the right a positron of about 20 million volts energy ($H\rho = 7.0 \times 10^4$ gauss-cm). Identification of the two tracks in the center is not possible. A negatron of about 15 million volts is shown at the left. This group represents early tracks which were broadened by the diffusion of the ions. The uniformity of this broadening for all the tracks shows that the particles entered the chamber at the same time.

From the fact that positrons occur in groups associated with other tracks it is concluded that they must be secondary particles ejected from an atomic nucleus. If we retain the view that a nucleus consists of protons and neutrons (and α-particles) and that a neutron represents a close combination of a proton and electron, then from the electromagnetic theory as to the origin of mass the simplest assumption would seem to be that an encounter between the incoming primary ray and a proton may take place in such a way as to expand the diameter of the proton to the same value as that possessed by the negatron. This process would release an energy of a billion electron-volts appearing as a secondary photon. As a second possibility the primary ray may disintegrate a neutron (or more than one) in the nucleus by the ejection either of a negatron or a positron with the result that a positive or a negative proton, as the case may be, remains in the nucleus in place of the neutron, the event occurring in this instance without the emission of a photon. This alternative, however, postulates the existence in the nucleus of a proton of negative charge, no evidence for which exists. The greater symmetry, however, between the positive and negative charges revealed by the discovery of the positron should prove a stimulus to search for evidence of the existence of negative protons. If the neutron should prove to be a fundamental particle of a new kind rather than a proton and negatron in close combination, the above hypotheses will have to be abandoned for the proton will then in all probability be represented as a complex particle consisting of a neutron and positron.

FIG. 4. A positron of about 200 million volts energy ($H\rho = 6.6\times 10^5$ gauss-cm) penetrates the 11 mm lead plate and emerges with about 125 million volts energy ($H\rho = 4.2\times 10^5$ gauss-cm). The assumption that the tracks represent a proton traversing the lead plate is inconsistent with the observed curvatures. The energies would then be, respectively, about 20 million and 8 million volts above and below the lead, energies too low to permit the proton to have a range sufficient to penetrate a plate of lead of 11 mm thickness.

While this paper was in preparation press reports have announced that P. M. S. Blackett and G. Occhialini in an extensive study of cosmic-ray tracks have also obtained evidence for the existence of light positive particles confirming our earlier report.

I wish to express my great indebtedness to Professor R. A. Millikan for suggesting this research and for many helpful discussions during its progress. The able assistance of Mr. Seth H. Neddermeyer is also appreciated.

17

Reprinted from *Nature* **129:**312 (1932)

POSSIBLE EXISTENCE OF A NEUTRON

J. Chadwick
Cavendish Laboratory
Cambridge, England

IT has been shown by Bothe and others that beryllium when bombarded by α-particles of polonium emits a radiation of great penetrating power, which has an absorption coefficient in lead of about 0·3 (cm.)$^{-1}$. Recently Mme. Curie-Joliot and M. Joliot found, when measuring the ionisation produced by this beryllium radiation in a vessel with a thin window, that the ionisation increased when matter containing hydrogen was placed in front of the window. The effect appeared to be due to the ejection of protons with velocities up to a maximum of nearly 3×10^9 cm. per sec. They suggested that the transference of energy to the proton was by a process similar to the Compton effect, and estimated that the beryllium radiation had a quantum energy of 50×10^6 electron volts.

I have made some experiments using the valve counter to examine the properties of this radiation excited in beryllium. The valve counter consists of a small ionisation chamber connected to an amplifier, and the sudden production of ions by the entry of a particle, such as a proton or α-particle, is recorded by the deflexion of an oscillograph. These experiments have shown that the radiation ejects particles from hydrogen, helium, lithium, beryllium, carbon, air, and argon. The particles ejected from hydrogen behave, as regards range and ionising power, like protons with speeds up to about $3{\cdot}2 \times 10^9$ cm. per sec. The particles from the other elements have a large ionising power, and appear to be in each case recoil atoms of the elements.

If we ascribe the ejection of the proton to a Compton recoil from a quantum of 52×10^6 electron volts, then the nitrogen recoil atom arising by a similar process should have an energy not greater than about 400,000 volts, should produce not more than about 10,000 ions, and have a range in air at N.T.P. of about 1·3 mm. Actually, some of the recoil atoms in nitrogen produce at least 30,000 ions. In collaboration with Dr. Feather, I have observed the recoil atoms in an expansion chamber, and their range, estimated visually, was sometimes as much as 3 mm. at N.T.P.

These results, and others I have obtained in the course of the work, are very difficult to explain on the assumption that the radiation from beryllium is a quantum radiation, if energy and momentum are to be conserved in the collisions. The difficulties disappear, however, if it be assumed that the radiation consists of particles of mass 1 and charge 0, or neutrons. The capture of the α-particle by the Be^9 nucleus may be supposed to result in the formation of a C^{12} nucleus and the emission of the neutron. From the energy relations of this process the velocity of the neutron emitted in the forward direction may well be about 3×10^9 cm. per sec. The collisions of this neutron with the atoms through which it passes give rise to the recoil atoms, and the observed energies of the recoil atoms are in fair agreement with this view. Moreover, I have observed that the protons ejected from hydrogen by the radiation emitted in the opposite direction to that of the exciting α-particle appear to have a much smaller range than those ejected by the forward radiation. This again receives a simple explanation on the neutron hypothesis.

If it be supposed that the radiation consists of quanta, then the capture of the α-particle by the Be^9 nucleus will form a C^{13} nucleus. The mass defect of C^{13} is known with sufficient accuracy to show that the energy of the quantum emitted in this process cannot be greater than about 14×10^6 volts. It is difficult to make such a quantum responsible for the effects observed.

It is to be expected that many of the effects of a neutron in passing through matter should resemble those of a quantum of high energy, and it is not easy to reach the final decision between the two hypotheses. Up to the present, all the evidence is in favour of the neutron, while the quantum hypothesis can only be upheld if the conservation of energy and momentum be relinquished at some point.

18

DISCUSSION OF MR. HEISENBERG'S REPORT

W. Pauli

This article was translated by F. Kertesz, Tennessee Technical Translators from pages 324–325 of Structure et Proprietes des Noyaux Atomiques, *Proceedings of the 7th Solvay Congress, Brussels, October, 1933, by permission of the publisher, Gauthier-Villars, Paris, 1934*

M. Pauli: As it is known, the difficulty resulting from the existence of a continuous β-ray spectrum lies in the fact that the mean lifetime of the nuclei that emit this radiation and that of the nuclei of the radioactive materials that are created by it possess well-defined values. From this it must be concluded that the state, and also the energy and mass of the nucleus that remains after the emission of a β-particle, are also well defined. I will not discuss the efforts that have been made to escape this conclusion, but I believe, in agreement with Mr. Bohr, that the explanation of these experimental facts will meet with insuperable difficulties.

In this connection the experiments may be interpreted in two ways. The interpretation supported by Mr. Bohr suggests that the laws of conservation of energy and momentum fail in the case of a nuclear process in which light particles play an essential role. I do not believe that this assumption is satisfactory or even plausible. In the first place, the electric charge is conserved in the process, and I do not see why the conservation of the charge should be more fundamental than the conservation of energy and of momentum. Then, the energy relations are the very ones that control several characteristic properties of β spectra (the existence of an upper limit and relation with the γ spectra, Heisenberg's stability criterion). If the conservation laws were not valid, these relations would imply that a β decay is always accompanied by a loss and never by a gain of energy; this conclusion implies an irreversibility of the process with respect to time, which does not seem acceptable to me.

In June 1931, at a conference in Pasadena, I proposed the following interpretation: The conservation laws remain valid, the expulsion of β-particles being accompanied by a very penetrating radiation of neutral particles, which has not been observed previously. The sum of the energies of the β particle and of the neutral particle (or neutral particles, because it is not known whether there is a single one or several) emitted by the nucleus in a single process will be equal to the energy that corresponds to the upper limit of the β spectrum. It goes without saying that we assume not only the conservation of energy, but also of momentum and of angular momentum, and the statistical character of all elementary processes.

As far as the properties of these neutral particles are concerned, the atomic weights of radioactive elements indicate to us, first of all, that their mass cannot exceed by very much the mass of the electron. In order to distinguish them from heavy neutrons, Mr. Fermi proposed the name "neutrino." It is possible that the specific mass of the neutrino might be equal to zero, and thus it should be able to propagate with the speed of light like photons. However, its penetrating power would probably greatly exceed that of photons having the same energy. I could accept the proposition that neutrinos possess a spin of ½ and that they follow the Fermi statistics, although this assumption is not supported by any direct experimental proof. We know nothing about the interaction of neutrinos with other material particles and with photons; the assumption that they possess a magnetic moment, as I suggested on another occasion (Dirac's theory leads to the prediction of the possible existence of magnetic neutral particles), does not seem to me to be well supported.

In this connection the experimental study of momentum balance in β decay represents a very important problem; major difficulties may be anticipated because of the low energy of the recoil nucleus.

19

Reprinted from *Phys.-Math. Soc. Japan Proc.*, ser. 3, **17**:48–57 (1935)

ON THE INTERACTION OF ELEMENTARY PARTICLES. I.

(The Meson)

H. Yukawa

§ 1. Introduction

At the present stage of the quantum theory little is known about the nature of interaction of elementary particles. Heisenberg considered the interaction of " Platzwechsel " between the neutron and the proton to be of importance to the nuclear structure.(1)

Recently Fermi treated the problem of β-disintegration on the hypothesis of " neutrino "(2). According to this theory, the neutron and the proton can interact by emitting and absorbing a pair of neutrino and electron. Unfortunately the interaction energy calculated on such assumption is much too small to account for the binding energies of neutrons and protons in the nucleus.(3)

To remove this defect, it seems natural to modify the theory of Heisenberg and Fermi in the following way. The transition of a heavy particle from neutron state to proton state is not always accompanied by the emission of light particles, i. e., a neutrino and an electron, but the energy liberated by the transition is taken up sometimes by another heavy particle, which in turn will be transformed from proton state into neutron state. If the probability of occurrence of the latter process is much larger than that of the former, the interaction between the neutron and the proton will be much larger than in the case of Fermi, whereas the probability of emission of light particles is not affected essentially.

Now such interaction between the elementary particles can be described by means of a field of force, just as the interaction between the charged particles is described by the electromagnetic field. The above considerations show that the interaction of heavy particles with this field is much larger than that of light particles with it.

---

(1) W. Heisenberg, Zeit f. Phys. **77**, 1 (1932); **78**, 156 (1932); **80**, 587 (1933). We shall denote the first of them by I.

(2) E. Fermi, ibid. **88**, 161 (1394).

(3) Ig. Tamm, Nature **133**, 981 (1934); D. Iwanenko, ibid. 981 (1934).

In the quantum theory this field should be accompanied by a new sort of quantum, just as the electromagnetic field is accompanied by the photon.

In this paper the possible natures of this field and the quantum accompanying it will be discussed briefly and also their bearing on the nuclear structure will be considered.

Besides such an exchange force and the oridinary electric and magnetic forces there may be other forces between the elementary particles, but we disregard the latter for the moment.

Fuller account will be made in the next paper.

§2. Field describing the interaction

In analogy with the scalar potential of the electromagnetic field, a function $U(x, y, z, t)$ is introducud to describe the field between the neutron and the proton. This function will satisfy an equation similar to the wave equation for the electromagnetic potential.

Now the eqnation

$$\left\{\Delta - \frac{1}{c^2}\frac{\partial^2}{\partial t^2}\right\}U=0 \tag{1}$$

has only static solution with central symmetry $\frac{1}{r}$, except the additive and the multiplicative constants. The potential of force between the neutron and the proton should, however, not be of Coulomb type, but decrease more rapidly with distance. It can be expressed, for example, by

$$+ \text{ or } -g^2\frac{e^{-\lambda r}}{r}, \tag{2}$$

where g is a constant with the dimension of electric charge, i. e., $\text{cm.}^{\frac{3}{2}}$ sec.^{-1} $\text{gr.}^{\frac{1}{2}}$ and λ with the dimention cm.^{-1}

Since this function is a static solution with central symmetry of the wave equation

$$\left\{\Delta - \frac{1}{c^2}\frac{\partial^2}{\partial t^2} - \lambda^2\right\}U=0, \tag{3}$$

let this equation be assumed to be the correct equation for U in vacuum. In the presence of the heavy particles, the U-field interacts with them and causes the transition from neutron state to proton state.

Now, if we introduce the matrices[(4)]

$$\tau_1=\begin{pmatrix}0 & 1\\ 1 & 0\end{pmatrix},\quad \tau_2=\begin{pmatrix}0 & -i\\ i & 0\end{pmatrix},\quad \tau_3=\begin{pmatrix}1 & 0\\ 0 & -1\end{pmatrix}$$

and denote the neutron state and the proton state by $\tau_3=1$ and $\tau_3=-1$ respectively, the wave equation is given by

$$\left\{\Delta-\frac{1}{c^2}\frac{\partial^2}{\partial t^2}-\lambda^2\right\}U=-4\pi g\tilde{\Psi}\frac{\tau_1-i\tau_2}{2}\Psi, \tag{4}$$

where Ψ denotes the wave function of the heavy particles, being a function of time, position, spin as well as τ_3', which takes the value either 1 or -1.

Next, the conjugate complex function $\tilde{U}(x, y, z, t)$, satisfying the equation

$$\left\{\Delta-\frac{1}{c^2}\frac{\partial^2}{\partial t^2}-\lambda^2\right\}\tilde{U}=-4\pi g\tilde{\Psi}\frac{\tau_1+i\tau_2}{2}\Psi, \tag{5}$$

is introduced, corresponding to the inverse transition from proton to neutron state.

Similar equation will hold for the vector function, which is the analogue of the vector potential of the electromagnetic field. However, we disregard it for the moment, as there's no correct relativistic theory for the heavv particles. Hence simple non-relativistic wave equation neglecting spin will be used for the heavy particle, in the following way

$$\left\{\frac{h^2}{4}\left(\frac{1+\tau_3}{M_N}+\frac{1-\tau_3}{M_P}\right)\Delta+ih\frac{\partial}{\partial t}-\frac{1+\tau_3}{2}M_Nc^2-\frac{1-\tau_3}{2}M_Pc^2 -g\left(\tilde{U}\frac{\tau_1-i\tau_2}{2}+U\frac{\tau_1+i\tau_2}{2}\right)\right\}\Psi=0, \tag{6}$$

where h is Planck's constant divided by 2π and M_N, M_P are the masses of the neutron and the proton respectively. The reason for taking the negative sign in front of g will be mentioned later.

The equation (6) corresponds to the Hamiltonian

$$H=\left(\frac{1+\tau_3}{4M_N}+\frac{1-\tau_3}{4M_P}\right)p^2+\frac{1+\tau_3}{2}M_Nc^2+\frac{1-\tau_3}{2}M_Pc^2 +g\left(\tilde{U}\frac{\tau_1-i\tau_2}{2}+U\frac{\tau_1+i\tau_2}{2}\right) \tag{7}$$

(4) Heisenberg, loc, cit. I.

where $\boldsymbol{p}$ is the momentum of the particle. If we put $M_N c^2 - M_P c^2 = D$ and $M_N + M_P = 2M$, the equation (7) becomes approximately

$$H = \frac{\boldsymbol{p}^2}{2M} + \frac{g}{2}\{\tilde{U}(\tau_1 - i\tau_2) + U(\tau_1 + i\tau_2)\} + \frac{D}{2}\tau_3, \tag{8}$$

where the constant term Mc^2 is omitted.

Now consider two heavy particles at points (x_1, y_1, z_1) and (x_2, y_2, z_2) respectively and assume their relative velocity to be small. The fields at (x_1, y_1, z_1) due to the particle at $(x_2 y_2, z_2)$ are, from (4) and (5),

$$\left.\begin{aligned} U(x_1, y_1, z_1) &= g\frac{e^{-\lambda r_{12}}}{r_{12}}\frac{(\tau_1^{(2)} - i\tau_2^{(2)})}{2} \\ &\text{and} \\ \tilde{U}(x_1, y_1, z_1) &= g\frac{e^{-\lambda r_{12}}}{r_{12}}\frac{(\tau_1^{(2)} + i\tau_2^{(2)})}{2}, \end{aligned}\right\} \tag{9}$$

where $(\tau_1^{(1)}, \tau_2^{(1)}, \tau_3^{(1)})$ and $(\tau_1^{(2)}, \tau_2^{(2)}, \tau_3^{(2)})$ are the matrices relating to the first and the second particles respectively, and r_{12} is the distance between them.

Hence the Hamiltonian for the system is given, in the absence of the external fields, by

$$\begin{aligned} H &= \frac{\boldsymbol{p}_1^2}{2M} + \frac{\boldsymbol{p}_2^2}{2M} + \frac{g^2}{4}\{(\tau_1^{(1)} - i\tau_2^{(1)})(\tau_1^{(2)} + i\tau_2^{(2)}) \\ &\quad + (\tau_1^{(1)} + i\tau_2^{(1)})(\tau_1^{(2)} - i\tau_2^{(2)})\}\frac{e^{-\lambda r_{12}}}{r_{12}} + (\tau_3^{(1)} + \tau_3^{(2)})D \\ &= \frac{\boldsymbol{p}_1^2}{2M} + \frac{\boldsymbol{p}_2^2}{2M} + \frac{g^2}{2}(\tau_1^{(1)}\tau_1^{(2)} + \tau_2^{(1)}\tau_2^{(2)})\frac{e^{-\lambda r_{12}}}{r_{12}} + (\tau_3^{(1)} + \tau_3^{(2)})D, \end{aligned} \tag{10}$$

where $\boldsymbol{p}_1, \boldsymbol{p}_2$ are the momenta of the particles.

This Hamiltonian is equivalent to Heisenberg's Hamiltonian (1),[5] if we take for "Platzwechselintegral"

$$J(r) = -g^2\frac{e^{-\lambda r}}{r}, \tag{11}$$

except that the interaction between the neutrons and the electrostatic repulsion between the protons are not taken into account. Heisenberg took the positive sign for $J(r)$, so that the spin of the lowest energy state of H^2 was O, whereas in our case, owing to the negative sign in front of g^2, the lowest energy state has the spin 1, which is required

(5) Heisenberg, I.

from the experiment.

Two constants g and λ appearing in the above equations should be determined by comparison with experiment. For example, using the Hamiltonian (10) for heavy particles, we can calculate the mass defect of H^2 and the probability of scattering of a neutron by a proton provided that the relative velocity is small compared with the light velocity.(6)

Rough estimation shows that the calculated values agree with the experimental results, if we take for λ the value between 10^{12}cm^{-1}. and 10^{13}cm^{-1}. and for g a few times of the elementary charge e, although no direct relation between g and e was suggested in the above considerations.

§ 3. Nature of the quanta accompanying the field

The U-field above considered should be quantized according to the general method of the quantum theory. Since the neutron and the proton both obey Fermi's statistics, the quanta accompanying the U-field should obey Bose's statistics and the quantization can be carried out on the line similar to that of the electromagnetic field.

The law of conservation of the electric charge demands that the quantum should have the charge either $+e$ or $-e$. The field quantity U corresponds to the operator which increases the number of negatively charged quanta and decreases the number of positively charged quanta by one respectively. $\tilde{U}$, which is the complex conjugate of U, corresponds to the inverse operator.

Next, denoting

$$p_x = -ih\frac{\partial}{\partial x}, \quad \text{etc.}, \quad W = ih\frac{\partial}{\partial t},$$

$$m_U c = \lambda h,$$

the wave equation for U in free space can be written in the form

$$\left\{p_x^2 + p_y^2 + p_z^2 - \frac{W^2}{c^2} + m_U c^2\right\} U = 0, \tag{12}$$

so that the quantum accompanying the field has the proper mass $m_U = \frac{\lambda h}{c}$.

(6) These calculations were made previously, according to the theory of Heisenberg, by Mr. Tomonaga, to whom the writer owes much. A little modification is necessary in our case. Detailed accounts will be made in the next paper.

Assuming $\lambda=5\times10^{12}\text{cm}^{-1}$., we obtain for m_U a value 2×10^2 times as large as the electron mass. As such a quantum with large mass and positive or negative charge has never been found by the experiment, the above theory seems to be on a wrong line. We can show, however, that, in the ordinary nuclear transformation, such a qnantum can not be emitted into outer space.

Let us consider, for example, the transition from a neutron state of energy W_N to a proton state of energy W_P, both of which include the proper energies. These states can be expressed by the wave functions

$$\Psi_N(x, y, z, t, 1)=u(x, y, z)e^{-iW_Nt/h}, \quad \Psi_N(x, y, z, t, -1)=0$$

and

$$\Psi_P(x, y, z, t, 1)=0, \quad \Psi_P(x, y, z, t, -1)=v(x, y, z)e^{-iW_Pt/h},$$

so that, on the right hand side of the equation (4), the term

$$-4\pi g\tilde{v}ue^{-it(W_N-W_P)/h}$$

appears.

Putting $U=U'(x, y, z)e^{i\omega t}$, we have from (4)

$$\left\{\Delta-\left(\lambda^2-\frac{\omega^2}{c^2}\right)\right\}U'=-4\pi g\tilde{v}u, \tag{13}$$

where $\omega=\frac{W_N-W_P}{h}$. Integrating this, we obtain a solution

$$U'(\boldsymbol{r})=g\iiint\frac{e^{-\mu|\boldsymbol{r}-\boldsymbol{r}'|}}{|\boldsymbol{r}-\boldsymbol{r}'|}\tilde{v}(\boldsymbol{r}')u(\boldsymbol{r}')dv', \tag{14}$$

where $\mu=\sqrt{\lambda^2-\frac{\omega^2}{c^2}}$.

If $\lambda>\frac{|\omega|}{c}$ or $m_Uc^2>|W_N-W_P|$, μ is real and the function $J(r)$ of Heisenberg has the form $-g^2\frac{e^{-\mu r}}{r}$, in which μ, however, depends on $|W_N-W_P|$, becoming smaller and smaller as the latter approaches m_Uc^2. This means that the range of interaction between a neutron and a proton increases as $|W_N-W_P|$ increases.

Now the scattering (elastic or inelastic) of a neutron by a nucleus can be considered as the result of the following double process: the neutron falls into a proton level in the nucleus and a proton in the latter jumps to a neutron state of positive kinetic energy, the total energy being conserved throughout the process. The above argument, then, shows that the probability of scattering may in some case increase with the velocity of the neutron.

According to the experiment of Bonner[7], the collision cross section of the neutron increases, in fact, with the velocity in the case of lead whereas it decreases in the case of carbon and hydrogen, the rate of decrease being slower in the former than in the latter. The origih of this effect is not clear, but the above considerations do not, at least, contradict it. For, if the binding energy of the proton in the nucleus becomes comparable with $m_U c^2$, the range of interaction of the neutron with the former will increase considerably with the velocity of the neutron, so that the cross section will decrease slower in such case than in the case of hydrogen, i. e., free proton. Now the binding energy of the proton in C^{12}, which is estimated from the difference of masses of C^{12} and B^{11}, is

$$12{,}0036-11{,}0110=0{,}9926.$$

This corresponds to a binding energy 0,0152 in mass unit, being thirty times the electron mass. Thus in the case of carbon we can expect the effect observed by Bonner. The arguments are only tentative, other explanations being, of course, not excluded.

Next if $\lambda<\frac{|w|}{c}$ or $m_U c^2<|W_N-W_P|$, μ becomes pure imaginary and U expresses a spherial undamped wave, implying that a quantum with energy greater than $m_U c^2$ can be emitted in outer space by the transition of the heavy particle from neutron state to proton state, provided that $|W_N-W_P|>m_U c^2$.

The velocity of U-wave is greater but the group velocity is smaller than the light velocity c, as in the case of the electron wave.

The reason why such massive quanta, if they ever exist, are not yet discovered may be ascribed to the fact that the mass m_U is so large that condition $|W_N-W_P|>m_U c^2$ is not fulfilled in ordinary nuclear transformation.

§4. Theory of β-disintegration

Hitherto we have considered only the interaction of U-quanta with heavy particles. Now, according to our theory, the quantum emitted when a heavy particle jumps from a neutron state to a proton state, can be absorbed by a light particle which will then in consequence of energy absorption rise from a neutrino state of negative energy to an

(7) T. W. Bonner, Phys. Rev. 45, 606 (1934).

electron state of positive energy. Thus an anti-neutrino and an electron are emitted simultaneously from the nucleus. Such intervention of a massive quantum does not alter essentially the probability of β-disintegration, which has been calculated on the hypothesis of direct coupling of a heavy particle and a light particle, just as, in the theory of internal conversion of γ-ray, the intervation of the proton does not affect the final result.[8] Our theory, therefore, does not differ essentially from Fermi's thory.

Fermi considered that an electron and a neutrino are emitted simultaneously from the radioactive nucleus, but this is formally equivalent to the assumption that a light particle jumps from a neutrino state of negative energy to an electron state of positive energy.

For, if the eigenfunctions of the electron and the neutrino be ψ_k, φ_k respectively, where $k=1, 2, 3, 4$, a term of the form

$$-4\pi g'\sum_{k=1}^{4}\tilde{\psi}_k\varphi_k \tag{15}$$

should be added to the right hand side of the equation (5) for $\tilde{U}$, where g' is a new constant with the same dimension as g.

Now the eigenfunctions of the neutrino state with energy and momentum just opposite to those of the state φ_k is given by $\varphi_k'=-\delta_{kl}\tilde{\varphi}_l$ and conversely $\varphi_k=\delta_{kl}\tilde{\varphi}_l'$, where

$$\delta=\begin{pmatrix} 0 & -1 & 0 & 0 \\ 1 & 0 & 0 & 0 \\ 0 & 0 & 0 & 1 \\ 0 & 0 & -1 & 0 \end{pmatrix},$$

so that (15) becomes

$$-4\pi g'\sum_{k,l=1}^{4}\tilde{\psi}_k\delta_{kl}\tilde{\varphi}_l'. \tag{16}$$

From equations (13) and (15), we obtain for the matrix element of the interaction energy of the heavy particle and the light particle an expression

$$gg'\int\cdots\int\tilde{v}(\boldsymbol{r}_1)u(\boldsymbol{r}_1)\sum_{k=1}^{4}\tilde{\psi}_k(\boldsymbol{r}_2)\varphi_k(\boldsymbol{r}_2)\frac{e^{-\lambda r_{12}}}{r_{12}}dv_1dv_2, \tag{17}$$

corresponding to the following double process: a heavy particle falls

(8) H. A. Taylor and N. F Mott, Proc. Roy. Soc. A, **138**, 665 (1932).

from the neutron state with the eigenfunction $u(\boldsymbol{r})$ into the proton state with the eigenfunction $v(\boldsymbol{r})$ and simultaneously a light particle jumps from the neutrino state $\varphi_k(\boldsymbol{r})$ of negative energy to the electron state $\psi_k(\boldsymbol{r})$ of positive energy. In (17) λ is taken instead of μ, since the difference of energies of the neutron state and the proton state, which is equal to the sum of the upper limit of the energy spectrum of β-rays and the proper energies of the electron and the neutrino, is always small compared with $m_U c^2$.

As λ is much larger than the wave numbers of the electron state and the neutrino state, the function $\frac{e^{-\lambda r_{12}}}{r_{12}}$ can be regarded approximately as a δ-function multiplied by $\frac{4\pi}{\lambda^2}$ for the integrations with respect to x_2, y_2, z_2. The factor $\frac{4\pi}{\lambda^2}$ comes from

$$\iiint \frac{e^{-\lambda r_{12}}}{r_{12}} dv_2 = \frac{4\pi}{\lambda^2}.$$

Hence (17) becomes

$$\frac{4\pi g g'}{\lambda^2} \iiint \tilde{v}(\boldsymbol{r}) u(\boldsymbol{r}) \sum_k \tilde{\psi}_k(\boldsymbol{r}) \varphi_k(\boldsymbol{r}) dv \tag{18}$$

or by (16)

$$\frac{4\pi g g'}{\lambda^2} \iiint \tilde{v}(\boldsymbol{r}) u(\boldsymbol{r}) \sum_{k,l} \tilde{\psi}(\boldsymbol{r}) \delta_{kl}' \tilde{\varphi}_l'(\boldsymbol{r}) dv, \tag{19}$$

which is the same as the expression (21) of Fermi, corresponding to the emission of a neutrino and an electron of positive energy states $\varphi_k'(\boldsymbol{r})$ and $\psi_k(\boldsymbol{r})$, except that the factor $\frac{4\pi g g'}{\lambda^2}$ is substituted for Fermi's g.

Thus the result is the same as that of Fermi's theory, in this approximation, if we take

$$\frac{4\pi g g'}{\lambda^2} = 4 \times 10^{-50} \text{cm}^3. \text{erg},$$

from which the constant g' can be determined. Taking, for example, $\lambda = 5 \times 10^{12}$ and $g = 2 \times 10^{-9}$, we obtain $g' \cong 4 \times 10^{-17}$, which is about 10^{-8} times as small as g.

This means that the interaction between the neutrino and the electron is much smaller than that between the neutron and the proton so that the neutrino will be far more penetrating than the neutron and consequently more difficult to observe. The difference of g and g' may be due to the difference of masses of heavy and light particles.

§5. Summary

The interaction of elementary particles are described by considering a hypothetical quantum which has the elementary charge and the proper mass and which obeys Bose's statistics. The interaction of such a quantum with the heavy particle should be far greater than that with the light particle in order to account for the large interaction of the neutron and the proton as well as the small probability of β-disintegration.

Such quanta, if they ever exist and approach the matter close enough to be absorbed, will deliver their charge and energy to the latter. If, then, the quanta with negative charge come out in excess, the matter will be charged to a negative potential.

These arguments, of course, of merely speculative character, agree with the view that the high speed positive particles in the cosmic rays are generated by the electrostatic field of the earth, which is charged to a negative potential.[9]

The massive quanta may also have some bearing on the shower produced by cosmic rays.

In conclusion the writer wishes to express his cordial thanks to Dr. Y. Nishina and Prof. S. Kikuchi for the encouragement throughout the course of the work.

Department of Physics,
Osaka Imperial University.

(9) G. H. Huxley, Nature **134**, 418, 571 (1934); Johnson, Phys. Rev. **45**, 569 (1934).

20

Reprinted from *Phys. Rev.* **51:**884–886 (1937)

Note on the Nature of Cosmic-Ray Particles

SETH H. NEDDERMEYER AND CARL D. ANDERSON
California Institute of Technology, Pasadena, California
(Received March 30, 1937)

MEASUREMENTS[1] of the energy loss of particles occurring in the cosmic-ray showers have shown that this loss is proportional to the incident energy and within the range of the measurements, up to about 400 Mev, is in approximate agreement with values calculated theoretically for electrons by Bethe and Heitler. These measurements were taken using a thin plate of lead (0.35 cm), and the observed individual losses were found to vary from an amount below experimental detection up to the whole initial energy of the particle, with a mean fractional loss of about 0.5. If these measurements are correct it is evident that in a much thicker layer of heavy material multiple losses should become much more important, and the probability of observing a particle loss less than a large fraction of its initial energy should be very small. For the purpose of testing this inference and also for checking our previous measurements[2] which had shown the presence of some particles less massive than protons but more penetrating than electrons obeying the Bethe-Heitler theory, we have taken about 6000 counter-tripped photographs with a 1 cm plate of platinum placed across the center of the cloud chamber. This plate is equivalent in electron thickness to 1.96 cm of lead, and to 1.86 cm of lead for a Z^2 absorption. The results of 55 measurements on particles in the range below 500 Mev are given in Fig. 1, and in Fig. 2 the distribution of particles is shown as a function of the fraction of energy lost. The shaded part of the diagram represents particles which either enter the chamber accompanied by other particles or else themselves produce showers in the bar of platinum. It is clear that the particles separate themselves into two rather well-defined groups, the one consisting largely of shower particles and exhibiting a high absorbability, the other consisting of particles entering singly which in general lose a relatively small fraction of their initial energy, although there are four cases in which the loss is more than 60 percent. A considerable part of the spread on the negative abscissa can be accounted for by errors; it seems likely, however, that the case plotted at the extreme left represents a particle moving upward. Particles of both signs are distributed over the whole diagram, and moreover, the initial energies of the particles of each group are distributed over the whole measured range.

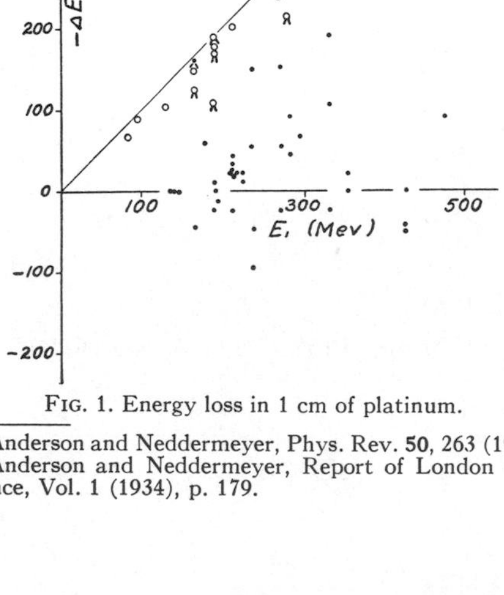

FIG. 1. Energy loss in 1 cm of platinum.

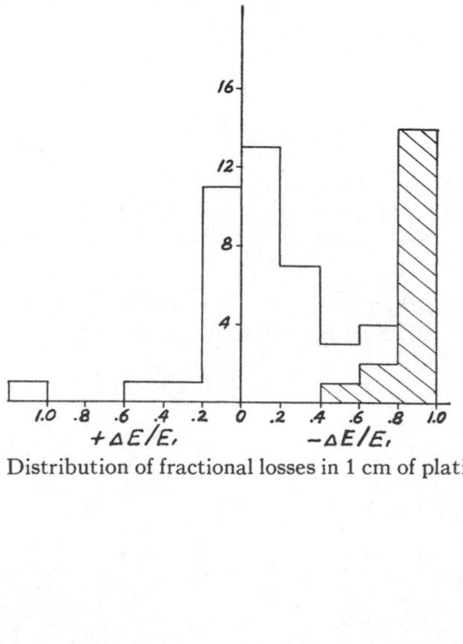

FIG. 2. Distribution of fractional losses in 1 cm of platinum.

[1] Anderson and Neddermeyer, Phys. Rev. **50**, 263 (1936).
[2] Anderson and Neddermeyer, Report of London Conference, Vol. 1 (1934), p. 179.

The chief source of error in these experiments lies not in the curvature measurements themselves, but in the track distortions produced by irregular motions of the gas in the chamber. The distortions are much larger when a thick plate is inside the chamber than when it is left unobstructed. These distortions are not sufficient to alter essentially the distribution of observed losses for the nonpenetrating group, but they could have a very serious effect in the part of the distribution representing small losses. This is especially true inasmuch as this group represents a small percentage of the total number of tracks, selected solely on the basis that they should exhibit a measurable curvature and at the same time be free from obvious distortion. The problem of measuring small energy losses is then evidently an extremely difficult one compared to that of measuring energy distributions in an unobstructed chamber. While it is possible in many cases to distinguish a distortion as such when a magnetic field is present, it is necessary to obtain independent criteria as to the reliability of the measurements; it is not a satisfactory procedure to try to do this simply by measuring curvatures of tracks taken with no field and comparing the curvature distribution thus found with the one obtained when the field is present. Observations made with no magnetic field indicate that serious distortions occur on about 5 percent of the photographs, and show that they are by no means a uniform function of the orientation and position of the track in the chamber. It is therefore not possible to correct for distortion in individual cases. When large distortions do occur, however, they are likely to obey one or both of the following correlations: (*a*) a curvature concave upward when the track makes a considerable angle with the vertical; (*b*) a curvature concave toward the center of the chamber. The observed percentages of measured single tracks obeying the correlations

TABLE I. *Correlations between track curvatures and positions and orientations of tracks.*

Designation (See text)	Percentage correlation		
	Observed	Expected	Observed (excluding apparent gains)
(*a*)	52	50	55
(*b*)	67	50	55
(*a*) and (*b*)	33	25	27
neither	15	25	18

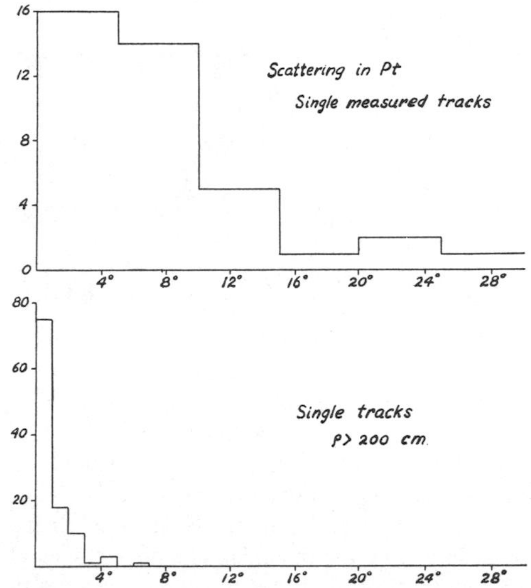

FIG. 3. Scattering distributions in 1 cm of platinum.

(*a*) and (*b*) are compared in Table I with the percentages expected if the observed curvatures have no relation to the positions and orientations of the tracks. If the 11 cases of apparent gains in energy are left out of consideration the observed percentages are brought somewhat closer to the expected values as shown in the last column.

A second independent check on the validity of the measurements can be obtained by measuring the scattering of the particles which show apparent curvatures, and comparing this with the scattering exhibited by those single tracks whose curvatures are just outside the range of measurability. In Fig. 3 are shown the distributions of scattering angles (the angles projected on the plane of the chamber) for the measured single tracks and for single tracks with a radius of curvature, $\rho \gtrsim 200$ cm (475 Mev). As it is scarcely conceivable that distortions could influence the scattering measurements by as much as 5°, these distributions constitute strong independent evidence that the measured tracks actually lie in the energy range indicated by the curvature determinations.

It has been known for a long time that there exist particles of both penetrating and nonpenetrating types. Crussard and Leprince-Ringuet[3] have recently made measurements of

[3] Crussard and Leprince-Ringuet, C. R. **204**, 240 (1937).

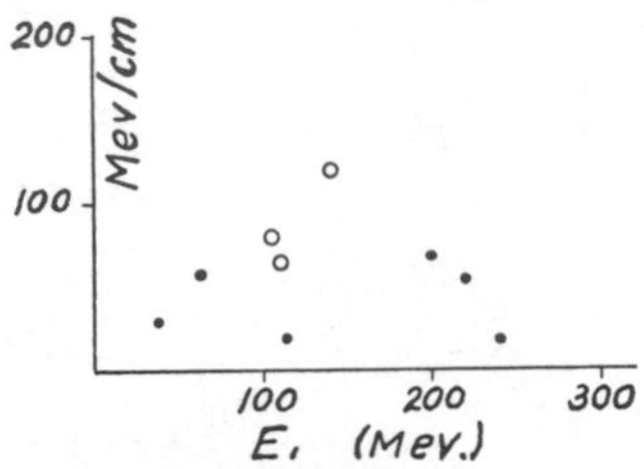

FIG. 4. Early measurements of energy loss in 0.7–1.5 cm of Pb. Dots indicate single particles; circles, shower particles.

energy loss in a range mainly above that covered in our experiments. They have concluded from their data that either the absorption law changes with energy or else that there is a difference in character among the particles. This same conclusion has already been twice stated by the writers.[4, 5] The present data appear to constitute the first experimental evidence for the existence of particles of both penetrating and nonpenetrating character in the energy range extending below 500 Mev. Moreover, the penetrating particles in this range do not ionize perceptibly more than the nonpenetrating ones, and cannot therefore be assumed to be of protonic mass. The lowest $H\rho$ among the penetrating group is 4.5×10^5 gauss cm. A proton of this curvature would ionize at least 25 times as strongly as a fast electron. It is interesting that our early measurements[2] of the energy loss in thicknesses of lead from 0.7 to 1.5 cm show a similar tendency to separate into two groups. They are reproduced in Fig. 4. If reinterpreted in the light of our present data they provide no evidence against high absorbability for electrons.

The nonpenetrating particles are readily interpreted as free positive and negative electrons. Interpretations of the penetrating ones encounter very great difficulties, but at present appear to be limited to the following hypotheses: (*a*) that an electron (+ or −) can possess some property other than its charge and mass which is capable of accounting for the absence of numerous large radiative losses in a heavy element; or (*b*) that there exist particles of unit charge, but with a mass (which may not have a unique value) larger than that of a normal free electron[6] and much smaller than that of a proton; this assumption would also account for the absence of numerous large radiative losses, as well as for the observed ionization. Inasmuch as charge and mass are the only parameters which characterize the electron in the quantum theory, assumption (*b*) seems to be the better working hypothesis. If the penetrating particles are to be distinguished from free electrons by a greater mass, and since no evidence for their existence in ordinary matter obtains, it seems likely that there must exist some very effective process for removing them.

The experimental fact that penetrating particles occur both with positive and negative charges suggests that they might be created in pairs by photons, and that they might be represented as higher mass states of ordinary electrons.

Independent evidence indicating the existence of particles of a new type has already been found, based on range, curvature and ionization relations; for example, Figs. 12 and 13 of our previous publication.[1] In particular the strongly ionizing particle of Fig. 13 cannot readily be explained except in terms of a particle of e/m greater than that of a proton. The large value of e/m apparently is not due to an e greater than the electronic charge since above the plate the particle ionizes imperceptibly differently from a fast electron, whereas below the plate its ionization definitely exceeds that of an electron of the same curvature in the magnetic field; the effects, however, are understandable on the assumption that the particle's mass is greater than that of a free electron. We should like to suggest, merely as a possibility, that the strongly ionizing particles of the type of Fig. 13, although they occur predominantly with positive charge, may be related with the penetrating group above.

We wish to express our gratitude to Professor Millikan for his helpful discussions and encouragement. These experiments have been made possible by the Baker Company, who very generously loaned us the bar of platinum; and by funds supplied by the Carnegie Institution of Washington.

Note added in proof: Excellent experimental evidence showing the existence of particles less massive than protons, but more penetrating than electrons obeying the Bethe-Heitler theory has just been reported by Street and Stevenson, Abstract no. 40, Meeting of American Physical Society, Apr. 29, 1937.

---

[4] Reference 2, p. 182.

[5] Reference 1, p. 268.

[6] The energies referred to throughout are, of course, calculated on the assumption of electronic mass. For a mass $m \lesssim 50 m_e$ the actual energy is very roughly $E = E_{el} - mc^2$ in the range of curvature here considered.

Part V

ENERGY IN THE ATOMIC NUCLEUS

Editor's Comments on Papers 21 Through 28

21 GAMOW
On the Quantum Theory of the Atomic Nucleus

22 CURIE and JOLIOT
I. Artificial Production of Radioactive Elements.
II. Chemical Proof of the Transmutation of the Elements

23 BETHE
An Attempt to Calculate the Number of Energy Levels of a Heavy Nucleus

24 FEENBERG and WIGNER
On the Structure of the Nuclei between Helium and Oxygen

25 HAHN and STRASSMAN
Excerpts from *On the Identification and Behavior of the Alkaline Earth Metals Resulting from the Irradiation of Uranium with Neutrons*

26 BOHR and WHEELER
The Mechanism of Nuclear Fission

27 BETHE
Energy Production in Stars

28 MÖSSBAUER
Nuclear Resonance Absorption of Gamma Radiation in Ir^{191}

From the earliest experimental investigations on radioactivity, it was realized that the various rays emitted by radioactive substances carry considerable energy with them,[1] indicating the existence of a substantial energy source within the nuclei of atoms of radioactive elements. This energy emission is a natural and uncontrollable process. At first, there was little indication how it might be controlled or manipulated. This had to await the development of further knowledge of the structure of the nucleus and its interaction with bombarding particles, as well as the invention of devices for accelerating charged particles to very high velocities.

In order to secure a theoretical hold on the energy of the nucleus, it was essential to construct a satisfactory theory for the emission of energetic particles from it. This was first attempted in a serious way by George Gamow (1904–1968). In an important 1928 article, he used the new quantum mechanics to investigate the emission of alpha particles. In particular, he used the existence of the tunnel effect in the sense that there is a finite probability for an alpha particle to emerge through the relatively high potential barrier surrounding the nucleus. His calculation verified theoretically the experimentally derived Geiger-Nuttall law, which says that the nuclear decay constant is proportional to the energy of the emitted alpha particles. We reproduce Gamow's 1928 article in translation as Paper 21.

For the first twenty years of research in radioactivity, great attention was paid to the properties of the alpha, beta, and gamma rays emitted by radioactive substances, as well as the properties of the decaying atoms. It was soon realized that the alpha particles emitted by some nuclei possess enough energy to serve as effective atomic projectiles with which to bombard other atoms and hopefully strike their nuclei and produce interesting reactions. It was this sort of use of alpha particles that led to Rutherford's theory of the nuclear atom. The development of particle-detecting devices like the zinc sulphide screen for recording scintillations, the C. T. R. Wilson (1869–1959) cloud chamber, and the Hans Geiger (1882–1945) counter made the observation of the activity of changed particles relatively easy. It was by the use of the scintillations method that Rutherford in 1919 discovered the disintegration of the nitrogen nucleus by capture of an alpha particle, with the emission of a proton. This was probably the first example of the use of atomic energy to effect the transmutation of a normally stable element. The nitrogen nucleus, gaining four mass units from the alpha particle and losing one with the emission of the proton, as well as gaining two elementary positive charges and losing one, takes on the structure of the nucleus of an oxygen isotope. This was the decisive indication of the role of energy in nuclear interactions and in the possible transformation of atomic physical and chemical properties. An important follow-up of this early work of Rutherford was the artificial production of radioactive elements and the definite chemical proof of the transmutation of elements by Irene Joliot-Curie (1897–1956) and Frederic Joliot-Curie (1900–1958) in 1934. They bombarded boron, magnesium, and aluminum with alpha particles from polonium and showed that the resulting products were radioactive. They also showed by chemical means that the nuclei so bombarded became nuclei of atoms differing from the original ones. This was apparently the first chemical proof of the transmutation of elements.

The fundamental article by Curie and Joliet is reproduced here in translation as Paper 22.

Any theoretical attempt to investigate the energy of the nucleus necessarily implies a theory of nuclear structure. This has been the subject of much effort during the time period covered by the present volume, and it is possible to include only a few examples of the many articles published in this field. It was true from the outset, based on the interaction between bombarded nuclei and elementary particles like neutrons and protons, that the nuclei of most atoms possess enormous numbers of energy levels. The early study of these was almost necessarily statistical. Such an investigation of the levels of a heavy nucleus was carried out by Hans A. Bethe (1906–) in 1936; in this he paid much attention to the relation between the energy levels and the capture of neutrons. This proved of importance in connection with the later development of nuclear reactors. Bethe's article is reproduced here as Paper 23.

At about the same time as Bethe's work, somewhat more precise calculations were made on light nuclei by Eugene Feenberg (1906–1977) and Eugene P. Wigner (1902–). They used essentially the Hartree-Fock method of the self-consistent field to calculate the energy levels in the nuclei between helium and oxygen. This involved somewhat arbitrary assumptions regarding the nature of the forces between neutrons and protons and neutrons and neutrons. The relation between nuclear binding energy and nuclear mass was evaluated, and rather good agreement found with the experimental values. The seminal 1937 article by Feenberg and Wigner is reproduced here as Paper 24.

The late 1930s saw continued interest in the experimental irradiation of heavy nuclei with various elementary particles, particularly neutrons, themselves products of nuclear bombardment. It was early assumed that when uranium is irradiated with neutrons, the result would be, in addition to the production of transuranium elements, various radium isotopes from the alpha particle decay of these transuranium atoms and the resulting U^{239}. In experiments carried out in 1939 to test this hypothesis by the German physical chemists Otto Hahn (1879–1968) and Fritz Strassmann (1902–), they obtained the surprising result that barium isotopes were actually produced. This was soon interpreted to mean that the bombardment of uranium with slow neutrons had fissioned the uranium nucleus, splitting it into components of roughly comparable nuclear mass. This was apparently the first indication of the possibility of what has come to be called nuclear fission and the basis for the practical exploitation of nuclear energy, since each fission of a nucleus is accompanied by the production of considerable kinetic energy in the fission fragments. The results of the experiments of Hahn and Strassmann were very shortly duplicated by other investigators, notably by the Austrian physicists Otto R. Frisch (1904–1979) and Lise Meitner (1878–1968). The fundamental article by Hahn and Strassmann is reproduced here in translation Paper 25.

The theorists were not long in applying their ideas to the structure of nuclei suggested by the discovery of Hahn and Strassmann. Bohr had already in 1936 devised a rather simple classical theory of nuclear structure, likening the nucleus to a liquid drop. This accounted for a number of observed nuclear properties. After the announcement of the results of Hahn and Strassmann, Bohr and John A. Wheeler (1911–   ) proceeded in 1939 to develop a theory of nuclear fission based on the liquid drop model. From rather simple considerations, they were able to predict theoretically from assumptions of mass changes on fission and the use of the Einstein mass-energy relation the amount of energy release involved in fission. For example, in the case of the fission of U^{239} into $_{46}Pd^{119}$ and $_{46}Pd^{120}$ the release comes out to be about 200 Mev for each fission. This article by Bohr and Wheeler undoubtedly encouraged further experimental work on nuclear fission. It was soon found that in the process of fission by slow neutrons, two or more neutrons are emitted for each fission; they are then available for further fission. This at once implied the possibility of a self-sustaining chain reaction. The first such reaction was carried out in Chicago in December 1942, providing the basis for nuclear engineering and the production of electrical energy by the use of so-called nuclear reactors. It also provided the basis for the construction of an atomic bomb. The seminal 1939 paper by Bohr and Wheeler is reproduced here as Paper 26.

In nuclear physics research in the 1920s, it was recognized that the loss of mass involved in the combination of two nuclei to form a new nucleus—for example, the fusion of two deuterium nuclei to form He^3 plus a neutron—could lead to the development of considerable kinetic energy. For the example cited, this would amount to 3.25 Mev. From a purely scientific point of view, this energy result of a possible nuclear reaction was seized upon in the 1930s to provide a plausible explanation for the source of the heat of the sun and other stars, thus solving a long-standing problem in astrophysics. Hans Bethe went into this problem in considerable detail in 1938. His preliminary article on the subject is reproduced here as Paper 27.

The nuclear fusion process or the thermonuclear process (it demands high temperatures to make it go in practice) has been hailed as the ultimate source of available energy production in the world of the future, but at present its practical exploitation in plausible nuclear reactors is fraught with difficult technical problems. It does work, of course, in the so-called hydrogen bomb.

We conclude this section with a reference to a low-energy

nuclear phenomenon that has proved of wide scientific use. The emission of a gamma ray by a nucleus involves the transition of the nucleus from an excited energy state to one of lower energy. The energy of the gamma ray will be equal to the difference between the energies of the two states in question minus the recoil energy of the nucleus, if it is free to recoil. This means that if the gamma ray is absorbed again by another nucleus, it will not have enough energy to promote the transition from the lower to the higher state corresponding to those from which it came. However, in 1959 German physicist Rudolf Ludwig Mössbauer (1929–   ) discovered that the nuclear recoil can be effectively reduced to zero by embedding the nucleus in the lattice of a solid, which because of its mass takes up the greater part of the recoil with practically zero velocity. In this way Mössbauer was able to produce gamma rays of precise frequency and wavelength. However, there is an indeterminacy ΔE in the energy level of the excited nuclear state due to its mean finite lifetime Δt, as given by the Heisenberg indeterminacy relations $\Delta E \cdot \Delta t \sim h/2\pi$. (See Paper 8 in this volume.) Hence in order to be used for resonance absorption, the emitted gamma ray must be modulated to take account of this indeterminacy. This is usually done by means of the Doppler effect. The discovery of the Mössbauer effect has led to the development of a new and valuable spectroscopic tool: nuclear gamma-ray resonance fluorescence spectroscopy. This has found wide application to genuine nuclear physics, structural chemistry, biological sciences, and solid-state physics. Mössbauer's seminal 1958 article is reproduced here in translation as Paper 28.

NOTE AND REFERENCE

1. R. Lindsay, ed., 1979, *Early Concepts of Energy in Atomic Physics,* Benchmark Papers on Energy, vol. 7., Stroudsburg, Pa.: Dowden, Hutchinson & Ross. See especially pages 316–340.

ANNOTATED BIBLIOGRAPHY

Bethe, H. A., 1947, *Elementary Nuclear Theory,* Wiley, New York.

Brink, D. M., 1965, *Nuclear Forces,* Pergamon Press, Oxford.

Korff, S. A., 1955, *Electron and Nuclear Counters,* 2nd ed., Van Nostrand, New York.

Thomson, J. J., 1937, *Recollections and Reflections,* Macmillan, New York. See in particular pages 416–420 for a discussion of the Wilson Cloud Chamber.

21

THE QUANTUM THEORY OF THE ATOMIC NUCLEUS

G. Gamow
Göttingen, Institute for Theoretical Physics

This article was translated by J. J. Pimajian, Oak Ridge National Laboratory, from Z. Phys. ***51****:204–212 (1928), by permission of the publisher, Springer-Verlag, New York. This translation appears as U.S. Atomic Energy Commission Report, AEC-tr-5915, August 1960. The figures have been reproduced from the original article.*

ABSTRACT

The attempt is made to examine more closely the processes of α-emission on the basis of wave mechanics and to obtain theoretically the experimentally established relationship between the decay constant and the energy of the α-particle.

1 It has been frequently conjectured[1] that the noncoulombic attractive forces play a very important role in the atomic nucleus. About the nature of these forces we can make many hypotheses.

The forces can be due to attractions between the magnetic moments of the separate nuclear elements of the forces due to electrical and magnetic polarization.

In any case, these forces decrease very rapidly with increasing distance from the nucleus, and only in the direct vicinity of the nucleus do they overcome the influence of the coulombic force.

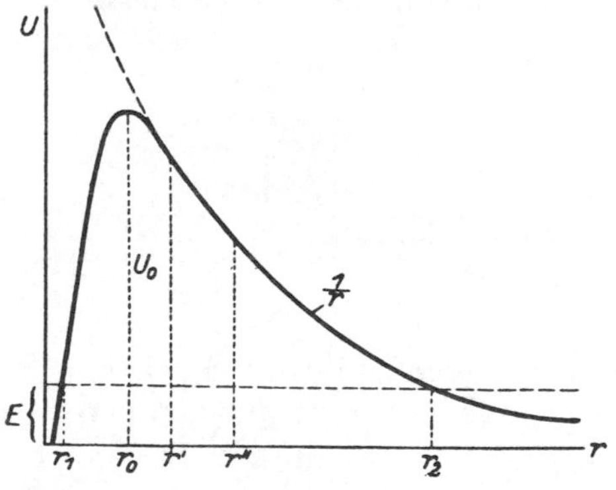

Fig. 1.

From the experiments on the scattering of α-particles, we can conclude that, for heavy elements, the attractive forces are not yet appreciable up to a distance of $\sim 10^{-12}$ cm. Thus, we can accept the diagram shown in Fig. 1 for the form of the potential energy.

Here, r'' signifies the distance up to which it has been experimentally shown that coulombic attraction alone exists. At r', the deviations begin (r' is unknown and perhaps much smaller than r''), and the U-curve has a maximum at U_0. For $r < r_0$, the attractive forces alone prevail; in this region a particle would revolve around the nuclear residue as a satellite.

This movement is, however, not stable; its energy is positive, and after some time the α-particle will be emitted (α-radiation). Here, however, we meet a principal difficulty.

In order to be emitted, the α-particle must overcome a potential barrier of height U_0 (Fig. 1); its energy must not be smaller than U_0. But the energy of the α-particle is much smaller, as has been proven experimentally. In a study of the scattering of RaC′ α-particles on uranium,[2] it is found that with such very fast particles, the coulombic law holds for the uranium nucleus up to a distance of 3.2×10^{-12} cm. On the other hand, the particles emitted from uranium itself have an energy that from the repulsion curve corresponds to a nuclear distance of 6.3×10^{-12} cm (r_2 in Fig. 1). If an α-particle that comes from within the nucleus is emitted, it must pass through the region between r_1 and r_2, where its kinetic energy would be negative. This is naturally an impossibility according to classical ideas.

In order to overcome this difficulty, Rutherford[3] hypothesized that the α-particle is neutral in the nucleus, since there it is said to include two electrons. First, at a fixed nuclear distance on the other side of the potential maximum, it loses, according to Rutherford, both its electrons, which fall back into the nucleus, and then the α-particle is emitted with the development of the coulombic repulsive force. However, this assumption appears to be very unnatural and would hardly correspond to the facts.

2 If we consider the problem from the standpoint of wave mechanics, then the above-mentioned difficulty is resolved automatically. In wave mechanics, namely, there is always a non-zero probability for a particle to move from one region to another region of the same energy, separated from the first by an arbitrary but finite potential barrier.[4]

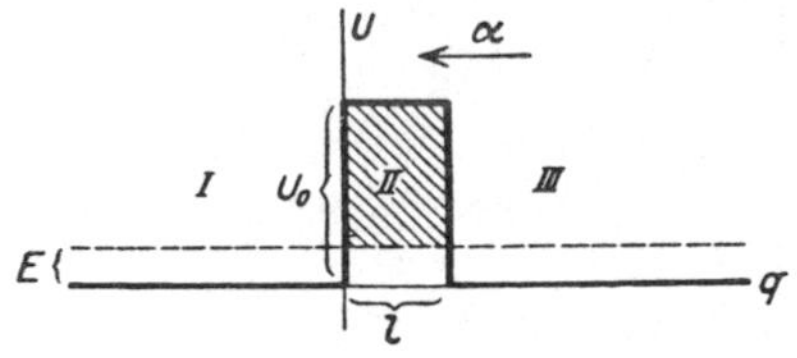

Fig. 2.

As we shall see later, the probability of such a transition is, to be sure, very small; that is, the higher the potential barrier to be overcome, the smaller the probability. In order to illustrate this fact, we will investigate a simple example.

We have a rectangular potential barrier and we intend to find the solution of the Schrödinger equation, which describes the passage of the particle from the right to the left. For the energy E we write the wave function ψ in the following form:

$$\psi = \Psi(q) \cdot e^{+\frac{2\pi i E}{h} t},$$

where $\psi(q)$, the amplitude, satisfies the equation

$$\frac{\partial^2 \psi}{\partial q^2} + \frac{8\pi^2 m}{h^2}(E - U)\Psi = 0. \tag{1}$$

For region I, we have the solution

$$\Psi_I = A \cos(kq + \alpha).$$

Here A and α are both arbitrary constants and k signifies

$$k = \frac{2\pi\sqrt{2m}}{h} \cdot \sqrt{E}. \tag{2a}$$

In region II the solution reads

$$\Psi_{II} = B_1 e^{-k'q} + B_2 e^{+k'q},$$

where k' is

$$k' = \frac{2\pi\sqrt{2m}}{h}\sqrt{U_0 - E}. \tag{2b}$$

At the boundary $q = 0$, these conditions apply:

$$\psi_I(0) = \psi_{II}(0) \quad \text{and} \quad \left[\frac{\partial \Psi_I}{\partial q}\right]_{q=0} = \left[\frac{\partial \Psi_{II}}{\partial q}\right]_{q=0},$$

from which we easily obtain

$$B_1 = \frac{A}{2\sin\vartheta}\sin(\alpha + \vartheta); \quad B_2 = -\frac{A}{2\sin\vartheta}\sin(\alpha - \vartheta),$$

where

$$\sin\vartheta = \frac{1}{\sqrt{1 + \left(\frac{k}{k'}\right)^2}}.$$

Hence the solution in region II reads:

$$\Psi_{II} = \frac{A}{2\sin\vartheta}[\sin(\alpha + \vartheta) \cdot e^{-k'q} - \sin(\alpha - \vartheta)\, e^{+k'q}].$$

Again, in region III we have:

$$\Psi_{III} = C\cos(kq + \beta).$$

At the boundary $q = l$, we have from the boundary restrictions:

$$\frac{A}{2\sin\vartheta}[\sin(\alpha + \vartheta)\, e^{-lk'} - \sin(\alpha - \vartheta)\, e^{+lk'}] = C\cos(kl + \beta)$$

and

$$\frac{A}{2\sin\vartheta} k'[-\sin(\alpha + \vartheta)\, e^{-lk'} - \sin(\alpha - \vartheta)\, e^{+lk'}] = -k\,C\cos(kl + \beta).$$

Thus,

$$C^2 = \frac{A^2}{4\sin^2\vartheta}\left\{\left[1+\left(\frac{k'}{k}\right)^2\right]\sin^2(\alpha-\vartheta)\,.\,e^{2lk'}\right.$$

$$-\left[1-\left(\frac{k'}{k}\right)^2\right]2\sin(\alpha-\vartheta)\sin(\alpha+\vartheta)$$

$$\left.+\left[1+\left(\frac{k'}{k}\right)^2\right]\sin^2(\alpha+\vartheta)\,e^{-2lk'}\right\}. \qquad (3)$$

The calculation of β is not of interest to us. The case where lk' is very large is of interest to us, so that we need to consider only the first term in (3).

Thus, we have the following solution:

left: right:

$$A\cos(kq+\alpha)\ \ldots\ A\frac{\sin(\alpha-\vartheta)}{2\sin\vartheta}\left[1+\left(\frac{k'}{k}\right)^2\right]^{\frac{1}{2}}.\,e^{+lk'}\cos(kq+\beta).$$

If we replace α by $\alpha-\frac{\pi}{2}$, multiply the obtained solution by i, and add both solutions, we thus obtain on the left

$$\Psi = Ae^{i(kq+\alpha)}. \qquad (4a)$$

However, on the right:

$$\Psi = \frac{A}{2\sin\vartheta}\left[1+\left(\frac{k'}{k}\right)^2\right]^{\frac{1}{2}}.\,e^{+lk'}\{\sin(\alpha-\vartheta)\cos(kq+\beta)$$

$$-i\cos(\alpha+\vartheta)\cos(kq+\beta')\}, \qquad (4b)$$

where β' is the new phase.

If we multiply this solution by $e^{2\pi i\frac{E}{h}t}$, then we obtain a traveling wave on the left for ψ; on the right, however, we obtain the complicated vibrational process with a very great amplitude ($e^{lk'}$), deviating slightly from the standing wave. That means nothing other than that the wave coming from the right is partially reflected and partially transmitted.

Thus, we see that the amplitude of the penetrating wave is smaller, the smaller the total E is; that is, the factor

$$e^{-lk'} = e^{-\frac{2\pi.\sqrt{2m}}{h}\sqrt{U_0-E}\,.\,l}$$

plays the most important role in this dependence.

3 At the present we can solve the problem for two symmetrical potential barriers (Fig. 3). We will look for two solutions.

A solution for positive q should be valid and for $q > q_0 + l$ gives the wave:

$$Ae^{i\left(\frac{2\pi E}{h}t-kq+\alpha\right)}.$$

The other solution applies for the negative q and gives for $q < -(q_0+l)$ the wave:

$$A\,.\,e^{i\left(\frac{2\pi}{h}Et+qx'-\alpha\right)}.$$

Then we cannot continuously join the two solutions at the limit $q = 0$ because we have here two boundary conditions to fulfill and only one constant α at our disposal.

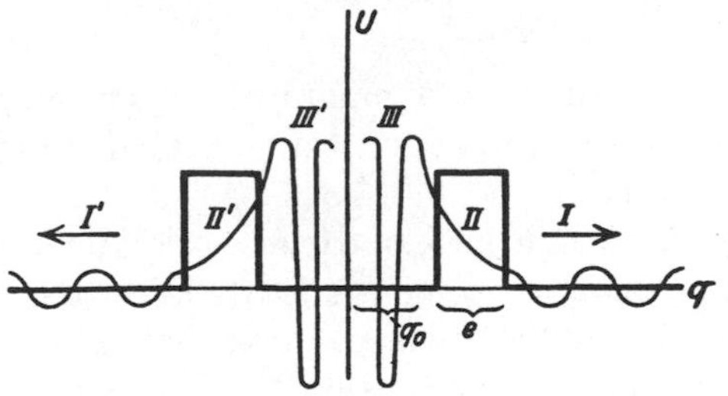

Fig. 3.

The physical reason for this impossibility is that the ψ function constructed out of these two solutions does not suffice for the conservation principle:

$$\frac{\partial}{\partial t}\int_{-(q_0+l)}^{+(q_0+l)} \psi\overline{\psi}\, dq = 2 \cdot \frac{-h}{4\pi i m}[\psi \operatorname{grad} \overline{\psi} - \overline{\psi} \operatorname{grad} \psi]_{\mathrm{I}}.$$

In order to overcome this difficulty, we must assume that the vibrations are damped and E is complex

$$E = E_0 + i\frac{h\lambda}{4\pi},$$

where E is the usual energy and λ the damping decrement (disintegration constant). Then, however, we see from the relations (2a) and (2b) that k and k' should also be complex; that is, that the amplitudes of our waves exponentially depend also on the coordinate q. For example, the amplitude of the traveling wave in the direction of the wave propagation will increase. That signifies, however, nothing further than that when the oscillation (vibration) of the wave is damped at the starting point, the amplitude of the emitted wave part must be greater. We can now choose α so that the boundary conditions are fulfilled. However, the rigorous solution does not interest us. When λ is small in comparison with E/h

$$\left(\frac{E}{h} \backsimeq \frac{10^{-5}}{10^{-27}\,\mathrm{sec}} = 10^{+22}\,\mathrm{sec}^{-1} \text{ and } \lambda = 10^{+5}\,\mathrm{sec}^{-1}\right),$$

then the change of $\psi(q)$ is very small and we can simply multiply the old solution by $e^{-\frac{\alpha}{2}t}$.

Then the conservation principle reads:

$$\frac{\partial}{\partial t} e^{-\lambda t}\int_{-(q_0+l)}^{+(q_0+l)} \Psi^{(q)}_{\mathrm{II,III}} \cdot \Psi^{(q)}_{\mathrm{II,III}}\, dq = -2 \cdot \frac{A^2 h}{4\pi i m} \cdot 2ik \cdot e^{-\lambda t},$$

from which follows:

$$\lambda = \frac{4\,h\,k\,\sin^2\vartheta}{\pi\,m\left[1+\left(\frac{k'}{k^0}\right)^2\right]2\,(l+q_0)\,\varkappa}\cdot e^{-\frac{4\,\pi\,l\sqrt{2\,m}}{h}\sqrt{U_0-E}}, \tag{5}$$

where x is a number of the order of magnitude 1. This formula gives the dependence of the disintegration constant on the disintegration energy for our simple nuclear model.

4 Now we can turn to the case of the real nucleus. We cannot solve the corresponding wave equation since we do not know the accurate potential variation in the vicinity of the nucleus. However, we can obtain a few results for the actual nucleus for our simple model without the actual knowledge of the potential variation. As usual, in the case of the central force, we shall look for the solution in polar coordinates—that is, in the form

$$\Psi = u\,(\theta, \varphi)\,\chi\,(r).$$

For u we obtain the spherical functions, and for X, the following differential equations must suffice:

$$\frac{\partial^2\chi}{\partial r^2} + \frac{2}{r}\frac{d\chi}{dr} + \frac{8\,\pi^2 m}{h^2}\left[E - U - \frac{h^2}{8\,\pi^2 m}\cdot\frac{n\,(n+1)}{r^2}\right]\chi = 0,$$

where n is the order of the spherical function. We can assume $n = 0$, since if $n > 0$, it would be as if the potential energy were increased, and consequently the damping becomes much smaller for these oscillations. The particle must first go over to the state $n = 0$, and only then can it be emitted.

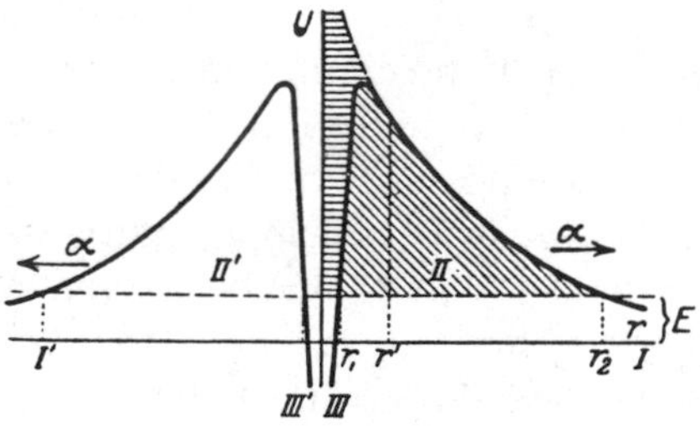

Fig. 4.

It is a good possibility that such transitions produce the γ-rays that always accompany α-emission. The probable form of U is reproduced in Fig. 4.

For larger values of r, we shall assume for X, the following solution:

$$\chi_{\mathrm{I}} = \frac{A}{r}\,e^{i\left(\frac{2\,\pi\,E}{h}t - k\,r\right)}.$$

Although one cannot obtain an accurate solution of the problem in this case, we surely can say that in the regions I and III, X will not decrease rapidly on the average (in the three-dimensional case as with $1/r$).

In region III, however, X will decrease exponentially, and indeed, we can, in analogy with our simple case, expect that the relationship between decrease of amplitude and E is approximately given by the following factor:

$$e^{-\frac{2\pi\sqrt{2m}}{h}\int_{r_1}^{r_2}\sqrt{U-E}\,dr}.$$

With the application of the conservation principles, we can again write the formula

$$\lambda = D \,.\, e^{-\frac{2\pi\sqrt{2m}}{h}\int_{r_1}^{r_2}\sqrt{U-E}\,dr}, \tag{6}$$

where D depends on the special properties of the nuclear model. We can disregard the dependence of D on E next to the exponential dependence of the second factor.

We can approximate the integral

$$\int_{r_1}^{r_2}\sqrt{U-E}\,dr$$

by

$$\int_0^{\frac{2Ze^2}{E}}\sqrt{\frac{2Ze^2}{r}-E}\,.\,dr.$$

The relative error that we make here becomes of the order of magnitude

$$\frac{\int_0^{r_1}\sqrt{\frac{1}{r}}\,dr}{\int_0^{r_2}\sqrt{\frac{1}{r}}\,dr} = \sqrt{\frac{r'}{r_2}}.$$

Since r'/r_2 is small, then this error does not become very large. Since with various radioactive elements, E does not vary much, we can write approximately

$$\lg\lambda = \lg D - \frac{4\pi\sqrt{2m}}{h}\left\{\int_0^{\frac{2Ze^2}{E_0}}\sqrt{\frac{2Ze^2}{r}-E_0}\,dr + \frac{\partial}{\partial E}\int_0^{\frac{2Ze^2}{E}}\sqrt{\frac{2Ze^2}{r}-E}\,dr\,.\,\Delta E\right\}$$

or

$$\lg\lambda = \mathrm{Const}_E + B_E\,.\,\Delta E,$$

where

$$B = -\frac{4\pi\sqrt{2m}}{h}\frac{\partial}{\partial E}\int_0^{\frac{2Ze^2}{E}}\sqrt{\frac{2Ze^2}{r} - E}\,dr = \frac{4\pi\sqrt{2m}}{2h}\int_0^{\frac{2Ze^2}{E}}\frac{dr}{\sqrt{\frac{2Ze^2}{r} - E}}\,dr.$$

We set

$$\left.\begin{aligned} &\varrho = \frac{E}{2Ze^2}r, \\ &\text{then} \\ &b = \frac{4\pi\sqrt{2m}\,.\,2Ze^2}{2hE^{3/2}}\int_0^1\frac{d\varrho}{\sqrt{\frac{1}{\varrho}-1}} = \frac{\pi^2\sqrt{2m}\,.\,2Ze^2}{hE^{3/2}}. \end{aligned}\right\} \quad (7)$$

Now we want to compare this formula with the experimental facts. It is known[5] that when the energy of the α-particles is plotted on the abscissa and the logarithm of the disintegration constants is plotted along the ordinate, all points for a given radioactive family lie on a straight line. For various families, one acquires parallel straight lines. The empirical formula reads

$$\lg\lambda = \text{Const} + bE,$$

where b is a constant common to all radioactive families.

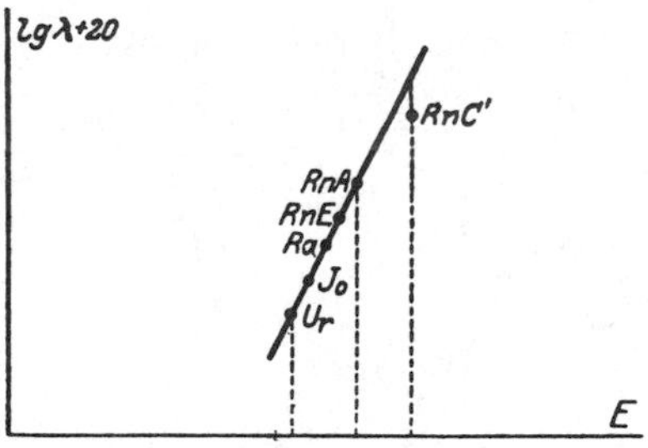

Fig. 5.

The experimental value of b (calculated from RaA and Ra) is $b_{exper} = 1.02 \times 10^{+7}$.

If we, however, insert the energy value in our formula for RaA, then the calculation gives $b_{theor} = 0.7 \times 10^{+7}$.[6]

The order of magnitude agreement indicates that the basic hypothesis of the theory may be right.

According to our theory, certain deviations from the linear law must exist; with increasing energy b must decrease—that is, $\log\lambda$ must decrease somewhat more slowly than E. The measurements of Jacobsen[7] agree with the above. These measurements indicate that RaC′, whose α-radiation is very energy rich (energetic), has a disintegration constant of 8.4×10^7, whereas a value of 5×10^7 is obtained from the linear law.

In conclusion, I would like to express to my friend N. Kotschin my thanks for the kind discussion of the mathematical problem. Also, I would like to heartily thank Professor Born for permission to work in his Institute.

NOTES

1. Frenkel, J., Z. Physik, 37, 243 (1926); Rutherford, E., Phil. Mag., 4, 580 (1927); Enskog, D., Z. Physik, 45, 852 (1927).
2. Rutherford, E., loc. cit., p. 581.
3. Rutherford, E., loc. cit., p. 584.
4. See, for example: Oppenheimer, Phys. Rev., 31, 66 (1928); Nordheim, Z. Physik, 46, 833 (1927).
5. Geiger and Nuttall, Phil. Mag., 23, 439 (1912); Swinne, Phys. Zeit., 13, 14 (1912).
6. For other elements we obtain approximately the same value, since Z for different radioactive elements is only slightly different.
7. Jacobsen, Phil. Mag., 47, 23 (1924).

22

I. ARTIFICIAL PRODUCTION OF RADIOACTIVE ELEMENTS
II. CHEMICAL PROOF OF THE TRANSMUTATION OF THE ELEMENTS

Irene Curie and F. Joliot

This article was translated expressly for this Benchmark volume by R. Bruce Lindsay, Brown University, from J. Phys. Radium, *ser. 7,* ***5:****153–156 (1934) by permission of the publisher, Les Editions de Physique, Orsay, France*

ABSTRACT

Boron, magnesium, and aluminum, after irradiation by alpha particles from polonium, manifest a lasting radioactivity, which shows itself, in the case of boron and aluminum, in the emission of positrons and, in the case of magnesium, in the emission of both negative electrons and positrons. Radioelements have thus been created by transmutation. Their decay follows an exponential law. The half-lives in the cases of boron, magnesium, and aluminum are found to be 14 minutes, 2.5 minutes, and 3.25 minutes, respectively. The decay is independent of the energy of the exciting alpha particles.

The radiation emitted by the irradiated aluminum and boron is composed exclusively of positrons without negative electrons and forms a continuous spectrum like the natural spectrum of beta rays from radioactive bodies. The maximum energy of the positron radiation is of the order 1.5×10^6 electron volts for boron and 3×10^6 electron volts for aluminum.

The positive and negative electrons from magnesium form two continuous spectra and correspond without doubt to the transmutation of two isotopes of magnesium.

The new radioactive elements are probably the nuclei ${}^{13}{}_{7}N$, ${}^{27}{}_{14}Si$, ${}^{28}{}_{13}Al$, and ${}^{30}{}_{15}P$, formed from the nuclei ${}^{10}{}_{5}B$, ${}^{24}{}_{12}Mg$ and ${}^{27}{}_{13}Al$.

We have chemically separated from boron and aluminum the radioactive elements that have been formed by irradiation. As predicted, they have the chemical properties of nitrogen and phosphorus, respectively. These experiments constitute the first chemical proof of artificial transmutation.

We propose to call the new radioelements radionitrogen, radiosilicon, radioaluminum, and radiophosphorus, respectively.

We have recently shown that certain light elements (beryllium, boron, and aluminum) emit positive electrons when bombarded with alpha particles from polonium.[1]

Beryllium emits positive and negative electrons of comparable energy, attaining several million electron volts. We have attributed the emission of these electrons to the "internal materialization" of gamma radiation of high energy (5×10^5 eV) emitted by beryllium:

$$ {}^{9}_{4}\mathrm{Be} + {}^{4}_{2}\mathrm{He} = {}^{12}_{6}\mathrm{C} + {}^{1}_{0}n + \overset{+}{\varepsilon} + \overset{-}{\varepsilon}. $$

Aluminum, on the contrary, emits only positive electrons whose energy can reach 3×10^6 eV. The negative electrons observed are all attributable to polonium (electrons emitted by internal conversion of gamma radiation), and their energy does not exceed 0.9×10^6 eV.

We have considered the positive electrons as real "transmutation" electrons, whose emission takes part in the nuclear transformation.

In the course of experiments made with a view to determining the minimum energy of alpha particles leading to the emission of positrons from aluminum, we have observed that the emission is not instantaneous but is produced only after some minutes of irradiation and persists some time after the cessation of the irradiation.[2]

We irradiate for some minutes an aluminum foil with alpha particles from a strong polonium source. When we remove the foil, it possesses an activity that decreases by one-half in 3.25 minutes. The emitted radiation, which can be observed by means of a counter or a Wilson cloud chamber, consists entirely of positive electrons.

Irradiated boron and magnesium also provide a lasting radioactivity, with periods of 14 minutes and 2.5 minutes, respectively. The decrease follows an exponential law. The curves in Fig. 1 represent the logarithmic variation of the intensities as a function of time.

We are therefore here in the presence of new radio elements and of a new type of radioactivity with the emission of positive electrons.

The initial intensity of the observed radiation increases with the time t of the irradiation up to a limiting value I_∞ following the law

$$ I = I_\infty \left(1 - e^{-\lambda t}\right), $$

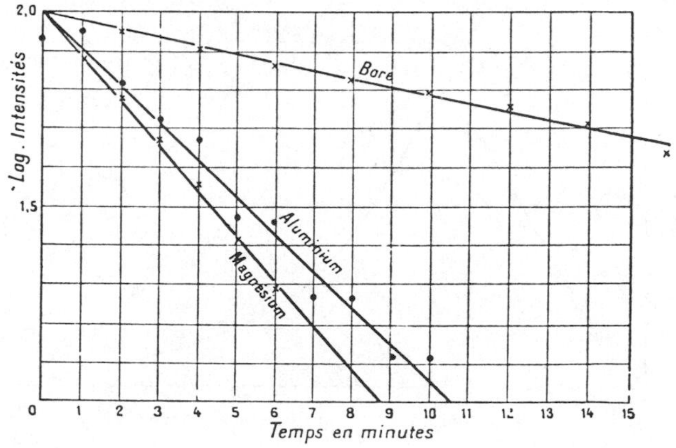

Fig. 1.

a law similar, for example, to that governing the accumulation of radon in radium. The activation limit is attained after some periods of the radioactive substances formed (around fifteen minutes for aluminum and an hour for boron).

If we activate to saturation using a 60 millicurie polonium source and if we immediately place the source close to a counter with a thin window, permitting the counting of electrons, the number of counts is of the same order (about 150 per minute) with boron, aluminum, and magnesium. Taking account of the soild angle used, we can calculate that this corresponds to a rate of creation of active nuclei of the order of 10^{-6} with respect to the number of incident alpha particles, a figure one would expect in a transmutation phenomenon.

In spite of the small difference in the rate of activation for boron, magnesium, and aluminum, the experiments are more difficult to perform with magnesium owing to the more rapid decrease in the activity in its case.

With the elements H, Li, C, Be, N, O, F, Na, Ca, Ni, Au, no effect has been observed. For certain of these elements, the phenomenon probably does not take place at all; for others, the period of decrease is perhaps too short or too long, thus necessitating a long irradiation to detect it.

The radiation emitted by aluminum is absorbed in large part by 1 gram/cm^2 of copper, that by boron and magnesium in 0.26 gram/cm^2.

The radiation from irradiated aluminum and boron, photographed in a Wilson cloud chamber in the presence of a magnetic field, shows itself to be wholly comprised of positrons, with complete absence of negative electrons. If there were actually any emission of negative electrons, it would have to be a very rare event; at any rate, the energy of such would have to be less than 2×10^5eV.

We present (Fig. 2, radiophosphorus and Fig. 3, radionitrogen) curves for the observed positive electrons.

For aluminum, the positrons are distributed in a continuous spectrum having an intensity maximum around 10^6 eV and an upper energy limit

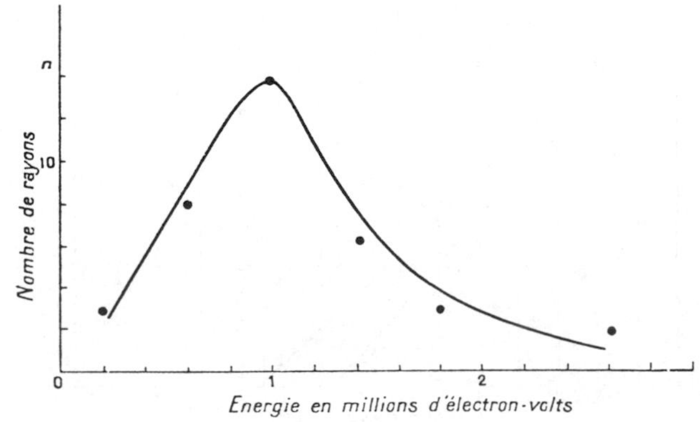

Fig. 2. — Radiophosphore.

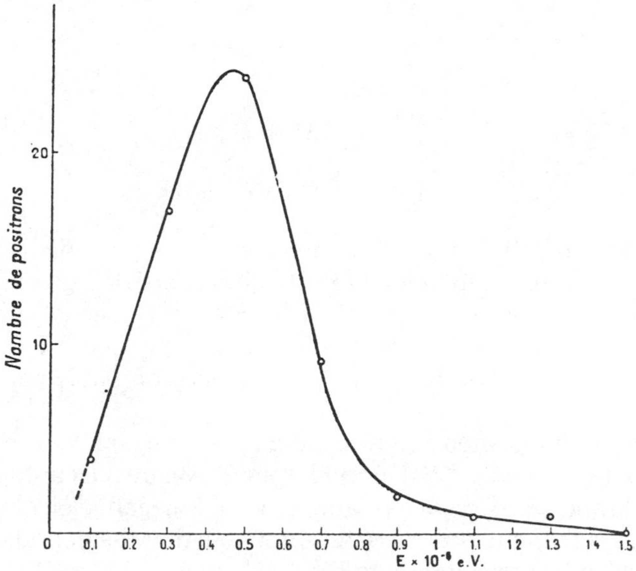

Fig. 3. — Radioazote.

around 3×10^6 eV. For boron, the intensity maximum is about 0.5×10^6 eV, and the upper limit is around 1.5×10^6 eV.

Irradiated magnesium emits both positive and negative electrons, forming two continuous spectra with maximum energy around 2.2×10^6 eV and 1.5×10^6 eV, respectively.

THE NATURE OF THE ELEMENTS PRODUCED

We can make two hypotheses concerning the nature of the elements found:

1. The active nuclei are unstable isotopes of certain light elements, or
2. They are known stable nuclei created in a state of particular excitation.

The first hypothesis appears to us to be the more reasonable one.

Here are the nuclear reactions that would agree with our first interpretation of the rate of emitted positive electrons.

The nuclei ${}^{10}{}_5B$, ${}^{24}{}_{12}Mg$, and ${}^{27}{}_{13}Al$ suffer a transformation with the capture of the alpha particle and emission of a neutron:

$$
\begin{aligned}
{}^{10}_{5}\mathrm{B} + {}^{4}_{2}\mathrm{He} &= {}^{13}_{7}\mathrm{N} + {}^{1}_{0}n \\
{}^{24}_{12}\mathrm{Mg} + {}^{4}_{2}\mathrm{He} &= {}^{27}_{14}\mathrm{Si} + {}^{1}_{0}n \\
{}^{27}_{13}\mathrm{Al} + {}^{4}_{2}\mathrm{He} &= {}^{30}_{15}\mathrm{P} + {}^{1}_{0}n .
\end{aligned}
$$

The isotopes ${}^{13}{}_7N$, ${}^{27}{}_{14}Si$, and ${}^{30}{}_{15}P$ are not known. They are probably unstable nuclei that decay with the emission of positrons, yielding the stable nuclei ${}^{12}{}_6C$, ${}^{27}{}_{13}Al$, ${}^{30}{}_{14}Si$, respectively.

Thus:

$$^{14}_{7}N = {}^{13}_{6}C + \overset{+}{\varepsilon}$$

$$^{27}_{14}Si = {}^{27}_{13}Al + \overset{+}{\varepsilon}$$

$$^{30}_{15}P = {}^{30}_{14}Si + \overset{+}{\varepsilon}.$$

It is very probable that fluorine $^{19}{}_{9}F$ and sodium $^{23}{}_{11}Na$, which emit protons and neutrons when irradiated with alpha particles, give rise to the unstable isotopes $^{22}{}_{11}Na$ and $^{26}{}_{13}Al$, respectively, which decay with the emission of positrons, yielding the stable nuclei $^{22}{}_{10}Ne$ and $^{26}{}_{12}Mg$, respectively. The period of the unstable elements is probably not favorable for observation.

The radioelement created in irradiated magnesium and emitting beta rays is probably a nucleus $^{28}{}_{13}Al$ formed from $^{25}{}_{12}Mg$ by the capture of an alpha particle and emission of a proton. Since the negative electrons are more numerous than the positrons, it is probable that the period of 2.25 minutes observed corresponds to this radioelement.

To preserve the laws of conservation of energy and spin, it is without doubt necessary to agree that the emission of positive electrons is accompanied by that of neutrinos, as in the case of the beta-ray spectra of the ordinary radioactive elements, or it may indeed be an "antineutrino" in the sense of Louis de Broglie.

The emission of a proton in the transmutation of aluminum, for example, should take place in accordance with the following reaction:

$$^{27}_{13}Al + {}^{4}_{2}He = {}^{30}_{14}Si + {}^{1}_{1}H.$$

The emission of a neutron followed by a positive electron would lead to the same nucleus, $^{30}{}_{14}Si$. In this second possible mode of transmutation, the emission of a proton would then be replaced by that of a neutron and a positron. The proton, the neutron, and the positron all being supposed to have spin ½, this transmutation is possible only on the condition of the emission of another particle of spin ½.

If our hypotheses concerning the nuclear transformations are correct, the minimum energy of the alpha particles necessary to create new radioelements should be the same as the minimum energy of excitation of neutrons.

Some experiments have been performed in exciting radiation from boron, magnesium, and aluminum by alpha particles slowed down by thin screens. The period of the radioelements formed does not depend on the energy of the incident alpha particles. We have not yet established any difference in the penetration of the radiation.

The minimum energy of alpha particles necessary for the excitation of neutrons is 2×10^6 eV for the case of boron (length of path, 4 cm) and 4 to 4.5×10^6 eV in the case of magnesium and aluminum (path length, 3 cm).

For magnesium and aluminum, the limit of emission of positive electrons has been found to be the same as that of neutrons to the limit of precision of the experiments. For boron we have observed excitation of positive electrons by alpha particles of energy around 3×10^6 eV (path length, 2 cm). It is very probable that only lack of intensity prevents us from observing them down to 2×10^6 eV.

CHEMICAL PROOF OF THE TRANSMUTATIONS

The phenomena of the artificial transmutation appear to us to have been established with certitude. Physical experiments have demonstrated the existence of these phenomena and have even permitted us to conclude that the particles producing the transmutation are captured in the nucleus analogously to the earlier experiments of Blackett, which have demonstrated the transmutation of nitrogen by alpha particles with the emission of a proton.

However, the reality of the transmutation has never, so far, been able to be verified chemically as in the case of the natural disintegration of the radioactive elements. This results from the fact that the small number of atoms produced in these artificial transmutations cannot be disclosed by any analytical chemical procedure, and consequently, their chemical properties cannot be recognized.

In the present work, on the contrary, on irradiating boron, magnesium, and aluminum, we obtain different nuclei in small quantities, to be sure, but of such amounts that their radioactive properties allow them to be identified.

We are prepared to test the hypotheses we have made as to the nature of these elements by determining their chemical character.[3]

TRANSMUTATION OF BORON

Solid boron is composed of very hard grains and does not lend itself readily to experiment. We have irradiated boron nitride (the nitrogen does not produce any effect). The boron nitride (BN) is decomposed in a hot soda solution, and the nitrogen is liberated as ammonia. This operation can be carried out in a few minutes. We confirm that the boron loses its activity. The ammonia that is released is recovered in a tube with thin walls and is measured with a Hoffmann electrometer on an ionization chamber closed by a thin aluminum foil. Taking account of the decrease in activity during the operation, we confirm that a large part of the activity is recovered in the tube. The radioelement, therefore, has the chemical properties of nitrogen. It is released either in the form of nitrogen itself or in the form of ammonia entrained with the inactive ammonia.

TRANSMUTATION OF ALUMINUM

A thin film of irradiated aluminum is dissolved in HCl. The aluminum salt is almost completely inactive when immediately dried. The released hydrogen can be recovered in a thin-walled tube, and we confirm that this has entrained the activity. It is necessary to work fast, in view of the rapid decrease of the activity. We can carry out the operation in three minutes.

The radioelement that ought to have the properties of phosphorus should go into the gaseous state in the form of a compound of phosphorus and hydrogen produced by the nascent hydrogen. If one dissolves aluminum in an oxidizing medium HCl + HNO_3, the activity remains with the aluminum.

However, the above operation would occur in the same manner if the radioelement had the properties of silicon, for in this case a compound of hydrogen and silicon would be formed.

One can also dissolve the irradiated aluminum, add sodium phosphate, and precipitate by means of a zirconium salt. The zirconium phosphate that precipitates in a weak acid solution entrains the activity. It is then indeed shown to be probable that an isotope of phosphorus has been formed.

For magnesium, we have not verified the chemical nature of the elements formed. These should be, as we have seen, isotopes of silicon and aluminum.

We propose to give the names radionitrogen, radioaluminum, radiosilicon, and radiophosphorus to the new radioactive elements, emitters of positrons, or beta rays.

The above experiments have provided the chemical proof that the element formed by transmutation is different from the initial element and that the alpha particle is captured in the nucleus.

CONCLUSIONS

It is probable that a certain number of nuclei (isotopes of known elements) are radioelements that do not exist in nature because of their instability but which we can create by transmutation of ordinary elements by means of different types of irradiating particles: protons, deuterons, alpha particles, neutrons, and perhaps still others. We can believe that these radioelements are not all emitters of positrons or negative electrons but that certain of them emit heavier particles. Perhaps indeed there also exist elements constituting real but so far unknown radioactive families.

The same radioactive element doubtless can be created by several different nuclear reactions. Cockcroft, Gilbert, and Walton, as a result of our experiments, have tried to create new radioelements by bombardment with protons. In irradiating carbon, they have obtained a radioelement that emits positrons and has a period of 10.5 minutes. They suppose they have created radioactive nitrogen, which we have created starting with boron. They explain the difference in period by means of a different state of excitation. This interpretation seems improbable to us, and we think that in their experiment, it is the case of another radioactive element.

The study of these phenomena cannot fail to make an important contribution to the problem of the stability of nuclei.

Furthermore, the energies and intensities obtainable by means of machines for accelerating particles permit us to envisage the time when we can create with such particles radioactive elements whose radiation intensity is far greater than that of naturally created elements. These may be useful for medical applications and other purposes. Such radioelements will be emitters of not only positrons but also gamma radiation through positron annihilation—that is, two quanta of 5×10^5 eV per emitted positron. They will then consitute powerful sources of penetrating radiation.

We thank Mme. Pierre Curie for the interest she has taken in this work and M. Preiswerk for the efficient help he has given us.

(Ms. received March 20, 1934)

NOTES

1. Irene Curie and F. Joliot, *Jour. de Physique et le Radium, 4,* 494, 1933.
2. Irene Curie and F. Joliot, *Comptes Rendus, 198,* 254, 1934.
3. Irene Curie and F. Joliot, *Comptes Rendus, 198,* 559, 1934.

23

Reprinted from *Phys. Rev.* **50:**332–341 (1936)

An Attempt to Calculate the Number of Energy Levels of a Heavy Nucleus

H. A. BETHE, *Cornell University*

(Received June 5, 1936)

Experiments on slow neutrons, and theoretical considerations of Bohr have shown that heavy nuclei possess an enormous number of energy levels which are very closely spaced if the nucleus is highly excited. A crude method is suggested for calculating the spacing between these levels. The method is statistical: The individual nuclear particles are supposed to move in a simple potential hole, and the energy of the complete nucleus is supposed to be the sum of the energies of the individual particles. A critical discussion of these assumptions is given in section 5. The problem then reduces itself to the calculation of the "entropy" of a Fermi gas containing a given number of particles A and having a given excitation energy Q above the zero point energy of the Fermi gas (cf. section 2 and 3). This calculation gives the total number of levels of the complete nucleus in a given energy interval irrespective of the angular momentum, which will, for most of the levels, be very large. For the theory of neutron capture, it is necessary to calculate the density of nuclear levels with a *given angular momentum* I (section 4). The spacing of nuclear levels is found to depend on the product of the mass number A and the excitation energy Q of the nucleus, and to be roughly given by

$$\Delta = 4.1 \cdot 10^6 x^4 e^{-x}/(2I+1) \text{ volts}$$
$$x = (AQ)^{\frac{1}{2}}/2.20,$$

Q being expressed in MV and I being the nuclear spin. For the capture of slow neutrons by nuclei of medium weight (A around 100), Δ is of the order 50 to 500 volts. The spacing between adjacent levels decreases rapidly with increasing atomic weight. For given atomic weight, the spacing of the nuclear levels responsible for neutron capture is wider if the capture leads to the formation of a radioactive nucleus than if a stable nucleus is formed. This explains the experimental fact that only moderately large cross sections are found for the capture of thermal neutrons leading to radioactive nuclei while the very largest cross sections are all connected with the formation of stable nuclei. The dependence of the spacing on various factors is discussed (section 6); the results seem to be in qualitative agreement with experiment.

1. Statement of Problem

BOHR[1] has given strong reasons for the existence of a very great number of closely spaced energy levels for a highly excited heavy nucleus. Breit and Wigner[2] and Bohr[1] have shown that the assumption of such levels leads automatically to a completely satisfactory explanation of all phenomena connected with slow neutrons, in particular the selective absorption, the high capture cross section and the large ratio of capture to scattering. Various investigators[3] have measured the position of the neutron resonance levels for several substances. The resonances are found to lie at neutron energies ranging from about 0.1 volt (Cd) to about 50 volts (I). These measurements indicate that the spacing between adjacent energy levels of the nuclei concerned in the energy region investigated is very small, maybe of the order of 100 volts or even less.

It is the purpose of this paper to give some fairly crude calculations leading to an estimate of this spacing. We consider a nucleus containing N neutrons and Z protons. The total number of particles (mass number) will be denoted by $A=N+Z$. The nucleus will have a certain ground state of energy U_0. We are interested in the energy levels of the nucleus which lie by a certain amount Q higher than the ground state, and we ask for the density of energy levels in this region, i.e., for the number of levels between Q and $Q+dQ$ which we may call $\rho(Q)dQ$. $1/\rho(Q)$ will then be the average spacing between neighboring levels.

We shall be particularly interested in such values of Q which are just sufficient to dissociate the given nucleus A into a neutron and a residual nucleus of atomic weight $A-1$. These energy levels will be important for the capture of slow neutrons by the nucleus $A-1$. In general, the "dissociation energy" Q, i.e., the energy set free when a neutron is captured by the nucleus $A-1$, will be of the order 8 MV. This figure applies if the packing fractions of the nuclei $A-1$ and A are about equal, and represents the excess of the

[1] Bohr, Nature **137** (1936).
[2] Breit and Wigner, Phys. Rev. **49**, 519 (1936).
[3] Frisch and Placzek, Nature **137**, 357 (1936); Weekes, Livingston and Bethe, Phys. Rev. **49**, 471 (1936); Rasetti, Fink, Goldsmith and Mitchell, Phys. Rev. **49**, 869 (1936); Collie, Nature **137**, 614 (1936); Fermi and Amaldi, Ricerca scient. **1**, No. 7–8 (1936).

neutron mass over one mass unit. In particular cases, Q will be lower or higher than 8 MV, but it will probably lie in the interval from 5 to 10 MV for most cases (see section 6).

2. Method of Calculation

In order to estimate the "density" $\rho(Q)$ of the levels of the nucleus as a whole, we shall start from the statistical model of the nucleus; in other words, from the individual-particle picture. We are fully aware of the crudeness of this assumption, but reasons will be given below (section 5) for the belief that the density of levels will come out fairly correctly from this picture, although the wave function of a particular state of the nucleus will differ greatly from that obtained by this picture.

In this model, we shall obtain, first of all, a certain set of energy levels for the individual neutrons and protons in the nucleus. The positions of these levels will depend on the potential which we assume to act on the particles. The ground level of the *nucleus as a whole* is then obtained by filling all the lowest "individual" states with particles and leaving all the higher individual states unoccupied. An excited level of the nucleus will be obtained by taking one particle out of one of the low individual states and putting it into one of the higher individual states, or else by leaving *two* of the "low" individual states empty and having two higher individual states occupied instead, etc. In fact, for excited levels about 8 MV above the ground level of the nucleus, we shall in general have a fairly large number (of the order $\sqrt{A}$) of the individual particles "excited." Since there is a great variety of "low" individual states which may be left empty, and an equally great variety of "high" states which may be occupied by a particle, it is obvious that there will be a very large number of different ways in which a certain excitation energy of the nucleus as a whole may be realized, particularly if the energy is sufficient to have many particles excited.

The problem of finding the number of these different modes of realization is identical with the problem of finding the "probability number" (entropy) of a Fermi gas whose energy is given and has a value larger than the minimum possible energy for the gas. The solution of this problem is well-known in Fermi statistics, and it is only necessary to go one step farther in the accuracy than is usually done because usually only the logarithm of $\rho(Q)$, i.e., the entropy, is wanted while we want to calculate $\rho(Q)$ itself correctly to quantities of the relative order $1/A$ or $1/\sqrt{A}$ where A is the total number of particles.

Since we know that Fermi statistics is applicable to our problem, the probability that a given *individual-particle* state of energy ϵ is occupied, will be

$$f(\epsilon)=1/(e^{\beta(\epsilon-\zeta)}+1) \qquad (1)$$

where β and ζ are two constants which have to be determined from the total number of particles N and the total energy U of the nucleus. As regards the number of particles, we have to consider neutrons and protons separately because both these numbers are given for a given nucleus. We shall, for the present, refer to neutrons alone and let U denote the total energy of the neutrons alone, U is then, of course, *not* given, but we must, later on, integrate over all possible distributions of the total nuclear energy between neutrons and protons.

The conditions determining β and ζ are then

$$N=\sum_i f(\epsilon_i), \quad U=\sum_i \epsilon_i f(\epsilon_i), \qquad (2)$$

the sums extending over all possible states of an individual neutron. We now assume that the neutrons and protons are contained in a box of volume

$$\Omega=(4\pi/3)R^3, \qquad (3)$$

where R is the nuclear radius. Then the number of neutron levels of energy between ϵ and $\epsilon+d\epsilon$ is given by the well-known formula

$$\varphi(\epsilon)d\epsilon=(2^{5/2}/3\pi)(MR^2/\hbar^2)^{\frac{3}{2}}\epsilon^{\frac{1}{2}}d\epsilon=(3/2)C\epsilon^{\frac{1}{2}}d\epsilon \qquad (4)$$

with
$$C=(2^{7/2}/9\pi)(MR^2/\hbar^2)^{\frac{3}{2}}. \qquad (4a)$$

Then (2) reduces to

$$N=(3/2)C\int f(\epsilon)\epsilon^{\frac{1}{2}}d\epsilon, \qquad (5)$$

$$U=(3/2)C\int f(\epsilon)\epsilon^{\frac{3}{2}}d\epsilon. \qquad (5a)$$

The number $\rho(U)dU$ of energy levels of the nucleus as a whole, between U and $U+dU$, is found very easily from Sommerfeld's[4] derivation of the Fermi statistics. We have

$$\log \sum_{N'} \int dU' \rho(U') e^{\beta\zeta N' - \beta U'} = \sum_i \log\left(1+e^{\beta(\zeta-\epsilon_i)}\right)$$

$$= (3/2)C \int \epsilon^{\frac{1}{2}} d\epsilon \log\left(1+e^{\beta(\zeta-\epsilon)}\right). \quad (6)$$

The notation is the same as in Sommerfeld's paper (Eqs. (3), (11)), except that his α has been denoted by $-\beta\zeta$. The left hand side of (6) is known to have a sharp maximum for $N'=N$ and $U'=U$. This fact makes it easy to determine $\rho(U)$ once the right hand side of (6) has been evaluated. $1/\rho(U)$ gives directly the desired spacing between the nuclear levels.

3. Evaluation of the Number of Levels

Since the excitation Q of the nucleus is small compared to the total kinetic energy U of all the nuclear particles, the "Fermi gas" is highly degenerate and the formulae known from the theory of metals may be applied. We have

$$N = C\zeta^{\frac{3}{2}}\left(1+(\pi^2/8)(\beta\zeta)^{-2}+\cdots\right), \quad (7)$$

$$U = \tfrac{3}{5}C\zeta^{5/2}\left(1+(5\pi^2/8)(\beta\zeta)^{-2}+\cdots\right). \quad (7a)$$

C may be regarded as a given constant, essentially determined by the nuclear radius. Therefor ζ is practically determined by the number of neutrons:

$$\zeta = (N/C)^{2/3}\left(1-(\pi^2/12)(\beta\zeta)^{-2}+\cdots\right). \quad (8)$$

ζ, in its turn, determines U except for a small term of the relative order $(\beta\zeta)^{-2}$. There will, therefore, be a large zero-point energy of the nucleus, plus a small additional (excitation) energy. The latter determines the constant β.

If the nucleus is in its ground state, β will be infinity. In this case, the "Fermi energy" becomes (cf. (8))

$$\zeta_0 = (N/C)^{2/3} \quad (9)$$

and the total kinetic energy of all neutrons (cf. (7a), "zero-point energy")

$$U_0 = \tfrac{3}{5}C\zeta_0^{5/2} = \tfrac{3}{5}N\zeta_0. \quad (9a)$$

From (8) and (9), we have

$$\zeta = \zeta_0\left(1-(\pi^2/12)(\beta\zeta_0)^{-2}+\cdots\right). \quad (10)$$

Inserting this into (7a), we find

$$U = \tfrac{3}{5}C\zeta_0^{5/2} + \tfrac{1}{4}\pi^2 C\beta^{-2}\zeta_0^{1/2}. \quad (10a)$$

Therefore the excitation energy is

$$Q = U - U_0 = \tfrac{1}{4}\pi^2 C\zeta_0^{1/2}\beta^{-2}, \quad (11)$$

wherefrom

$$\beta^{-2} = 4Q/\pi^2 C\zeta_0^{\frac{1}{2}} \quad (11a)$$

or, by inserting (7)

$$\beta^{-1} = (2/\pi)(\zeta_0 Q/N)^{\frac{1}{2}}. \quad (12)$$

β^{-1} is the average excitation energy of the individual particles (width of the "tail" of the Fermi distribution).

Numerically, ζ_0 turns out to be somewhat, but not very much, larger than Q. Therefore β^{-1} is of the order $QN^{-\frac{1}{2}}$. In other words, the excitation energy is, in our model, shared between $N^{\frac{1}{2}}$ particles.

The right-hand side of (6) can be transformed by partial integration; we obtain exactly (i.e., not only for large β)

$$(3/2)C\int \epsilon^{\frac{1}{2}} d\epsilon \log\left(1+e^{\beta(\zeta-\epsilon)}\right)$$
$$= \beta C\int \epsilon^{\frac{3}{2}} d\epsilon/\left(e^{\beta(\zeta-\epsilon)}+1\right) = \tfrac{2}{3}\beta U. \quad (13)$$

Therefore

$$\Phi = \log \sum_{N'} \int dU' \rho(U') e^{\beta\zeta(N'-N)-\beta(U'-U)}$$
$$= (5/3)\beta U - \beta\zeta N. \quad (14)$$

Inserting (7) (7a) we get

$$\Phi = \beta C\zeta^{5/2}\tfrac{1}{2}\pi^2(\beta\zeta)^{-2} = \tfrac{1}{2}\pi^2 C\beta^{-1}\zeta^{\frac{1}{2}} \quad (14a)$$

and with (11a) (9)

$$\Phi = \pi(NQ/\zeta_0)^{\frac{1}{2}}. \quad (15)$$

This formula contains the fundamental result of our calculations. The following calculations, down to the end of section 4, represent only refinements.

In order to obtain ρ from (14), (15) we remark that the argument of the logarithm on the left-hand side in (14) is essentially $\rho(U)$ because the integrand has a sharp maximum for $U'=U$. Therefore we put

$$\rho(Q) = \lambda(N, Q)e^{\Phi(Q)}, \quad (16)$$

where λ is a slowly varying function of Q. We shall determine λ by carrying out the summation

[4] Sommerfeld, Zeits. f. Physik **47**, 1 (1927).

in (14), regarding λ as constant over the range of N' and U' involved. We find then from (14)

$$\lambda(N,Q)\int dQ' \exp\left[\beta\zeta(N'-N) -\beta(U'-U)+\Phi(Q')-\Phi(Q)\right]=1, \quad (17)$$

having replaced the integration variable U' by Q'.

In order to evaluate (17), we consider the exponent

$$f(\beta',\zeta')=\beta\zeta(N'-N)-\beta(U'-U)+(5/3)\beta'U' -\beta'\zeta'N'-(5/3)\beta U+\beta\zeta N=((5/3)U'-N'\zeta')(\beta'-\beta) +\tfrac{2}{3}\beta(U'-U)-N'\beta(\zeta'-\zeta), \quad (18)$$

where β' and ζ' are the parameters corresponding to $N'U'$. Using the formulae (7), (7a) repeatedly, we find by an elementary calculation

$$(5/3)U'-N'\zeta'=\tfrac{1}{2}\pi^2 C\beta'^{-2}\zeta'^{\frac{1}{2}}, \quad (18a)$$

$$\tfrac{2}{3}(U'-U)-N'(\zeta'-\zeta)=-\tfrac{3}{4}C\zeta'^{\frac{1}{2}}(\zeta'-\zeta)^2 +\tfrac{1}{4}\pi^2 C\zeta'^{\frac{1}{2}}(\beta'^{-2}-\beta^{-2}), \quad (18b)$$

leaving out some terms of the relative order $(\beta\zeta)^{-2}$. Inserting into (18), we have

$$f=-C\beta\zeta^{\frac{1}{2}}\left[\tfrac{3}{4}(\zeta'-\zeta)^2+\tfrac{1}{4}\pi^2(\beta'-\beta)^2(\beta\beta')^{-2}\right]. \quad (19)$$

The exponential in (17) behaves thus as $e^{-a(\zeta'-\zeta)^2-b(\beta'-\beta)^2}$ as should be expected. The differences $\zeta'-\zeta$ and $\beta'-\beta$ may be expressed in terms of $N'-N$ and $Q'-Q$, respectively, using (9) and (12). Neglecting again some small terms, we get

$$f=-\frac{\pi}{2}\left(\frac{N}{\zeta_0 Q}\right)^{\frac{1}{2}}\left[\tfrac{1}{3}(N'-N)^2\frac{\zeta}{N}+\frac{1}{4}\frac{(Q'-Q)^2}{Q}\right]. \quad (19a)$$

This may be inserted into (17) and the integrations carried out. Then

$$\frac{1}{\lambda(N,Q)}=\int_{-\infty}^{+\infty}dQ'\int_{-\infty}^{\infty}dN' \exp\left[-\frac{\pi}{6}\left(\frac{N}{\zeta Q}\right)^{\frac{1}{2}}\frac{\zeta}{N}(N'-N)^2 -\frac{\pi}{8}\left(\frac{N}{\zeta Q}\right)^{\frac{1}{2}}\frac{(Q'-Q)^2}{Q}\right]=\frac{\pi}{48^{-\frac{1}{2}}\pi}\left(\frac{\zeta Q}{N}\right)^{\frac{1}{2}}\left(\frac{N}{\zeta}\right)^{\frac{1}{2}}Q^{\frac{1}{2}}=48^{\frac{1}{2}}Q. \quad (20)$$

Therefore we find finally (cf. (16), (15))

$$\rho(Q)dQ=48^{-\frac{1}{2}}e^{\pi(NQ/\zeta)^{\frac{1}{2}}}dQ/Q \quad (21)$$

for the number of states of the system composed of all the neutrons, having a total excitation energy between Q and $Q+dQ$. A similar expression holds for the protons. Therefore the total number of levels of the nucleus as a whole per dQ is

$$\rho(Q)=\int dQ_1 \exp\left(\pi(NQ_1/\zeta_1)^{\frac{1}{2}} +\pi(ZQ_2/\zeta_2)^{\frac{1}{2}}\right)/48Q_1Q_2, \quad (22)$$

where $Q_2=Q-Q_1$ is the excitation energy of the protons.

The integration is facilitated by the fact that, for all existing nuclei, practically $N/\zeta_1=Z/\zeta_2$. The Fermi energies $\zeta_1\zeta_2$ are given by (cf. 9)

$$\zeta_1=(N/C)^{\frac{2}{3}}, \qquad \zeta_2=(Z/C)^{\frac{2}{3}}, \quad (23)$$

so that

$$a_1=\pi(N/\zeta_1)^{\frac{1}{2}}=\pi C^{\frac{1}{3}}N^{1/6}, \quad a_2=\pi(Z/\zeta_2)^{\frac{1}{2}}=\pi C^{\frac{1}{3}}Z^{1/6}. \quad (24a)$$

The ratio a_2/a_1 is, even for uranium, only $(146/92)^{1/6}=1.08$ and for other nuclei even closer to unity. We put therefore

$$a_1=a_2=a=\pi C^{\frac{1}{3}}(\tfrac{1}{2}A)^{1/6} \quad (24)$$

and have

$$\rho(Q)=\int dQ_1 e^{a(Q_1^{\frac{1}{2}}+(Q-Q_1)^{\frac{1}{2}})}/48Q_1(Q-Q_1). \quad (25)$$

The exponential has a sharp maximum for $Q_1=\frac{1}{2}Q$. We may write

$$Q_1^{\frac{1}{2}}+(Q-Q_1)^{\frac{1}{2}}=\sqrt{2}Q^{\frac{1}{2}}(1-\tfrac{1}{8}(2Q_1-Q/Q)^2+\cdots), \quad (25a)$$

while the denominator may simply be replaced by $12Q^2$. The integration gives then

$$\rho(Q)=\tfrac{1}{12}(\pi/a)^{\frac{1}{2}}2^{\frac{1}{4}}Q^{-5/4}e^{\sqrt{2}aQ^{\frac{1}{2}}}. \quad (26)$$

From its definition (24a), the constant a may be written

$$a=\pi(A/2\zeta_0)^{\frac{1}{2}}, \quad (27a)$$

where ζ_0 is the average Fermi energy for protons and neutrons. Then (26) reduces to

$$\rho(Q)=\tfrac{1}{12}\sqrt{2}\zeta_0^{\frac{1}{4}}Q^{-5/4}A^{-\frac{1}{4}}e^{\pi(AQ/\zeta_0)^{\frac{1}{2}}}. \quad (27)$$

The value of ζ_0 follows from (9) and (4a):

$$\zeta_0=(A/2C)^{\frac{2}{3}}=(3^{4/3}\pi^{\frac{2}{3}}/8)(\hbar^2A^{\frac{2}{3}}/MR^2). \quad (28)$$

Now the nuclear volume is proportional to the atomic weight A, so that

$$R=r_0A^{1/3}, \quad (29)$$

where r_0 may be calculated from the experimental data on α-radioactivity. Assuming $R=9\cdot10^{-13}$ cm for the average radius of radioactive nuclei, corresponding to A about 222, we have

$$r_0=1.48\cdot10^{-13}\ \text{cm} \quad (29a)$$

and $\quad \zeta_0=(3^{4/3}\pi^{\frac{2}{3}}/8)(\hbar^2/Mr_0^2)=21.5\ \text{MV}. \quad (30)$

Putting now

$$x=\pi(AQ/\zeta_0)^{\frac{1}{2}}=(AQ/2.20)^{\frac{1}{2}} \quad (31)$$

(Q in MV), we have for the spacing between neighboring levels

$$1/\rho(Q)=12\cdot(2\pi)^{-\frac{1}{2}}Qx^{\frac{1}{2}}e^{-x}. \quad (32)$$

For medium atomic weight, let us say $A=110$, and for $Q=8$ MV, we have $x=20$ and the spacing (32) becomes

$$5\cdot8\cdot10^6\cdot4.5\cdot2\cdot10^{-9}=0.4\ \text{volt}. \quad (32a)$$

This is obviously too small compared to the experimental spacing between neutron resonance levels. The reasons will be explained in the following section.

4. States with Given Angular Momentum

Most of the nuclear energy levels calculated in the preceding section will have very large angular momenta. Already the angular momentum of an individual particle in a heavy nucleus is apt to run up to about 6 (see Eq. (44a), below), and if the momenta of a fairly large number of particles are added, extremely large momenta for the nucleus as a whole may result. On the other hand, only small momenta of the nucleus A are of any importance for the capture of slow neutrons by the nucleus $A-1$. The nucleus $A-1$ will, in its ground state, have a certain total angular momentum ("nuclear spin"), say I_0. The neutron must have orbital momentum $l=0$ in order to be captured by the nucleus $A-1$, and therefore total momentum $j=\frac{1}{2}$, considering its spin. The resulting state of the "compound nucleus" A must therefore have an angular momentum $I_0\pm\frac{1}{2}$. Of all the levels of the nucleus A, we should therefore consider only those with angular momenta $I_0\pm\frac{1}{2}$; and I_0 will be small compared to the "average angular momentum" of all possible states of the nucleus A. Only a small fraction of the levels considered in the preceding section will fulfill this condition.[5]

We want to calculate the probability that a nuclear level has a given angular momentum I. In order to do this, we assume that each individual particle in the nucleus has the same angular momentum j. (The value of j will be calculated later; also we shall then consider the variation of j among the particles.) The resultant of all individual particle momenta will be the moment of the nucleus, I. Following the usual procedure, we consider the *components* of the angular momenta in a given direction z. Let m_i be the z-component of the momentum of particle i, and M the z-component of the total momentum. Then

$$M=\sum_i m_i, \tag{33}$$

[5] The importance of the angular momentum for the number of nuclear states was first pointed out to me by Dr. Placzek, to whom I am indebted for this suggestion.

We want to know the probability of a given resultant M, each value of m_i from $-j$ to $+j$ being equally probable. Provided the number n of particles is large, the probability for a given M is given by the "Gauss formula"

$$p(M)=(\tfrac{2}{3}\pi nj(j+1))^{-\frac{1}{2}}e^{-3M^2/2nj(j+1)}. \tag{34}$$

To prove this, we make use of three facts:

(1) If $p_n(M)$ is the probability that n momenta have the resultant M in the z direction, we must have the "addition theorem"

$$p_{r+s}(M)=\sum_{M_1}p_r(M_1)p_s(M-M_1)$$
$$=\int_{-\infty}^{\infty}p_r(M_1)p_s(M-M_1)dM_1. \tag{34a}$$

If this is to be generally true for arbitrary values of r, s and M, then p must be of the form

$$p_n(M)=c_ne^{-\alpha M^2/n}, \tag{35a}$$

where α is a constant independent of n while c_n depends on n.

(2) The average of M^2, *viz.*

$$\overline{M^2}=\sum_M M^2p_n(M), \tag{36}$$

must be given by

$$\overline{M^2}=\sum_{ik}\overline{m_im_k}=\sum_i\overline{m_i^2}+\sum_{i\neq k}\overline{m_i}\,\overline{m_k}=n\overline{m^2}, \tag{36a}$$

since two different m's (m_i and m_k) are independent of each other, and the average value of an individual m is zero. Now

$$\overline{m^2}=\sum_{m=-j}^{j}m^2/(2j+1)=\tfrac{1}{3}j(j+1). \tag{36b}$$

On the other hand, we have from (35a)

$$\overline{M^2}=\int p_n(M)M^2dM/\int p_ndM=n/2\alpha. \tag{37}$$

Comparing (36a), (36b), (37), we find

$$\alpha=3/2j(j+1). \tag{37a}$$

(3) The total probability must be unity:

$$\int_{-\infty}^{\infty}p_n(M)dM=1. \tag{38}$$

This fixes the constant c_n in (35a).

The number of states of given total angular momentum I is, as is well known, equal to the number of states with $M=I$, minus the number of states with $M=I+1$. Since $\rho(Q)dQ$ is the total number of nuclear states in the energy interval dQ, the number of states with a given M is $\rho(Q)p(M)dQ$ and the number of energy levels with a given I therefore

$$\rho(Q)[p(I)-p(I+1)]dQ. \tag{39}$$

Since p varies slowly with its argument M, this gives

$$\rho(Q, I) = \rho(Q)(dp/dM)_{I+\frac{1}{2}}. \quad (40)$$

Here we insert (34), carry out the differentiation and then put the exponential equal to unity, since for all cases of interest $I^2 \ll nj(j+1)$. We obtain thus

$$\rho(Q, I) = \rho(Q)(2I+1)(8\pi)^{-\frac{1}{2}}(3/nj(j+1))^{3/2}. \quad (41)$$

We have now to compute $nj(j+1) \approx n(j+\frac{1}{2})^2$. We have therefore to know the average angular momentum j of the individual particles, and the number n of the particles which contribute to the resultant angular momentum of the nucleus. We know that in the ground state of a nucleus the resultant momentum is almost zero, because the momenta of the various individual particles nearly cancel each other ("closed shells"). Therefore the angular momentum of an excited state of a nucleus comes from the "tail of the Fermi distribution," i.e., from those particles whose energy is larger than the Fermi energy ζ, and from the empty states of energy smaller than ζ.

For j we have thus to take the average angular momentum of an individual particle whose energy is near ζ. We therefore calculate the number of quantum states (per unit energy) of given orbital momentum l for an individual particle of kinetic energy ζ which is enclosed in a deep spherical potential well of radius R. This problem is similar to the problem of the first appearance of an electron of orbital momentum l in the periodic system which was treated by Fermi.[6] The problem can be solved by separating the wave equation in polar coordinates, and then using the WKB (Wentzel Kramers Brillouin) method for treating the radial wave equation. This leads to the well known "quantum condition"

$$\int^R dr[(2M/\hbar^2)\epsilon - (l+\tfrac{1}{2})^2/r^2]^{\frac{1}{2}} = (n+\tfrac{1}{2})\pi, \quad (42)$$

where n is an integer and ϵ the energy of the particle. Since there is one quantum state for each integral value of n, the number of states of orbital momentum l in the energy interval $d\epsilon$ is

$$\begin{aligned} w(\epsilon)d\epsilon &= \frac{d\epsilon}{\pi}\frac{d}{d\epsilon}\int^R dr\left[\frac{2M}{\hbar^2}\epsilon - \frac{(l+\frac{1}{2})^2}{r^2}\right]^{\frac{1}{2}} \\ &= \frac{Md\epsilon}{\pi\hbar^2}\int^R \frac{rdr}{[2M\epsilon\hbar^{-2}r^2-(l+\frac{1}{2})^2]^{\frac{1}{2}}} \\ &= \frac{d\epsilon}{2\pi\epsilon}[2M\epsilon\hbar^{-2}R^2-(l+\tfrac{1}{2})^2]^{\frac{1}{2}}. \quad (43) \end{aligned}$$

The number of states with a given j is equal to the number of states with $l=j-\frac{1}{2}$, plus the number of states with $l=j+\frac{1}{2}$. Therefore it will be approximately proportional to

$$w'(j) = [2M\zeta R^2\hbar^{-2} - (j+\tfrac{1}{2})]^{\frac{1}{2}}, \quad (43a)$$

[6] Fermi, Zeits. f. Physik **48**, 73 (1928).

leaving out factors independent of j and putting the particle energy equal to ζ. The number of *particles* of angular momentum j will be proportional to $(2j+1)w'(j)$ because each level j has the statistical weight $2j+1$. Therefore the average value of $(j+\frac{1}{2})^2$ is

$$\overline{(j+\tfrac{1}{2})^2} = \int(2j+1)dj(j+\tfrac{1}{2})^2w'(j)/\int(2j+1)djw'(j) = (2/5)\cdot 2MR^2\hbar^{-2}\zeta. \quad (44)$$

Inserting the value of ζ from (30) and R from (29), we find

$$\overline{(j+\tfrac{1}{2})^2} = (3^{4/3}\pi^{\frac{2}{3}}/10)A^{\frac{2}{3}} = 0.93A^{\frac{2}{3}}. \quad (44a)$$

There remains the computation of n, which is the number of neutrons and protons having energies larger than ζ, plus the number of unoccupied states of energy smaller than ζ. Obviously, n is about four times the number of neutrons with energy larger than ζ. According to the Fermi distribution, this number is given by

$$\begin{aligned} \tfrac{1}{4}n &= (3/2)C\int_\zeta^\infty \frac{\epsilon^{\frac{1}{2}}d\epsilon}{e^{\beta(\epsilon-\zeta)}+1} = 3/2\frac{C\zeta^{\frac{1}{2}}}{\beta}(1-\tfrac{1}{2}+\tfrac{1}{3}-\cdots) \\ &= 3/2\frac{C\zeta^{\frac{1}{2}}}{\beta}\log 2 \quad (45) \end{aligned}$$

or, inserting the value of β from (11a), and of C from (9), we have

$$n = \frac{12}{\pi}\log 2\left(\frac{QN}{\zeta}\right)^{\frac{1}{2}} = \frac{6\sqrt{2}\log 2}{\pi}\left(\frac{QA}{\zeta}\right)^{\frac{1}{2}} = 1.87\left(\frac{QA}{\zeta}\right)^{\frac{1}{2}}. \quad (46)$$

For $A=100$, $Q=8$ MV and $\zeta=20$ MV, this would be about 12.

With this value for n, and (44a) for $j(j+1)$, we find for the number of nuclear levels with angular momentum I:

$$\rho(Q, I) = (2I+1)\rho(Q)(5/\log 2)^{\frac{3}{2}} \cdot 2^{-9/4}\cdot 3^{-2}(\zeta/Q)^{3/4}A^{-7/4}. \quad (47)$$

Inserting $\rho(Q)$ from (27), we have

$$\rho(Q, I) = 2^{-15/4}\cdot 3^{-3}(5/\log 2)^{\frac{3}{2}}(2I+1)\zeta_0^{-1} \times(\zeta_0/QA)^2 e^{\pi(AQ/\zeta_0)^{\frac{1}{2}}} \quad (48)$$

or, introducing the abbreviation $x=\pi(AQ/\zeta_0)^{\frac{1}{2}}$ from (31):

$$\rho(Q, I) = \frac{\pi^4\cdot 2^{\frac{1}{4}}}{432}\left(\frac{5}{\log 2}\right)^{\frac{3}{2}}(2I+1)\zeta^{-1}x^{-4}e^x. \quad (49)$$

The spacing between two levels of spin I is the reciprocal of this, *viz.*

$$\Delta = \frac{432}{2^{\frac{1}{4}}\pi^4}\left(\frac{\log 2}{5}\right)^{\frac{3}{2}}\frac{\zeta}{2I+1}x^4e^{-x} \quad (50)$$

or $$\Delta = \Delta_0/(2I+1) \quad (50a)$$

with $$\Delta_0 = 4.1\cdot 10^6 x^4 e^{-x} \text{ volts} \quad (50b)$$

and $$x = (AQ/2.20)^{\frac{1}{2}}, \quad (Q \text{ in MV}). \quad (50c)$$

In applications to the capture of slow neutrons, it should be remembered that levels with $I = I_0 \pm \frac{1}{2}$ of the "final" nucleus are effective if I_0 is the "spin" of the capturing nucleus.

5. Criticism of the Method Used

We have assumed the energy levels of the nucleus as a whole to be given by the sum of the energies of the individual particles. In other words, we have taken into account the interaction between the particles only insofar as it can be expressed in terms of a potential acting on each particle (Hartree method). This method would, of course, lead to a hopelessly wrong result if we wanted to deduce the actual characteristics (wave function) of each nuclear level from it. It might, however, give fairly correct results for the number of levels in a given energy interval.

The interaction between particles will thoroughly mix the wave functions of the various nuclear levels obtained from the Hartree approximation. A group of levels which would have approximately the same energy in the Hartree approximation will, by the interaction between the particles, be drawn out into a spectrum extending over a wide energy range, probably several MV. Conversely, the wave functions of the *actual* nuclear states in a given energy interval will be linear combinations of Hartree wave functions belonging to much lower as well as much higher Hartree levels.

From these considerations, it might seem that the general behavior of the density of levels, as a function of the energy, might *in the average* be not very greatly changed by the interaction between the particles. There is, however, one fact which will somewhat invalidate this conclusion: The lowest level of the nucleus lies certainly lower than the corresponding Hartree level. It might seem that Q should be taken as the energy of a nuclear level as compared to that of the lowest *Hartree* level. Then Q would be smaller than the actual excitation energy counted from the *true* ground state of the nucleus, which we may call Q'. If we inserted Q' into our formulae we should then obtain too small a spacing between the nuclear levels.

However, we believe that this error is compensated by the fact that we are considering only nuclear levels of low angular momentum. From experience, and from some calculations made recently on light nuclei,[7] we know that the nuclear levels of low angular momentum usually lie lowest, those with high momentum highest among the levels arising from a given configuration. It may be expected that the levels of low momentum arising from "excited configurations" are depressed by the same amount, as compared to their position in the Hartree approximation, as the ground level. Thus we may expect our formulae to give us about the correct density of levels of low angular momentum if we insert for Q the actual energy of excitation above the ground state. Of course, the formulae would give us too high a density of levels of high momenta.

Another reason why we believe our formulae to be not too far wrong, is the fact that the actual levels in a given energy interval will mostly arise from Hartree levels of higher energy, simply because there is a rapid increase of the density of Hartree levels with increasing energy. Most of the levels will therefore be related to the corresponding Hartree levels in a similar way as the ground state.

6. Discussion of the Spacing of Nuclear Levels

The spacing of nuclear energy levels depends, according to formula (50), only on the product of the mass number A of the nucleus and the excitation energy Q. Light nuclei, and heavy nuclei at energies just above the ground level, should possess very few quantum states while highly excited heavy nuclei ought to have an enormous number of closely spaced levels.[1]

Table I gives the spacing between the levels with $I=0$ for various values of the product AQ. E.g., if a nucleus of atomic weight 112(Cd) captures a neutron with the evolution of about

Table I. *Spacing Δ_0 of nuclear energy levels of zero angular momentum in volts.**

QA (in MV)	100	200	400	600	800	1000	1200	1500	1800
Δ_0 (in Volts)	10^7	$2.4\cdot10^6$	$1.9\cdot10^5$	$2.1\cdot10^4$	2800	450	85	8.5	1.0

* For angular momentum I, the spacing would be $\Delta = \Delta_0/(2I+1)$.

[7] Bethe and Bacher, Rev. Mod. Phys. 8, 82 (1936) (quoted as B), §36, Wigner and Feenberg, to appear shortly in the Phys. Rev., and unfinished calculations of Bethe and Rose.

9 MV energy, QA would be 1000 and the spacing of S levels of that nucleus only 450 volts. For higher angular momenta, the spacing between levels becomes even less, e.g., for $I=3$ it would be only 1/7 of the previous value, i.e., 60 volts. Since two values of the angular momentum ($I=I_0\pm\frac{1}{2}$, see section 4) lead to neutron capture, the distance between neutron resonance levels would be only one half of the values given. The distance between the highly excited levels of fairly heavy nuclei is thus very small indeed.

For smaller charge, the spacing between levels becomes very much larger. If we let Q be again of the order 9 MV, the spacing of levels for a nucleus such as Fe($A=55$) will be of the order of ten thousand volts, and for really light nuclei such as O($A=16$) of the order of a million volts. This explains why simple capture of neutrons is practically never found with any great intensity for really light nuclei. It would also mean that the lowest neutron resonance level for a nucleus of atomic weight around 50 will, in the average, lie at very much higher energy than for atomic weights of the order 100.

Another cause for an increased spacing of levels would be a smaller value of the excitation energy Q. In connection with the capture of slow neutrons, this would mean that the energy set free in the capture process would have to be smaller than 9 MV. This energy is given by

$$Q=M_{A-1}+M_n-M_A, \qquad (51)$$

where M_{A-1}, M_n and M_A denote the (exact) masses of the capturing nucleus, the neutron and the product nucleus respectively, in energy units. In the average, nuclei of medium atomic weight have packing fractions of $-1/1000$, so that M_A-M_{A-1} will be about 0.999 mass unit. The neutron mass being almost 1.009, we find *in the average* the above-mentioned figure $Q=9$ MV (For details, see below).

However, deviations from this figure are to be expected if either of the nuclei $A-1$ or A is exceptionally stable or unstable. The greatest variations will in general come in through the product nucleus A. If this nucleus is radioactive, it is obviously less stable than if it is not. Thus it is to be expected that the energy evolution Q is smaller (in the average) if a radioactive nucleus A is produced, than otherwise. Therefore the spacing of neutron energy levels will be larger in the case of the production of a radioactive nucleus, and the first resonance level will lie at a higher neutron energy. Now it is known from the Breit-Wigner theory[2] of neutron capture that the capture probability for thermal neutrons is, cet. par., the larger the lower the energy of the first resonance level. *The probability of capture of thermal neutrons will, therefore, in the average be smaller if the capture leads to a radioactive nucleus than if it leads to a stable nucleus.*

This fact has been known for some time experimentally and has puzzled investigators to some extent. All the very large cross sections (of the order 10^{-21} cm^2 and more, e.g., Cd, Sm) for the absorption of slow neutrons are connected with the formation of stable nuclei whereas the cross section for the formation of radioactive nuclei are in general only moderately large (about 10^{-22} cm^2 or smaller). Even more marked differences should be found in the positions of the lowest resonance level for neutrons: This level should lie, in the average, at higher neutron energies for the formation of radioactive nuclei than for capture processes leading to stable nuclei.

Apart from irregularities for the individual nuclei, the value of the "dissociation energy" Q will depend on charge and mass number of the nuclei A and $A-1$. We know that generally nuclei with even charge and even mass are most stable, such with odd mass number less stable, and nuclei of odd charge and even mass number unstable to the extent of being radioactive.[8] Therefore we have to distinguish three cases:

(1) The capturing nucleus $A-1$ has *odd charge.* Its mass must then be also odd. Then the capture of a neutron will certainly lead to a radioactive nucleus of odd charge and even mass. The energy Q evolved will be comparatively small, the spacing between the neutron levels fairly large and the capture cross section for temperature neutrons only moderately large.

(2) The capturing nucleus $A-1$ has *even charge and even mass.* The nucleus produced will then have even charge and odd mass. It may be radioactive or stable. But in any case, it will have relatively higher energy in its ground state than the capturing nucleus. Therefore the energy evolved will again be comparatively small, and probably of the same order as in case (1), irrespective of whether the nucleus A is radioactive or not.

(3) The capturing nucleus $A-1$ has *even charge and odd mass.* The nucleus produced will then be of the most stable

[8] For a discussion and explanation of this fact, see B, §10.

type, i.e. even charge and even mass. The energy Q evolved will be exceptionally large, the lowest resonance level will lie at very low neutron energy and the cross section for temperature neutrons will be exceedingly large.

From these considerations it would seem that very strong capture of temperature neutrons can only be due to nuclei of even charge and odd mass number. We suggest that the neutron absorption levels at very low neutron energies which have been observed[9] for Cd, Sm and Hg are due to the abundant "odd isotopes" Cd^{111} or Cd^{113}, Sm^{147} or Sm^{179}, and Hg^{199} or Hg^{201}, respectively.

Another factor which will also increase the capture in case 3 is the factor $2I+1$ in the number of energy levels per unit energy (cf. 49). The spin I of the "resonance level" of the product nucleus A may be either $I_0-\frac{1}{2}$ or $I_0+\frac{1}{2}$ if I_0 is the angular momentum of the original nucleus $A-1$ (cf. beginning of section 4). The number of levels of the nucleus A suitable for the capture of neutrons will therefore be proportional to $2(2I_0+1)$, i.e., the larger the greater the spin of the original nucleus. Now all nuclei with even charge and even mass (class 2 above) seem to have spin $I_0=0$, while nuclei with odd mass (classes 1 and 3) have spins different from zero and therefore are more likely to capture slow neutrons.

We shall now try to compute roughly the actual values of Q to be expected and the differences in the Q values between the above-mentioned cases 1, 2 and 3. An estimate of the latter may perhaps be based on the average energy of the β-particles obtained from neutron captures of class 1. The lifetimes of the radioactive nuclei obtained from neutron capture vary from about 20 sec. to some hours or days, at least for the radioactivities known at present. According to the Sargent rule, this corresponds to energies from about 1 to 3 MV. In the average, we find therefore that nuclei of even mass number and odd nuclear charge have energies by 2 MV greater than their neighboring isobars of even charge into which they transform by emitting a β-ray. Now we may safely assume that the average packing fraction of nuclei of *odd* mass number is independent of whether their nuclear charge is even or odd, because firstly there seems to be experimentally no difference between the number of species and the abundance of these two types of nuclei, and secondly there is no theoretical reason for assuming any difference (cf. B, §10). Consequently, the energy Q evolved in the capture of neutrons by nuclei of odd weight and even charge (class 3 above) ought to be about 2 MV more than for nuclei of odd weight and odd charge (class 1). For nuclei of class 2 (even weight, even charge) we may expect about the same Q's as for class 1, because in both cases the transition goes from a more stable to a less stable type of nucleus.

A rough estimate of the *average* value of Q, i.e. the mean between the cases (2) and (3) above, may be obtained from an empirical formula for the average nuclear mass defects as a function of mass number A and charge Z, such as that derived by Weizsäcker[10] or by the author (B, §30). The energy (excess of the exact mass value over the mass number) of the most stable nucleus of atomic weight A is approximately given by (B, Eq. (186))

$$E(A)=-6.6_5A+14.2A^{2/3}+0.156A^{5/3}\cdot 135/(134+A^{2/3}), \quad (52)$$

the unit of energy being a thousandth of a mass unit. The difference in energy between the nuclei A and $A-1$ is therefore (cf. B (186b))

$$E(A)-E(A-1)=-6.6_5+9.5A^{-1/3}+0.156A^{2/3}\cdot 135(223+A^{2/3})/(134+A^{2/3})^2. \quad (52a)$$

The mass of the neutron may be calculated from the following data:

(1) The mass spectroscopic comparison of the deuteron and the proton by Bainbridge and Jordan,[11] giving

$$2H-D=0.00153\pm 0.00004 \text{ mass unit.}$$

(2) The binding energy of the deuteron as measured by Feather[12]

$$H+n-D=2.22\pm 0.06 \text{ MV}=0.00238.$$

(3) The mass of the deuteron as derived from

[9] Rasetti, Fink, Goldsmith and Mitchell, Phys. Rev. **49**, 869, 1936. Placzek and Frisch (private communication). Amaldi and Fermi, Ricerca Scientifica **1**, 11–12 (1936).

[10] Weizsäcker, Zeits. f. Physik **96**, 431 (1935).

[11] Bainbridge and Jordan, Bull. Am. Phys. Soc., 1936, Washington meeting, report 123.

[12] Feather, Nature **136**, 467 (1935). A correction of 40,000 volts has been applied to Feather's value because of the range energy relation (B, p. 123).

TABLE II. *Mass excesses of nuclei and energy evolved in neutron capture.*

A	20	50	100	150	200	240
$E(A)-E(A-1)$ (mass units)	−1.45	−1.0	0	0.75	1.4	1.9
$\bar{Q}$(MV)	9.5	9.1	8.2	7.5	6.9	6.4

disintegration data[13]

$$D=2.01445.$$

A combination of these three data gives

$$n=H+0.00085=\tfrac{1}{2}D+0.00161=1.00884. \quad (53)$$

Thus the energy $\bar{Q}$ evolved in the neutron capture will be 8.84 thousandths of a mass unit, minus the energy difference (52a). Converted into MV, this gives

$$\bar{Q}=14.4-8.8A^{-1/3}-0.145A^{2/3}\cdot 135(223+A^{2/3})/(134+A^{2/3})^2. \quad (54)$$

Table II gives the average difference between the mass excess of neighboring isotopes, $E(A)-E(A-1)$, according to the semi-empirical formula (52a), in thousandths of a mass unit, and the average energy evolved in the capture of a neutron, $\bar{Q}$, in MV. The values in the table are of course only averages, and in individual cases large deviations ought to be expected. Moreover, it seems from the observed energies of radioactive α-particles that the mass excess of heavy nuclei increases actually somewhat faster with increasing A, so that $\bar{Q}$ for $A=200$ or more may actually be about 0.5 MV smaller than indicated in the table.

Accepting the energies given in the table for the *average* energy evolved in neutron capture, we should expect values by about 1 MV *higher* for the capture by nuclei of even charge and odd mass number (class 3 above) and about 1 MV *lower* than the values of the table for the other cases (classes 1 and 2 above). Thus the odd isotopes of Cd would probably correspond to Q values slightly over 9 MV, giving for AQ a value somewhat over 1000. With a spin of $I_0=\frac{1}{2}$ (B, Table 19) for both the "odd" isotopes of Cd, we should thus expect a spacing of the nuclear levels of about 50 volts (cf. Table I, divide by $2(2I_0+1)$). For Ag, we would expect $Q=7$ MV, approximately, $AQ=800$, and, with $I_0=3/2$, a spacing of about 300 volts between neighboring levels. These figures seem reasonable, although perhaps a little high.

Turning now to the very heavy nuclei (A around 200), we should expect very many very closely spaced levels. The increase in the density of levels due to the increased number of particles A in the nucleus is, however, partly offset by the decrease in the energy evolved in the neutron capture,[14] as shown in Table II. Even so, we should expect values of QA of the order 1400 for $A=200$ (Hg, cf. Table II). This would correspond to a spacing of about 20 volts between levels with $I=0$. Now Hg^{201} has a spin of 3/2; therefore the average spacing between the resonance levels of neutrons captured by Hg^{201} ought to be about two volts. An element of such high atomic number should have an almost continuous absorption spectrum for slow neutrons if the element belongs to class 3, i.e., has even charge and odd mass.

For elements of odd charge (or of even charge and even mass), the increase in the density of levels for high mass number should be less marked. Suppose the Q value is 1 MV less than the "average" given in Table II; which may easily happen. Then AQ is reduced to 1200 for $A=200$, i.e., not much more than for elements of medium atomic weight such as Cd. The distance between adjacent levels would, accordingly, be of the order of 10 volts. This is compatible with the observed resonance level of Au (2.5 volts.)

I wish to express my thanks to Dr. G. Placzek and Dr. L. Nordheim for interesting discussions and valuable suggestions. My thanks are also due to Dr. M. E. Rose for a critical revision of the manuscript.

[13] Cockcroft and Lewis, Proc. Roy. Soc. **A154**, 261 (1936).

[14] The importance of this factor was first pointed out to me by Dr. Nordheim.

24

Reprinted from *Phys. Rev.* **51**:95-106 (1937)

On the Structure of the Nuclei Between Helium and Oxygen

E. Feenberg* and E. Wigner†
University of Wisconsin, Madison, Wisconsin

(Received October 10, 1936)

In order to test the present assumptions on nuclear forces, the theory is applied to the nuclei in which the first p shell of protons and neutrons is being built up, i.e., to the nuclei with masses between 5 and 16. The Hartree-Fock approximation is used for the numerical calculations, but the more qualitative results are independent to a large degree of the approximations used. The angular momenta of the ground states appear to be given correctly by the theory. Although the wave functions used do not correspond to preformed alpha-particles, the first-order energies exhibit a marked four-shell structure. The experimental energy difference between the nuclear pairs $(N, N+1)$, $(N+1, N)$ may possibly be somewhat larger than the difference in the electrostatic energies. Support for the use of spin exchange (Heisenberg) forces to account for the singlet-triplet separation in the deuteron is found in the singlet-triplet separation inferred from the Li^6-He^6 normal state energy difference. However, the $B^{10}-Be^{10}$ and $N^{14}-C^{14}$ normal state differences do not fit the simple theory which is adequate for the singlet-triplet spacing in the two and six particle problems.

I. Introduction

It is generally accepted now that the explanation of the binding energies and scattering properties of the nuclei n, H^1, H^2, H^3, He^3, He^4 requires several kinds of forces.[1] The forces which are generally assumed at present between a proton and a neutron ("between unlike particles") are: (1) a "Majorana force" involving an exchange P of the Cartesian coordinates of the two particles and (2) a "Heisenberg force" involving the product of a Cartesian coordinate exchange P and a spin coordinate exchange Q. The Heisenberg force has about $\frac{1}{4}$ the depth of the Majorana force and the same range of action. For all present calculations the exact dependence of the potential on distance seems to be relatively unimportant and we shall use, for the sake of convenience, the usual

$$(1-g)A_{\nu\pi}e^{-r^2/r_0^2}P=(1-g)A_{\nu\pi}e^{-\alpha r^2}P \quad \text{(1a)}$$

for the Majorana force and

$$gA_{\nu\pi}e^{-\alpha r^2}PQ \quad \text{(1b)}$$

for the Heisenberg force between unlike particles. Here $A_{\nu\pi}=72\ mc^2$, $r_0=2.25\cdot 10^{-13}$ cm, $g=0.22$ and $\alpha=16$ if r is measured in units of $\hbar/c(Mm)^{\frac{1}{2}}=8.97\cdot 10^{-13}$ cm.

The forces between like particles are less well known. Between pairs of protons and pairs of neutrons one assumes a potential with a depth $A_{\nu\nu}=41\ mc^2$ and the same width as between unlike particles.[1] This force is either assumed to involve an exchange P of the Cartesian coordinates

$$A_{\nu\nu}e^{-\alpha r^2}P \quad \text{(2a)}$$

or else the scalar product of their spin operators:

$$-\tfrac{1}{3}A_{\nu\nu}e^{-\alpha r^2}(\sigma_1\cdot\sigma_2). \quad \text{(2b)}$$

The latter possibility can be considered, because of Dirac's identity for antisymmetric wave functions[2]

$$-\tfrac{1}{3}(\sigma_1\cdot\sigma_2)=\tfrac{1}{3}+\tfrac{2}{3}P_{12}, \quad \text{(3)}$$

as the sum of an ordinary and an exchange force. It is undecided, at present, which of the two forms of interaction deserves preference. In fact it has been proposed[3] to assume interactions which would give equal attractions between pairs of like and unlike particles in the singlet state. We obtain a problem having this property by assuming for the interaction between like particles the same forces as between unlike particles:[4]

$$(1-g)A_{\nu\pi}e^{-\alpha r^2}P+gA_{\nu\pi}e^{-\alpha r^2}PQ. \quad \text{(2c)}$$

* Now at the Institute for Advanced Study.
† Now at Princeton University.
[1] E. Feenberg and J. K. Knipp, Phys. Rev. **48**, 906 (1935). E. Feenberg and S. S. Share, Phys. Rev. **50**, 253 (1936).
[2] Cf. P. A. M. Dirac, *Quantum Mechanics* (Oxford, 1935), §19, §61; J. H. Van Vleck, Phys. Rev. **48**, 367 (1935).
[3] G. Breit and E. U. Condon, private communications.
[4] The assumption that the forces between all kinds of particles are the same was first put forward by L. A. Young, Phys. Rev. **47**, 972 (1935).

For like particles the second part of this expression can be written, because of the antisymmetry of the wave function in the coordinates of the protons (or neutrons), as an ordinary repulsive force $-gA_{\nu\pi}e^{-\alpha r^2}$. We shall consider this "all forces equal model" because it is at least a useful approximation, the properties of which in some cases, can be discussed with greater ease[5] than the properties of a model with forces (2*a*) or (2*b*). In addition to the forces (2), there are between protons the electrostatic forces and also the ordinary spin-orbit and spin-spin interactions are generally assumed to exist. These latter interactions seem to play only a minor role in nuclear problems.

Of course, it is most probable that the interaction of the constituents of nuclei cannot be described at all rigorously by a Schrödinger equation, the variables of which are the Cartesian and spin coordinates of the protons and neutrons only. A similar description is impossible, strictly speaking, for the extranuclear electrons also. The cause of the interaction is probably some field in both cases—the electromagnetic field of light quanta in the second case and, perhaps, the electron-neutrino field in the first case. The elimination of the field variables is possible always in a certain approximation only and breaks down in a higher order approximation. This is manifested by the phenomena of spontaneous emission and line width for the extranuclear electrons and by the β disintegration, for example, in the case of nuclear constituents. Nevertheless, the usual Schrödinger equation gives a practically perfect description of atomic states and it may be hoped that a similar equation exists for the nuclear constituents also.

We shall attempt here to find experimental criteria which can be used to answer the following three questions:

(1) Whether or not the difference between proton-proton and neutron-neutron interaction is only the Coulomb force.

(2) Whether (2*a*), (2*b*) or (2*c*) is the more correct form for this interaction.

(3) Whether or not the assumption is correct that the neutron-proton interaction operator is a linear combination of Majorana and Heisenberg terms.

Before examining these questions we apply the Hartree-Fock approximation method to the light nuclei between and including Li^6 and O^{16} and perform the same calculation for the position of the terms arising from the lowest configuration which Slater[6] has made for atomic spectra.

II. Hartree-Fock Calculations

It is well known that the Hartree-Fock method gives only very roughly correct solutions of the nuclear wave equation. This has been pointed out by Weizsäcker, Flügge and Heisenberg[7] for the older models (no interaction between like particles). It is true, however, also for the forces (1) and (2), though to a somewhat lesser degree. Bethe and Bacher[8] performed similar calculations for both light and heavy nuclei using the newer model with the result that the method gives practically no binding energy for the observed nuclear densities. Since the Schrödinger equation without doubt has solutions with much lower characteristic values at the same densities (conglomerates of slightly compressed alpha-particles), the result obtained by Bethe and Bacher must be interpreted as revealing the inaccuracy of the Hartree approximation in nuclear problems. Considering the great similarity between the nuclear and metallic wave equations and the importance of the correlation energy in the latter,[9] this is not surprising.

Table I. *Binding energies of elements with masses between 4 and 16.*

n_1		$n_2=0$	1	2	3	4	5	6
0	He	55	—	56				
1	Li		62	76	89			
2	Be			109	113	126		
3	B			110	125	147	153	
4	C				141	176	186	202
5	N					180	201	222
6	O						215	246

[5] It is equivalent to "approximation (2)" of the paper of one of the present writers in this issue.

[6] Cf. E. U. Condon and G. Shortley, *The Theory of Atomic Spectra* (Cambridge, 1936). In the Hartree approximation the nuclei of the group He^6-O^{16} are obtained by successive additions to the *p* shell which is completed at O^{16}. This has been first pointed out by J. H. Bartlett, Nature **130**, 165 (1932). Cf. also his letter in the Phys. Rev. **41**, 370 (1932) and G. Gamow, Zeits. f. Physik **89**, 592 (1934) and especially W. M. Elsasser, J. de phys. **4**, 549 (1933); **5**, 389 and 635 (1934).

[7] C. F. v. Weizsäcker, Zeits. f. Physik **96**, 431 (1935); S. Flügge, p. 459; W. Heisenberg, p. 473.

[8] H. A. Bethe and R. F. Bacher, Rev. Mod. Phys. **8**, 82 (1936).

[9] It is more than half of the binding energy in Na.

Nevertheless, it can be expected that in the case of light nuclei the order of the terms will be given correctly by such a calculation. This is true for the corresponding atomic spectra[6] although the calculated ratios of the term differences show marked deviations from the experimental values.

Table I gives the elements with which we shall be concerned; n_1+2 is the number of protons, n_2+2 the number of neutrons. The figures in the table are the binding energies[10] in units of mc^2. As long as we neglect the Coulomb forces, the constitutions of elements symmetric with respect to the main diagonal of the table are identical. The theory of holes[11] gives us furthermore the constitution of a nucleus from its mirror image with respect to the other diagonal.

In this section we shall neglect the Heisenberg forces. This makes the Hamiltonian operate on the Cartesian coordinates only and both protons and neutrons will have a "multiplicity"; each resultant spin angular momentum will be a good quantum number. In addition to these, we have the total azimuthal quantum number. The possible terms for Be^9, for example, can be determined by combining every term of the configuration s^2p^2 of the protons with every term arising from the configuration s^2p^3 of the neutrons.[12] The former terms are ^1S, ^1D, ^3P, the latter ones ^2P, ^2D, ^4S. Taken together these two configurations yield for the whole nucleus the terms ^{12}P, ^{12}D, ^{14}S, ^{12}P, ^{12}D, ^{12}F, ^{12}S, ^{12}P, ^{12}D, ^{12}F, ^{12}G, ^{14}D, ^{32}S, ^{32}P, ^{32}D, ^{32}P, ^{32}D, ^{32}F, ^{34}P. The first index represents the multiplicity of the protons, the second the multiplicity of the neutrons. (The Heisenberg forces will introduce an interaction between proton spin and neutron spin and split many of these terms into several new ones.)

Since the number of terms is quite high, we first want a quick orientation as to which of the terms will be the lowest ones. It is clear that most of the terms just enumerated lie high in the continuous spectrum. We shall be interested in the low terms only.

For this first orientation, we shall assume the interaction between like particles to be the same as that between unlike particles. This amounts to assuming (1*a*) to be valid for all pairs of particles, since the interaction (1*b*) involving the spin is omitted in this section. We shall call this model the "equal orbital forces model." It does not constitute a good approximation to either (1*a*) and (2*a*) or to (1*a*) and (2*b*), but it is useful in obtaining a first orientation.

In the equal orbital forces model, the Hamiltonian is symmetric in all particles and acts on the Cartesian coordinates only. Four particles can be in the same orbit, namely two protons and two neutrons. Hence, not only those representations of the symmetric group will occur[13] which occur in atomic spectra, but we shall have representations with 1, 2 and also 3, 4 as addends. The representations play a very great role in the calculation of the potential energy, because the interaction operator contains a permutation of the particles.

The kinetic energy is the same for all wave functions arising from the same configuration.

If $\psi_1, \psi_2, \cdots \psi_s$ belong to a certain representation D of the symmetric (permutation) group of $n=n_1+n_2$ particles, we can calculate the matrix elements of the potential energy

$$V_{\kappa\kappa}=(\psi_\kappa(x_1\cdots x_n),\ \sum_{\alpha\beta} J(x_\alpha-x_\beta)P_{\alpha\beta}\psi_\kappa(x_1\cdots x_n)), \quad (4)$$

which is equal to

$$V_{\kappa\kappa}=\sum_{\alpha\beta}\sum_{\lambda} D(\alpha\beta)_{\lambda\kappa}(\psi_\kappa,\ J(x_\alpha-x_\beta)\psi_\lambda) \quad (5)$$

because of the relations

$$P_{\alpha\beta}\psi_\kappa(x_1\cdots x_n)+\sum_{\lambda} D(\alpha\beta)_{\lambda\kappa}\psi_\lambda(x_1\cdots x_n). \quad (6)$$

Only those terms will be great in the sum (5) for which $\kappa=\lambda$; in the other terms the positive and negative regions will about cancel each other. This canceling would be exact if the range of the forces were very long: in this case J could be taken out of the integral as a constant $J(0)$ and the remaining integral would vanish for $\kappa\neq\lambda$ and be 1 for $\kappa=\lambda$. The whole $V_{\kappa\kappa}$ is for very

[10] Computed from the mass values given by Bethe and Livingston at the Cornell Nuclear Conference, July, 1936. The C^{14} value is taken from a paper by T. W. Bonner (same conference).

[11] Cf. W. Heisenberg, Ann. d. Physik **10**, 888 (1931); G. Shortley, Phys. Rev. **43**, 451 (1933).

[12] Cf. reference 8, Table XIV for a complete list of terms.

[13] Cf. E. Wigner, *Gruppentheorie* etc. (Braunschweig, 1931), Chap. 13. The "all orbital forces equal" model is equivalent to approximation (1) of the paper mentioned in reference 5.

TABLE II. *The terms of a p^n configuration which correspond to a definite partition.*

p^2		p^3		
2	1+1	3	2+1	1+1+1
$S\ D$	P	$F\ P$	$D\ P$	S
1/1	−1/1	1/1	0/2	−1/1

p^4				p^5			
4	3+1	2+2	2+1+1	4+1	3+2	3+1+1	2+2+1
$G\ D\ S$	$F\ D\ P$	$D\ S$	P	$P\ D\ F\ G$	$F\ D\ P$	$D\ S$	P
1/1	1/3	0/2	−1/3	2/4	1/5	0/6	−1/5

p^6				
4+2	3+3	4+1+1	3+2+1	2+2+2
$G\ F\ D\ D\ S$	$F\ P$	$F\ P$	$D\ P$	S
3/9	1/5	2/10	0/16	−1/5

long range forces

$$V_{\kappa\kappa}=J(0)\sum_{\alpha\beta}D(\alpha\beta)_{\kappa\kappa}. \tag{7}$$

The last sum has to be extended over all $n(n-1)/2$ transpositions and is equal to $n(n-1)/2$ times the character $\chi(T)$ corresponding a transposition, divided by the dimension $\chi(E)=s$ of the representation:

$$V_{\kappa\kappa}=J(0)(n(n-1)/2)\chi(T)/\chi(E). \tag{8}$$

This expression is independent of the detailed shape of the wave function and depends only on the representation D to which the wave function belongs. The formulas for the characters of representations[14] show that it is greatest for that representation which contains as many 4's as possible and in addition to these only one other addend. The terms corresponding to this representation will be called the "low terms."

[14] I. Schur, Berl. Ber. (1908), p. 664. We found the following formula most suitable for the calculation of characters:

$$\{\lambda_1, \lambda_2, \cdots\lambda_\rho\}=\{\lambda_1-1, \lambda_2, \cdots, \lambda_\rho\} + \{\lambda_1, \lambda_2-1, \cdots\lambda_\rho\}+\cdots+\{\lambda_1, \lambda_2, \cdots\lambda_\rho-1\}.$$

Here the symbol $\{\lambda_1, \lambda_2, \cdots\lambda_\rho\}$ is the character of a permutation in the representation corresponding to the partition $\lambda_1+\lambda_2+\cdots+\lambda_\rho=n$. The left side gives the character of the representation of the permutation group of n elements, the right side contains characters of representations of the permutation group of $n-1$ elements. The formula holds for every permutation and can be used as a recursive formula. On the right side, all symbols must be omitted in which one number in the bracket is greater than the preceding. If the last figure in a bracket is a zero, it can be dropped. Thus, e.g., for a transposition,

$$\{4+1+1\}=\{3+1+1\}+\{4+0+1\}+\{4+1+0\} =\{3+1+1\}+\{4+1\}.$$

The values of these symbols can be obtained by a further application of the recursive formula, since, for a transposition, all symbols can be reduced finally to $\{2\}$ and $\{1+1\}$ which are +1 and −1, respectively. They are, for a transposition, 0 and 2, so that the character corresponding to a transposition in $\{4+1+1\}$ is 2.

TABLE III. *Interaction energies between like particles.*

INTERACTION	p^2 or π^2			p^3 or π^3		
	^1S	^1D	^3P	^2P	^2D	^4S
(2*a*)	$L+2K$	$L-K$	$-L+3K$	$5K$	$3K$	$-3L+9K$
(2*b*)	$L+2K$	$L-K$	$(-L+3K)/3$	$L+2K$	L	$-L+3K$
ORDINARY FORCE	$L+2K$	$L-K$	$L-3K$	$3L-4K$	$3L-6K$	$3L-9K$

INTERACTION	p^4 or π^4	p^5 or π^5	p^6 or π^6
(2*a*)	ADD $-L+8K$ to p^2	$-2L+16K$	$-3L+24K$
(2*b*)	ADD $L+2K$ to p^2	$2L+4K$	$3L+6K$
ORDINARY FORCE	ADD $5L-10K$ to p^2	$10L-20K$	$15L-30K$

It is easy to determine the terms of a p^n configuration which correspond to a definite partition[15] and the result is given in Table II. The last row contains $\chi(T)/\chi(E)$. The four s particles form a closed shell and contribute the same amount of energy to every term of a configuration.

We proceed now to the secular equations, using (2*a*) and (2*b*) for the interaction between like particles. The kinetic energy has been omitted, because it is the same for all terms of the same configuration. In the calculation of the potential energy, the closed s shell has been omitted for the same reason and the wave function for the p particles only used. To illustrate, for Li^7 the configuration is $p\pi^2$, the Greek letter π corresponding to the neutron, p to the proton. First the wave functions for the different π^2 terms ^1S, ^1D, ^3P are written down. Then, for example, to calculate the ^{12}P term, wave functions with the azimuthal quantum number 1 are constructed from the proton wave function and the ^1S neutron function, and from the proton function and the ^1D neutron function. The spin functions can always be omitted when calculating in this way.

With this choice of wave functions, the interaction between like particles has only diagonal elements, these being the sum of the energies of the proton term and the neutron term used. The matrix elements for the proton-neutron

[15] The simplest method for this is described in Section 4 of reference 3. (Cf. Table I there.)

interaction must be calculated separately for each case.

It is well known that all potential energy integrals for the p^n configuration can be expressed in terms of the following two,

$$
\begin{aligned}
L&=\int\cdots\int(x/r)^2R_p(r)^2(x'/r')^2R_p(r')^2\\
&\qquad\times J(|r-r'|)d\tau d\tau',\\
K&=\int\cdots\int(xy/r^2)R_p(r)^2(x'y'/r'^2)\\
&\qquad\times R_p(r')^2J(|r-r'|)d\tau d\tau'
\end{aligned}
\tag{9}
$$

by means of the identity

$$
L-2K=\int\cdots\int(x/r)^2R_p(r)^2(y'/r')^2\times R_p(r')^2J(|r-r'|)d\tau d\tau', \tag{10}
$$

where $(x/r)R_p(r)$ denotes the p wave function. If there is any ambiguity we shall use L and K with the index $\nu\pi$ if J signifies the interaction between unlike particles, with an index $\nu\nu$ if it is the interaction between like particles, and with an index c if J is the electrostatic potential.

The matrix elements for the interaction energies between like particles are tabulated in Table III.

Table IV contains the matrix elements

TABLE IV. *Matrix elements of the interaction between unlike particles.*[16]

Li^6

$({}^2P{}^2P){}^{22}S$ $L+2K$, $({}^2P{}^2P){}^{22}D$ $L-K$, $({}^2P{}^2P){}^{22}P$ $-L+3K$

Li^7

$({}^2P{}^1D){}^{21}F$ $2L-2K$, $({}^2P{}^3P){}^{23}P$ $L+2K$, $({}^2P{}^3P){}^{23}D$ L,
$({}^2P{}^1D){}^{21}D$ $-L+4K$, $({}^2P{}^3P){}^{23}S$ $-2L+6K$,

	$({}^2P{}^1S){}^{21}P$	$({}^2P{}^1D){}^{21}P$
$({}^2P{}^1S){}^{21}P$	$2(L+2K)/3$	$(20)^{\frac12}(L-K)/3$
$({}^2P{}^1D){}^{21}P$	$(20)^{\frac12}(L-K)/3$	$(L+14K)/3$

Be^8

	$({}^1S{}^1S){}^{11}S_+$	$({}^1D{}^1D){}^{11}S_+$
$({}^1S{}^1S){}^{11}S_+$	$4(L+2K)/3$	$(80)^{\frac12}(L-K)/3$
$({}^1D{}^1D){}^{11}S_+$	$(80)^{\frac12}(L-K)/3$	$2(L+14K)/3$

	$({}^1S{}^1D){}^{11}D_+$	$({}^1D{}^1D)\ D_+$
$({}^1S{}^1D){}^{11}D_+$	$4(2L+K)/3$	$(56)^{\frac12}\ L-K)/3$
$({}^1D{}^1D){}^{11}D_+$	$(56)^{\frac12}(L-K)/3$	$(-2L+23K)/3$

$({}^1D{}^1D){}^{11}G_+$ $4L-4K$.

Li^8

	$({}^2P{}^2P){}^{22}P$	$({}^2P{}^2D){}^{22}P$
$({}^2P{}^2P){}^{22}P$	$(L+7K)/2$	$(15)^{\frac12}(L-K)/2$
$({}^2P{}^2D){}^{22}P$	$(15)^{\frac12}(L-K)/2$	$(-L+13K)/2$

	$({}^2P{}^2D){}^{22}D$	$({}^2P{}^2P){}^{22}D$
$({}^2P{}^2D){}^{22}D$	$(L+7K)/2$	$3^{\frac12}(L-K)/2$
$({}^2P{}^2P){}^{22}D$	$3^{\frac12}(L-K)/2$	$(3L+K)/2$

$({}^2P{}^2D){}^{22}F$ $2L-K$.

Be^9

	$({}^1S{}^2P){}^{12}P$	$({}^1D{}^2P){}^{12}P$	$({}^1D{}^2D){}^{12}P$
$({}^1S{}^2P){}^{12}P$	$2L+4K$	0	$(20/3)^{\frac12}(L-K)$
$({}^1D{}^2P){}^{12}P$	0	$(L+17K)/2$	$-(49/12)^{\frac12}(L-K)$
$({}^1D{}^2D){}^{12}P$	$(20/3)^{\frac12}(L-K)$	$-(49/12)^{\frac12}(L-K)$	$(-L+23K)/2$

	$({}^1S{}^2D){}^{12}D$	$({}^1D{}^2D){}^{12}D$	$({}^1D{}^2P){}^{12}D$
$({}^1S{}^2D){}^{12}D$	$2L+4K$	0	$2(L-K)$
$({}^1D{}^2D){}^{12}D$	0	$(L+17K)/2$	$(7/4)^{\frac12}(L-K)$
$({}^1D{}^2P){}^{12}D$	$2(L-K)$	$(7/4)^{\frac12}(L-K)$	$(3L+11K)/2$

	$({}^1D{}^2P){}^{12}F$	$({}^1D{}^1D){}^{12}F$
$({}^1D{}^2P){}^{12}F$	$3L+K$	$-2^{\frac12}(L-K)$
$({}^1D{}^2D){}^{12}F$	$-2^{\frac12}(L-K)$	$2L+4K$

$({}^1D{}^2D){}^{12}G$ $4L-2K$.

Be^{10}

	$({}^1S{}^1S){}^{11}S$	$({}^1D{}^1D){}^{11}S$
$({}^1S{}^1S){}^{11}S$	$8(L+2K)/3$	$-(80)^{\frac12}(L-K)/3$
$({}^1D{}^1D){}^{11}S$	$-(80)^{\frac12}(L-K)/3$	$(-8L+50K)/3$

	$({}^1S{}^1D-{}^1D{}^1S){}^{11}D$	$({}^1S{}^1D+{}^1D{}^1S){}^{11}D$	$({}^1D{}^1D){}^{11}D$
$({}^1S{}^1D-{}^1D{}^1S){}^{11}D$	$4L+4K$	0	0
$({}^1S{}^1D+{}^1D{}^1S){}^{11}D$	0	$(4L+20K)/3$	$(56)^{\frac12}(L-K)/3$
$({}^1D{}^1D){}^{11}D$	0	$(56)^{\frac12}(L-K)/3$	$(5L+28K)/3$

$({}^1D{}^1D){}^{11}F$ $4L+4K$, $({}^1D{}^1D){}^{11}G$ $4L$.

B^{10}

	$({}^2P{}^2P){}^{22}S_+$	$({}^2D{}^2D){}^{22}S_+$
$({}^2P{}^2P){}^{22}S_+$	$2L+9K$	$(15)^{\frac12}(L-K)$
$({}^2D{}^2D){}^{22}S_+$	$(15)^{\frac12}(L-K)$	$15K$

	$({}^2P{}^2P){}^{22}D_+$	$({}^2P{}^2D+{}^2D{}^2P){}^{22}D_+$	$({}^2D{}^2D){}^{22}D_+$
$({}^2P{}^2P){}^{22}D_+$	$(7L+9K)/2$	0	$(21)^{\frac12}(L-K)/2$
$({}^2P{}^2D+{}^2D{}^2P){}^{22}D_+$	0	$5L+5K$	0
$({}^2D{}^2D){}^{22}D_+$	$(21)^{\frac12}(L-K)/2$	0	$(3L+21K)/2$

$({}^2D{}^2P+{}^2P{}^2D){}^{22}F_+$ $5L+2K$, $({}^2D{}^2D){}^{22}G_+$ $5L$.

The $+$ sign signifies that the wave function remains unchanged if the proton coordinates are interchanged with the neutron coordinates. This quantum number exists only for elements with equal numbers of protons and neutrons and also in this case only if the multiplicity for both is the same. If the multiplicities for protons and neutrons are different, the $+$ term coincides with the $-$ term and the degeneracy is doubled.

[16] The matrix elements for Li^6 are given by Bethe and Bacher, reference 8, §36.

for the proton-neutron interaction. The symbol $({}^1D{}^3P){}^{13}F$ denotes the wave function with the azimuthal quantum number 3 which can be constructed out of the ${}^1D$ protonic and the ${}^3P$ neutronic wave functions.

Table IV contains all matrix elements for Li^6 and Li^7, but for the other nuclei only those which are necessary for the calculation of the "low terms" (as determined by the equal orbital forces approximation). The matrix elements $\bar{V}_{\kappa\lambda}$ for the nucleus with $6-n_1$ protons and $6-n_2$ neutrons are obtained from the corresponding matrix elements $V_{\kappa\lambda}$ of the nucleus with n_1 protons and n_2 neutrons by means of the relation

$$\bar{V}_{\kappa\lambda}=V_{\kappa\lambda}+(6-n_1-n_2)(2L+4K)\delta_{\kappa\lambda}, \quad (11)$$

which can be derived from the theory of holes. Thus we have all the matrix elements for the Majorana forces.

The next task is the calculation of the integrals L and K. Because of

$$\int\cdots\int (xy'-x'y)^2/r^2r'^2\cdot R_p(r)^2R_p(r')^2 \times J(|r-r'|)d\tau d\tau'=2L-6K>0 \quad (12)$$

we know that at any rate $3K<L$. The actual calculation of the ratio and absolute values of L and K is given in the appendix. It is clear that these quantities depend principally on the radius r_p of the p shell and on the range r_0 of the nuclear forces. If r_0/r_p is very large, K will vanish. This corresponds to the approximation made in Eqs. (7) and (8). The appendix shows that K, though not zero, is 7 to 11 times smaller than L.

For the calculation of energies of terms which do not appear as solutions of secular equations, the interaction energy of like particles has to be added directly to the matrix element of Table IV. In the other cases, the like particle interaction matrix elements must be added to the diagonal elements of the secular determinant of Table IV. It must be remembered that the L and K in Table III are not identical with the corresponding quantities in Table IV. If, however, we suppose that the two different L's are equal and set $K=0$, the result must be that given by Eq. (8). This shows that in the equal orbital forces model, the differences between the "low terms" are due to the finite value of K. With the other models the numerical results for the spacing of the "low terms" in a multiplet depend essentially on the value of K and hardly at all on the L/K ratio.

The actual solutions of the secular equations are given in Table V in units of mc^2 for the terms which are "low" in the equal orbital forces approximation. Although the term differences are independent of the choice of like particle interaction, we do not believe them to be correct (because of the use of the Hartree method). However, we think that the order of the terms will be right. The situation is the same in the corresponding atomic spectra.[6]

The symbols (*i*) and (*ii*) designate different methods of making the numerical calculations which are fully described in the appendix. Models (2*a*) and (2*b*) (and consequently also model (2*c*)) give the same order and spacing of the "low terms" within a multiplet. The spacing is also practically independent of the method of calculation, because K has nearly the same value in both methods. Fig. 1 gives the calculated total energy (method (*ii*)) of the lowest term and the experimental binding energy. Only the most stable nucleus with a given total mass is shown on the diagram. To obtain the theoretical curve a linear function $c(n_1+n_2)+d$ is added to the computed total energy and the constants c and d determined to fit the experimental points at He^4 and O^{16}. The observed "4 shell" structure is clearly evident in the computed curve, although the wave functions used do not correspond to preformed alpha-particles. In general the agreement is somewhat worse than for the similar calculation in atomic spectra.[17]

For the comparison of the calculated excited levels with experiment there is available only a limited amount of experimental material. P. Savel[18] and W. Bothe[19] have investigated the gamma-radiation associated with some nuclear transmutations. Unfortunately, Savel's results do not allow any conclusions to be drawn as to the character of the excited states. Bothe seems to have found three states in the C^{12} nucleus, the positions of which agree fairly well with our "low" terms. However an interpretation of Bothe's terms on these lines hardly seems

[17] R. Peierls, Zeits. f. Physik **59**, 738 (1929).
[18] P. Savel, Ann. d. Physik **4**, 88 (1935).
[19] W. Bothe, Zeits. f. Physik **100**, 273 (1936).

TABLE V. *Calculated p shell potential energies in units* mc^2.

TERM	(2a) (i)	(2a) (ii)	(2b) (i)	(2b) (ii)
Li^6				
^{22}S	−18.6	−12.7	−18.6	−12.7
^{22}D	−14.3	−8.4	−14.3	−8.4
^{22}P	11.5	5.5	11.5	5.5
For N^{14} add	−137.1	−105.7	−195.9	−133.6
Li^7				
^{12}P	−44.0	−28.4	−44.0	−28.4
^{12}F	−37.8	−22.1	−37.8	−22.1
For C^{13} or N^{13} add	−102.8	−79.3	−146.9	−100.2
Li^8				
^{22}P	−39.2	−27.4	−46.5	−30.9
^{22}D	−36.2	−23.9	−43.5	−27.4
^{22}F	−32.8	−22.4	−40.1	−25.9
For B^{12} add	−68.5	−52.9	−97.9	−66.8
Be^8				
^{11}S	−88.1	−56.8	−88.1	−56.8
^{11}D	−84.3	−53.0	−84.3	−53.0
^{11}G	−75.6	−44.1	−75.6	−44.1
For C^{12} add	−68.5	−52.9	−97.9	−66.8
Be^9 or B^9				
^{12}P	−83.1	−55.8	−90.5	−59.3
^{12}D	−80.7	−53.3	−88.1	−56.8
^{12}F	−77.0	−49.5	−84.4	−53.0
^{12}G	−72.0	−44.4	−79.4	−47.9
For B^{11} or C^{11} add	−34.3	−26.4	−49.0	−33.4

TERM	(2a) (i)	(2a) (ii)	(2b) (i)	(2b) (ii)
Be^{10} or C^{10}				
^{11}S	−90.0	−62.6	−104.7	−69.6
^{11}D	−87.6	−60.2	−102.3	−67.2
^{11}D	−86.8	−59.4	−101.5	−66.4
^{11}F	−84.1	−56.7	−98.8	−63.7
^{11}G	−78.4	−50.9	−93.1	−57.9
B^{10}				
^{22}S	−96.9	−67.6	−111.6	−74.6
^{22}D	−92.9	−63.5	−107.6	−70.5
^{22}D	−92.5	−63.0	−107.2	−70.0
^{22}F	−88.7	−59.2	−103.4	−66.2
^{22}G	−84.1	−54.5	−98.8	−61.5
C^{14} or O^{14}				
^{11}S	−148.9	−113.7	−207.7	−141.6
^{11}D	−146.2	−111.0	−205.0	−138.9
N^{15} or O^{15}				
^{12}P	−171.4	−132.2	−244.9	−167.1
O^{16}				
^{11}S	−205.6	−158.6	−293.8	−200.5

feasible, because the transition from ^{11}G to ^{11}S should be so much less probable than the transition to the ^{11}D level that the gamma-ray corresponding to the former transition should be unobservable. Hence, if our results are correct, at least one of his lines must have another interpretation. A possible interpretation is that two of the lines observed by Bothe come from the transitions $^{11}G-^{11}D$, $^{11}D-^{11}S$ and the third from another excited level which happens to be near ^{11}G. Such an excited level might arise from an excited configuration. This interpretation gains in plausibility from the fact that N^{15} has an excited state[20, 21] with an excitation energy of 10.5 mc^2. Since the normal state configuration of N^{15} has only one term, all the excited states must be ascribed to configurations in which one or more particles are excited. The occurrence of short and long range groups of particles together in transmutations in which C^{12} is produced confirms the existence of an excited level about 10 mc^2 above the normal state.[20, 21, 22] Cockcroft and Lewis[21] find two excited states in B^{11}, one of which falls between the levels ^{12}G, ^{12}F and the other between the levels ^{12}F, ^{12}D of the theoretical calculation. Bonner and Brubaker[22] observe three excited states in B^{10} with the energy 8.4 mc^2 available to produce excitation. The theoretical calculation shows three triplet levels and one singlet level in this range.

If we assume in analogy with H^2 that there

[20] E. O. Lawrence, E. McMillan and M. C. Henderson, Phys. Rev. **47**, 273 (1935).

[21] J. D. Cockcroft and W. B. Lewis, Proc. Roy. Soc. **A154**, 261 (1935).

[22] T. W. Bonner and W. M. Brubaker, Phys. Rev. **50**, 308 (1936).

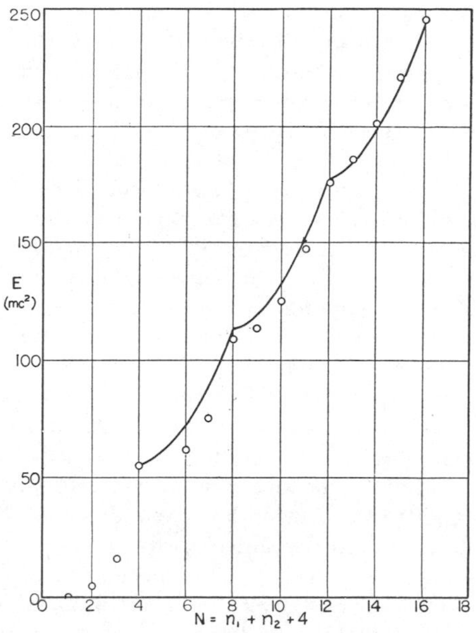

FIG. 1. Binding energy against nuclear mass. The calculated values lie on the solid curve; the circles mark experimental points.

are forces which tend to direct the proton spin parallel to the neutron spin (the Heisenberg forces will be shown to have such an effect), we obtain Table VI for the total angular momenta of the ground states. The ambiguity, in the cases for which several possible values of the angular momentum are given, is caused by the lack of knowledge concerning forces coupling the spin to the orbit. Experimentally the following momenta are known: Li^6-1, $Li^7-3/2$, $C^{12}-0$, $N^{14}-1$, $O^{16}-0$. These agree with the theory.

III. THE HEISENBERG FORCE BETWEEN UNLIKE PARTICLES

The comparatively large singlet-triplet separation in H^2 (~ 4 mc^2) points to an interaction between protons and neutrons which involves the spin in some way. The simplest assumption for such an interaction—though not the only possible one—is the Heisenberg force (1*b*) which involves the exchange of both Cartesian and spin coordinates of the interacting particles.

It is characteristic of this interaction that it is invariant with respect to the rotation of all spin coordinates—though it is not invariant with respect to the rotation of the proton spin alone any more, as were the Majorana forces. Consequently, it will leave the total spin angular momentum an exact quantum number, not, however, the proton spin and the neutron spin separately.

By Dirac's identity the exchange of the spin coordinates can be written

$$Q_{12} = \tfrac{1}{2} + \tfrac{1}{2}(\sigma_1 \cdot \sigma_2). \qquad (13)$$

The first part gives matrix elements which are, apart from the factor $g/(1-g)$, just half as great as the matrix elements of the Majorana force, thus effectively increasing this by the factor $(1-\frac{1}{2}g)/(1-g)$. The matrix elements of the second part vanish by the selection rules if both wave functions have a proton spin 0 or a neutron spin 0 or if the proton spins or the neutron spins differ by 2 or more. In these cases the addition of the Heisenberg force amounts only to an increase of the Majorana force by the factor $(1\ \frac{1}{2}g)/(1\ \ g)$. For most purposes it is more practical to unite the first part of the Heisenberg force with the Majorana force and consider the rest as an interaction involving P, the interchange of the Cartesian coordinates, and the scalar product of the spin operators. The effect of the first part of the Heisenberg force is numerically much greater than the second in most cases. It has been included in Tables IV and V and in Fig. 1 by calculating with a modified form of (1*a*) in which g is replaced by $\frac{1}{2}g$.

The splitting of a term with proton multiplicity s and neutron multiplicity σ is caused by the second part only. There are new components with the total spin angular momenta $S=|s-\sigma|$, $|s-\sigma|+1$, $\cdots$, $s+\sigma$ the displacement of the term with the total spin S being proportional to

$$s(s+1)+\sigma(\sigma+1)-S(S+1). \qquad (14)$$

The matrix elements of the second part of (1*b*) for the ground state when they do not vanish

TABLE VI. *Angular momenta for normal states.*

Li^6	Li^7	Li^8	Be^8	Be^9	B^9	Be^{10}	B^{10}	C^{10}
1	$\frac{1}{2},\frac{3}{2}$	0,1,2	0	$\frac{1}{2},\frac{3}{2}$	$\frac{1}{2},\frac{3}{2}$	0	1	0
B^{11}	C^{11}	B^{12}	C^{12}	C^{13}	N^{13}	N^{14}	N^{15}	O^{16}
$\frac{1}{2},\frac{3}{2}$	$\frac{1}{2},\frac{3}{2}$	0,1,2	0	$\frac{1}{2},\frac{3}{2}$	$\frac{1}{2},\frac{3}{2}$	1	$\frac{1}{2},\frac{3}{2}$	0

(in all these cases $s=\sigma=\frac{1}{2}$) are given in Table VII for the triplet state only, since the elements for the singlet state can be obtained by multiplication with -3. Here L and K are identical with the same symbols in Table IV. These matrix elements must be multiplied by the factor $g/2(1-\frac{1}{2}g)$ and added to the corresponding effective Majorana matrix elements in Table IV.

The numerical calculation (method (*i*)) yields a splitting of 9 mc^2 for both Li^6 and B^{10} (and thus also for N^{14}), the triplet term being always the lowest. Method (*ii*) gives a splitting of 6.3 mc^2 for Li^6 and N^{14} and of 6.6 mc^2 for B^{10}. For Li^8 both methods give very little splitting so that it is impossible, in this case, to tell what should be the normal state.

Since the azimuthal quantum number and the resultant spin angular momentum both remain good quantum numbers, the Heisenberg force does not introduce a splitting of the ^3P or similar terms into the fine structure components $^3P_0$, $^3P_1$, $^3P_2$. It has been suggested by D. R. Inglis (at the summer symposium at Ann Arbor, as yet unpublished) that the spin orbit coupling should have the same sign (giving the larger j the lower energy) for both neutrons and protons in nuclei. A relativistic (Thomas) term results from the acceleration of the particles by the nuclear forces. According to Inglis this is the preponderating spin orbit coupling term in nuclei.

IV. The Coulomb Energy

A further question of interest is that concerning the difference between proton-proton and neutron-neutron interaction. The simplest assumption is that the only difference is the electrostatic interaction between protons.

According to this assumption the binding energy and the whole spectrum of two nuclei should differ only in the Coulomb energy if the number of neutrons in the first element is equal to the number of protons in the second element and *vice versa*. Known pairs of this type are H^3-He^3, Be^9-B^9, $B^{11}-C^{11}$, $C^{13}-N^{13}$, $N^{15}-O^{15}$, $O^{17}-F^{17}$.

Table VII. *Matrix elements of the second part of the Heisenberg force.*

Li^6	$(^2P^2P)^{22}S$ $L+2K$	
Li^8	$(^2P^2P)^3P$	$(^2P^2D)^3P$
$(^2P^2P)^3P$	$(-L+3K)/2$	$(15)^{\frac{1}{2}}(L-3K)/6$
$(^2P^2D)^3P$	$(15)^{\frac{1}{2}}(L-3K)/6$	$(-5L+11K)/6$
B^{10}	$(^2P^2P)^3S$	$(^2D^2D)^3S$
$(^2P^2P)^3S$	$(2L-K)$	$(15)^{\frac{1}{2}}(-L+3K)/3$
$(^2D^2D)^3S$	$(15)^{\frac{1}{2}}(-L+3K)/3$	$(8L-5K)/3$

For He^3 there already exists a simple calculation of the Coulomb energy[1] and a more accurate calculation giving substantially the same result has been made by S. S. Share.[23] The computed Coulomb energy for He^3 appears to be about 15 percent smaller than the experimental H^3-He^3 energy difference. For the other pairs we have performed the calculation, adding the electrostatic energy to the like particle interaction. The details of the computation are given in the appendix. For all cases in which the positron spectrum has been observed the experimental energy difference is obtained as a sum of three terms: (1) the upper limit of the positron spectrum (in electron mass units, taking the inspection value for the upper limit),[24] (2) the n^1-H^1 mass difference, (3) the mass of two electrons. Table VIII exhibits the electrostatic and experimental differences. The agreement is quite good. Nevertheless the experimental values are always somewhat greater than the theoretical ones. Considering the rather large uncertainty of these calculations, one cannot claim with certainty at present that the neutron-neutron interaction is stronger than the proton-proton interaction.

V. Comparison of Isobaric Nuclei

In this section we wish to make a few remarks concerning the spectra of isobars and the possibility of deciding between the alternatives (2*a*), (2*b*) and (2*c*) for the interaction between like particles.

It is clear from Table V that for the same values of the depth and range of the potentials, (2*a*) gives much less binding than (2*b*). Hence using the statistical method one obtains a smaller discrepancy with (2*b*) than with (2*a*). This cannot be considered, however, to be decisive evidence in favor of (2*b*), since it is possible that

[23] S. S. Share, Phys. Rev. **50**, 488 (1936).

[24] W. A. Fowler, L. A. Delsasso and C. C. Lauritsen, Phys. Rev. **49**, 561 (1936); F. N. D. Kurie, J. R. Richardson and H. C. Paxton, Phys. Rev. **48**, 167 (1935); L. Meitner, Nature **22**, 420 (1934).

TABLE VIII. *Electrostatic and measured energy differences.*

	H^3-He^3	Be^9-B^9	$B^{11}-C^{11}$	$C^{13}-N^{13}$	$N^{15}-O^{15}$	$O^{17}-F^{17}$
Electrostat.	1.37	3.11	4.06	4.62	5.54	5.62
Experim.	1.57	3.20	4.90	5.10	6.00	6.80

the correlation energy can account for the greater difference between the observed energies and the results of the statistical calculation with (2*a*). The discrepancy for (2*c*) is even greater than for (2*a*), but (2*c*) also cannot be excluded on this basis.[25]

In all cases only a small fraction of the binding energy is due to the interaction between like particles. This makes it very difficult to draw conclusions concerning these forces from experimental term values. If we consider, however, the Schrödinger equations of two isobars in the equal orbital forces approximation, their characteristic values will be absolutely equal. Nevertheless, the energy values will not necessarily be equal even in this approximation, because a characteristic value which is allowed for one isobar, may be forbidden for the other by Pauli's principle. Thus the low terms of Be^8 arise from the partition 4+4 of the equal orbital force approximation, but this gives no allowed terms for Li^8, the low terms of which come from the 4+3+1 partitions. These are allowed for Be^8 also, but correspond to excited states. For Be^{10} and B^{10}, however, the low states come in both cases from the 4+4+2 partition, in the first case giving singlet-singlet, in the second doublet-doublet terms. These coincide exactly in the equal orbital forces approximation and their similarity can be seen even in Table V. If one now introduces the difference between (2) and (1*a*) as a perturbation, the Be^{10}, B^{10} energy separation will appear as a difference of two small quantities.

Experimentally the binding energy of Be^{10} is greater than that of B^{10} by 0.8 mc^2. From Table VIII we estimate that the difference between the proton-proton and the neutron-neutron interactions (Coulomb energy) has decreased the binding energy by 3.2 mc^2. Thus the binding energy due to nuclear forces alone is greater for B^{10} than for Be^{10} by 2.4 mc^2. The corresponding quantity for the $N^{14}-C^{14}$ pair is $5.1-0.8=4.3$ mc^2. *Such small differences are possible only if the constants in the interactions between like and unlike particles are essentially equal.*

With (2*a*) or (2*b*) the difference between the Majorana energies of the B^{10} and Be^{10} ground states is seen from Table V to be 5.0 mc^2 (method (*ii*)). This must be increased by the amount 1.6 mc^2 by which the normal state of B^{10} is lowered when account is taken of the second part of the Heisenberg force between unlike particles. This gives a total difference of 6.6 mc^2. Using (2*c*) it can be shown that the ground state of Be^{10} coincides with the singlet state arising from ^{22}S in B and hence the difference between the binding energies of Be^{10} and B^{10} is 6.6 mc^2 for all three models. The agreement between models (2*a*), (2*b*) and (2*c*) in this case is not due to numerical coincidence, but can be shown to be true for all values of L and K simply by comparing the like particle contributions to the secular equations of the two systems for each of the three different models.

We see that the experimental difference between the binding energies of Be^{10} and B^{10} is much smaller than the theoretical difference. For C^{14} and N^{14} the situation is similar. Both sets of low terms belong to the partition 4+4+2 and hence coincide in the equal orbital forces approximation. Here also the energy difference is independent of the model. From Table V (method (*ii*)) the $C^{14}-N^{14}$ energy difference is $4.7+1.6=6.3$ mc^2, while experimentally the difference is only 4.3 mc^2.

In view of these discrepancies it is surprising to find excellent agreement between the computed and experimental Li^6-He^6 energy differences. Bjerge and Broström[26] report the value 7.5 ± 1.0 mc^2 for the maximum kinetic energy of the electrons produced in the transformation of He^6 into Li^6. Since the additional Coulomb energy in Li^6 is almost exactly balanced by the greater mass of the constituents of He^6, we

[25] H. A. Bethe and R. F. Bacher, reference 8, §6, have shown that a symmetrical Hamiltonian gives stable nuclei consisting of protons or of neutrons alone if one does not take into account the Pauli principle. However the Pauli principle causes the effective interaction between like particles to be smaller than that between unlike particles, even though the fundamental Hamiltonian does not distinguish between like and unlike particles. A good example in this connection is Li^8 which is much less stable than Be^8.

[26] T. Bjerge, Nature **138**, 400 (1936), and T. Bjerge and K. J. Broström, Nature **138**, 400 (1936).

obtain 7.5 ± 1.0 mc^2 for the experimental singlet-triplet splitting in Li^6. The computed value is 6.3 mc^2 (method (*ii*)) or 9.0 mc^2 (method (*i*)).

The secular equations for the states of Be^8 corresponding to the $4+3+1$ partitions are given in Table IX.

We obtain -1.0 mc^2 for the $(Li^8)\ {}^3P - (Be^8)\ {}^{13}P$ energy difference using model (2*a*) and -3.4 mc^2 using model (2*b*). These numbers, added to the experimental normal state energy difference, yield the results 21 mc^2 (2*a*) or 24.4 mc^2 (2*b*) for the excitation energy of the ${}^{13}P$ term in Be^8. This term could be observed directly since it is stabilized by the spin conservation law.[27] A similar result can be deduced for C^{12} also, by comparison with B^{12}, since the experimental binding energy difference is nearly the same. It should be noted that method (*ii*) gives a $Li^8 - Be^8$ normal state spacing in fair agreement with experiment. Here, as well as in the problems of the "four shell" structure and the singlet-triplet separation, method (*ii*) gives better agreement with the experimental facts than method (*i*).

Appendix

We assume that each particle has a separate wave function (Hartree-Fock model) and, furthermore, that this has the form

$$R_s(r) = ce^{-r^2/r_s^2} = ce^{-\alpha\sigma r^2} \tag{15}$$

for the s particles and

$$(x/r)R_p(r), \quad (y/r)R_p(r), \quad (z/r)R_p(r), \qquad R_p(r) = c' r e^{-r^2/r_p^2} = c' e^{-\tau\alpha r^2} \tag{16}$$

for the p particles, where α is the reciprocal square of the range of the forces and is defined in (1).

It is well known that if a determinantal wave function for the whole system is built up out of these functions and the energy calculated, choosing the parameters σ and τ so as to make the energy a minimum, the absolute value of this minimum will be very small and the corresponding radii greater than the radius of the nucleus. The reason for this discrepancy is that it is impossible to express the finer statistical correlations between the positions of the particles assuming a separate wave function for each particle. In order to take into account at least approximately the correlation energy, a semi-empirical method is used, modifying the Hamiltonian in such a way that it gives the experimental binding energy for O^{16}.

[27] The spin conservation law is practically rigorous for nuclei if the singlet-triplet splitting is produced by a force of the form (1*b*). We should mention at this occasion that it is not sufficient for the validity of such a conservation law that the forces acting on the spin be small. It is necessary, in addition to this, that the separation of states with different spin should not be small also. For atomic spectra, this follows from the fact that the singlet-triplet separation is of electrostatic nature, while the spin forces are magnetic. For nuclei, if (1*b*) is correct, the singlet-triplet separation is due to the Heisenberg force, while the spin forces still are, under this assumption, of magnetic nature. Thus if (1*b*) is correct, the situation is much better for the spin law in nuclei than in the external shells. It is invalidated, however, by any force, coupling the spin to orbit, as e.g. that assumed by Inglis (cf. Section III, end).

Table IX. *Continuation of Table V. Be^8.*

	$({}^1S{}^3P)\,{}^{13}P$	$({}^1D{}^3P)\,{}^{13}P$
$({}^1S{}^3P)\,{}^{13}P$	$4(L+2K)/3$	$-(20)^{\frac{1}{2}}(L-K)/3$
$({}^1D{}^3P)\,{}^{13}P$	$-(20)^{\frac{1}{2}}(L-K)/3$	$-(4L-25K)/3$

$({}^1D{}^3P)\,{}^{13}D \quad 2L+3K, \qquad ({}^1D{}^3P)\,{}^{13}F \quad 2L$

The kinetic energy per particle is $(3/2)\alpha\sigma$ and $(5/2)\alpha\tau$ for the s and p particles, respectively. For a system with N particles the total kinetic energy is taken to be

$$(1-1/N)(6\sigma - 10\tau + (5/2)N\tau)\alpha. \tag{17}$$

The factor $(1-1/N)$ serves to eliminate the spurious kinetic energy of the center of gravity which arises from the use of single particle wave functions with coördinates measured from a fixed point.[28] For the Coulomb energy we obtain

$$CE(s) = (\alpha\sigma/8)^{\frac{1}{2}}, \quad \text{(within the } s \text{ shell)},$$
$$CE(sp) = (1/3)(\sigma\tau)^{\frac{1}{2}}(\sigma+\tau)^{-\frac{3}{2}}(3\tau + 2\sigma - 2\tau^2\sigma(\sigma+\tau)^{-2})\alpha^{\frac{1}{2}}, \tag{18}$$

(between the s shell and each proton in the p shell),

$$L_c = (49/120)(\alpha\tau/2)^{\frac{1}{2}}, \qquad K_c = (1/40)(\alpha\tau/2)^{\frac{1}{2}}. \tag{19}$$

If we use the relation $A_{\nu\nu} \sim (1-2g)A_{\nu\pi}$ the total potential energy in the s shell is

$$6(1-g)A_{\nu\pi}B(\sigma), \quad B(\sigma) = (\sigma/\sigma+1)^{\frac{3}{2}}, \tag{20}$$

for all three forms of the like particle interaction.

The total interaction energy between the s shell and a single particle in the p shell (including both like and unlike particle forces) has the form

$$\begin{aligned} &3(1-g)A_{\nu\pi}D + (1-2g)A_{\nu\pi}(D-C), && (1a+2a), \\ &3(1-g)A_{\nu\pi}D, && (1a+2b), \\ &3(1-g)A_{\nu\pi}D + (1+g)A_{\nu\pi}(D-C), && (1a+2c), \end{aligned} \tag{21}$$

with

$$\begin{aligned} D(\sigma,\tau) &= 16\sigma^{\frac{3}{2}}(\tau/(\sigma+\tau))^{5/2}(\sigma+\tau+2)^{-5/2}, \\ C(\sigma,\tau) &= (2\sigma)^{\frac{3}{2}}\tau^{5/2}(1+2\sigma)(2\sigma\tau+\sigma+\tau)^{-5/2}. \end{aligned} \tag{22}$$

Within the p shell

$$\begin{aligned} &L_{\nu\pi} = A_{\nu\pi}(1-g/2)L^0, \quad L_{\nu\nu} = A_{\nu\pi}(1-2g)L^0, \quad \text{(for } 2a \text{ and } 2b\text{)}, \\ &K_{\nu\pi} = A_{\nu\pi}(1-g/2)K^0, \quad K_{\nu\nu} = A_{\nu\pi}(1-2g)K^0, \\ &L^0 = (\tau/\tau+1)^{\frac{3}{2}}\{1-(4\tau+1)/4(\tau+1)^2\} \\ &K^0 = (\tau/\tau+1)^{\frac{3}{2}}/4(\tau+1)^2. \end{aligned} \tag{23}$$

In computing Eqs. (21) and (22) the Majorana interaction between unlike particles is given the depth $(1-g/2)A_{\nu\pi}$ in order to take into account part of the Heisenberg interaction (as discussed in Section III).

It appears to be sensible to try to compensate for the correlation energy by either multiplying the potential

[28] S. Flügge, reference 7.

energy by a suitable constant,[7, 8] or by allowing the particles to interact at a greater distance than they really do. With (2*b*) we found that multiplying the potentials by 1.36 yields a function for the sum of all energies which has a minimum value of $-246\ mc^2$ at $\sigma=0.80$ (taking $\tau=\sigma$ since very little improvement is obtained from varying both τ and σ independently). The second method (method (*i*)) is believed to be somewhat more reliable. With (2*b*) the radius of action must be increased by the factor 1.3 to make the minimum value of $E(O^{16})$ agree with experiment. The minimum is assumed at $\sigma=1.01$ (again taking $\sigma=\tau$ for the same reason as before). To obtain the numerical results discussed in the text we use the values

$$\begin{array}{ll} L_{\nu\pi}=15.74, & K_{\nu\pi}=1.41, \\ L_{\nu\nu}=10.05, & K_{\nu\nu}=0.90, \\ L_c=0.892, & K_c=0.055, \quad \text{Method } (i). \\ CE(s)=1.09, & CE(ps)=1.63, \\ \multicolumn{2}{l}{3(1-g)D(\sigma,\sigma)A_{\nu\pi}=14.95,} \\ \multicolumn{2}{l}{6(1-g)B(\sigma)A_{\nu\pi}=120.00.} \end{array} \tag{24}$$

In computing the kinetic and Coulomb energies α is replaced by $\alpha/(1.3)^2$.

The straight-forward application of this procedure to the models (2*a*) and (2*c*) is not very satisfactory, because an increase in the strength of the potentials or in the radius of action which serves to fit the experimental O^{16} energy will give far too much energy to the lighter elements.

Finally it was found most satisfactory to make the calculations without modifying either the range or depth of the potentials. With $\sigma=1.01/(1.3)^2$ the kinetic and Coulomb energies are exactly as in method (*i*). The other matrix elements have the values

$$\begin{array}{ll} L_{\nu\pi}=9.83, & K_{\nu\pi}=1.44, \\ L_{\nu\nu}=6.19, & K_{\nu\nu}=0.90, \quad \text{Method } (ii). \\ \multicolumn{2}{l}{3(1-g)D(\sigma,\sigma)A_{\nu\pi}=12.13,} \\ \multicolumn{2}{l}{6(1-g)B(\sigma)A_{\nu\pi}=77.37.} \end{array} \tag{25}$$

The correlation energy is introduced by adding to the total computed energy a linear function of the number of particles. Since the general linear function contains two parameters it is possible in this way to fit exactly the measured binding energies of two different nuclei, i.e., He^4 and O^{16}. Then the energies of all the others are uniquely determined.

25

ON THE IDENTIFICATION AND BEHAVIOR OF THE ALKALINE EARTH METALS RESULTING FROM THE IRRADIATION OF URANIUM WITH NEUTRONS

O. Hahn and F. Strassmann
(Kaiser Wilhelm Institut fur Chemie, Berlin, Dahlem)

This article was translated expressly for this Benchmark volume by R. Bruce Lindsay, Brown University, from Naturwissenschaften ***27:****11–15 (1939), by permission of the publisher, Springer-Verlag, Heidelberg*

In a previous communication,[1] which appeared a short time ago in this journal, it was reported that in the irradiation of uranium with neutrons, besides the transuranium elements 93 to 96 already described by Meitner, Hahn, and Strassmann, a lot of other transformation products result. These apparently owe their formation to a successive alpha-particle decay of the transiently produced uranium 239. Such a decay must lead to the production of an element with atomic number 88 (radium) from that with atomic number 92 (uranium). In the communication in question, we set forth in wholly tentative fashion decay schemes for three radium isotopes and gave their roughly estimated half-lives. We also described their transformation products: three isomeric actinium isotopes, which apparently decay into thorium isotopes.

At the same time, we pointed out the at first unexpected observation that the radium isotopes mentioned arise by bombardment with not only fast but also with slow neutrons. The conclusion that it was a question of radium isotopes in the case of the initial members of the three new isomeric series followed from the fact that these substances could be separated with barium salts and show all the characteristics of that element. All other known elements beginning with uranium and the transuranium elements—namely, thorium out to actinium—have different chemical properties from barium and can easily be separated from it. The same holds true for the elements below radium, such as bismuth, lead, polonium, and ekacesium.

Hence if one leaves barium out of consideration, there remains only radium.

In what follows, we shall briefly describe the separation of the isotope mixtures and the production of the individual members. From the activity development of the individual isotopes, we obtain their half-lives and the products resulting from their decay. In this communication, the latter will not be described in detail, since because of the very complicated phenomena—it is a question of at least three and probably four series for every three substances—the half-lives of all the successive products have not been determined to our complete satisfaction. Barium served naturally as the carrier substance for the "radium isotopes." The precipitation of barium as barium sulfate seemed to be the most appropriate; after the chromate, it is the least soluble barium salt. However, on the basis of previous experience and after some preliminary trials, the separation of the "radium isotopes" with barium sulfate was abandoned. In addition to small quantities of uranium, these precipitates carry away with them not inconsiderable quantities of actinium and thorium isotopes and accordingly also the supposed transformation products of the "radium isotopes" and hence permit no pure representation of the initial members. Hence in place of the sulfate precipitate, highly insoluble barium chloride in strong hydrochloric acid was used as the precipitation material, a method that worked best.

In the formation of radium isotopes by the bombardment of uranium with slow neutrons (a production not easy to understand from the standpoint of energy), a particularly thorough determination of the character of the new artificial radioactive elements was indispensable. In the separation of the individual artificial groups of elements from the solution of irradiated uranium, we always found, in addition to the large group of the transuranium elements, an activity in the alkaline earths (carrier substance barium), in the rare earths (carrier substance lanthanum), and in the elements of the fourth group of the periodic system (carrier substance zirconium). The barium precipitates were first exhaustively investigated, since they obviously contained the initial members of the observed isomeric series. It is to be shown that the transuranium elements, as well as uranium, protactinium, thorium, and actinium, can always be simply and completely separated from the activity emanating from the barium.

[*Editor's Note:* The above is followed by a detailed discussion of the chemistry of the separation of the various products resulting from the neutron bombardment of uranium together with results of measurements of their activity. The decay curves of the three so-called radium isotopes are presented for various irradiation times. It is not necessary to reproduce a translation of this part of the paper here, particularly since it is clear that during the progress of working up these results, the authors were growing increasingly suspicious that the isotopes in question were not, after all, those of radium but rather of barium. The latter part of the paper, reproduced next, is a discussion of the probability of this conclusion.]

We must now mention some more recent investigations, which, because of the strange results, we have been hesitant about publishing. In order to establish beyond the slightest doubt the proof of the chemical nature of the initial members of the series that have been separated out with barium and that we have designated in the foregoing as "radium isotopes," we have undertaken fractional crystallization and fractional precipitation with the active barium salts, a method well known for the enrichment (or the reverse) of radium in barium salts.

Barium bromide strongly enriches the radium by fractional crystallization; barium bromate with not too rapid development of the crystals works even better. Bromium chloride enriches less strongly than the bromide; barium carbonate goes somewhat in the opposite direction. Corresponding investigations, which we have made with our active barium preparations purified from consecutive products, have proved negative without exception. The activity remained equally distributed over all barium fractions within limits of experimental error. A couple of fractionalization experiments were then made with the radium isotope thorium X and the radium isotope mesothorium. They turned out precisely as we would have expected from all previous experience with radium. We then applied the indicator method to a mixture of purified Ra IV (with long half-life) with pure radium-free $MsTh_1$. The mixture with barium bromide as carrier substance was fractionally crystallized. The $MsTh_1$ was enriched, whereas the Ra IV was not; its activity remained the same for the same barium content of the fractions. We then come to the conclusion: Our "radium isotopes" have the properties of barium. As chemists we must truly say that as far as the new body is concerned, it is not radium but barium; no other elements besides radium and barium come into question.

Finally, we have run an indicator test with our pure precipitate Ac II (half-life around 2.5 hours) and the pure actinium isotope $MsTh_2$. If our "radium isotopes" are not radium, the "actinium isotopes" are not actinium but should be lanthanum. Following Mme. Curie,[2] we have undertaken a fractionalization of lanthanim oxalate, which both active substances contain, from a nitric acid solution. The $MsTh_2$ was strongly enriched in the final fraction, as Mme. Curie had found. In the case of "Ac II," there was no noticeable enrichment at the end. In agreement with I. Curie and P. Savitch,[3] in connection with their admittedly nonhomogeneous 3.5-hour substance, we found accordingly that the rare-earth metal arising from our active alkaline earth metal by emission of beta rays is not actinium. We expect to test more precisely the finding of Curie and Savitch that they enriched the activity in lanthanum, which accordingly speaks against an identity with lanthanum. It may even be possible that an enrichment in the mixture they were working with was simulated.

We have not yet tested whether the end products of our series resulting from the "Ac-La preparations" and designated as "Thorium" turn out to be cerium.

As far as the transuranium elements are concerned, they are indeed chemically related to their lower homologs rhenium, osmium, indium, and platinum but are not identical with them. Whether they are chemically equivalent with the still lower homologs masurium, ruthenium, rhodium, and palladium has not yet been tested. Until this has been done, one ought not to think of them. However, it is of interest that the sum of the mass numbers for barium and masurium—138 + 101—gives 239!

As chemists we must, on the basis of the investigations just summarized, redesignate the originally adopted scheme and in place of Ra, Ac, Th set Ba, La, and Ce. As nuclear chemists closely allied with physics, we cannot yet come to a firm decision on this conclusion, which contradicts all previous experience in nuclear physics. It can still be possible that we have been deceived by a series of strange accidents.

It is our intention to carry out further indicator investigations with the new transformation products, and in particular to carry out a common fractionation of the radium isotopes arising from thorium that has been bombarded with high-velocity neutrons, as studied by Meitner, Strassmann, and Hahn,[4] and compare them with our alkaline earth metals arising from uranium. In places at which strong artificial radioactive sources are available, this program could eventually be carried out more easily.

In conclusion we thank Frl. Cl. Lieber and Frl. I. Bohne for their efficient help in carrying out the countless precipitations and measurements.

NOTES

1. O. Hahn and F. Strassmann, *Naturwissenschaften, 26,* 756, 1938.
2. Mme. Pierre Curie, *J. Chim. Physique, 27,* 1, 1930.
3. I. Curie and P. Savitch, *Comptes Rendus, 206,* 1643, 1938.
4. L. Meitner, F. Strassmann, and O. Hahn, Z. Physik, **109,** 538, 1938.

26

Reprinted from *Phys. Rev.* **56**:426–450 (1939)

The Mechanism of Nuclear Fission

NIELS BOHR
University of Copenhagen, Copenhagen, Denmark, and The Institute for Advanced Study, Princeton, New Jersey

AND

JOHN ARCHIBALD WHEELER
Princeton University, Princeton, New Jersey

On the basis of the liquid drop model of atomic nuclei, an account is given of the mechanism of nuclear fission. In particular, conclusions are drawn regarding the variation from nucleus to nucleus of the critical energy required for fission, and regarding the dependence of fission cross section for a given nucleus on energy of the exciting agency. A detailed discussion of the observations is presented on the basis of the theoretical considerations. Theory and experiment fit together in a reasonable way to give a satisfactory picture of nuclear fission.

INTRODUCTION

THE discovery by Fermi and his collaborators that neutrons can be captured by heavy nuclei to form new radioactive isotopes led especially in the case of uranium to the interesting finding of nuclei of higher mass and charge number than hitherto known. The pursuit of these investigations, particularly through the work of Meitner, Hahn, and Strassmann as well as Curie and Savitch, brought to light a number of unsuspected and startling results and finally led Hahn and Strassmann[1] to the discovery that from uranium elements of much smaller atomic weight and charge are also formed.

The new type of nuclear reaction thus discovered was given the name "fission" by Meitner and Frisch,[2] who on the basis of the liquid drop model of nuclei emphasized the analogy of the process concerned with the division of a fluid sphere into two smaller droplets as the result of a deformation caused by an external disturbance. In this connection they also drew attention to the fact that just for the heaviest nuclei the mutual repulsion of the electrical charges will to a large extent annul the effect of the short range nuclear forces, analogous to that of surface tension, in opposing a change of shape of the nucleus. To produce a critical deformation will therefore require only a comparatively small energy, and by the subsequent division of the nucleus a very large amount of energy will be set free.

Just the enormous energy release in the fission process has, as is well known, made it possible to observe these processes directly, partly by the great ionizing power of the nuclear fragments, first observed by Frisch[3] and shortly afterwards independently by a number of others, partly by the penetrating power of these fragments which allows in the most efficient way the separation from the uranium of the new nuclei formed by the fission.[4] These products are above all characterized by their specific beta-ray activities which allow their chemical and spectrographic identification. In addition, however, it has been found that the fission process is accompanied by an emission of neutrons, some of which seem to be directly associated with the fission, others associated with the subsequent beta-ray transformations of the nuclear fragments.

In accordance with the general picture of nuclear reactions developed in the course of the last few years, we must assume that any nuclear transformation initiated by collisions or irradiation takes place in two steps, of which the first is the formation of a highly excited compound nucleus with a comparatively long lifetime, while

---

[1] O. Hahn and F. Strassmann, Naturwiss. **27**, 11 (1939); see, also, P. Abelson, Phys. Rev. **55**, 418 (1939).

[2] L. Meitner and O. R. Frisch, Nature **143**, 239 (1939).

[3] O. R. Frisch, Nature **143**, 276 (1939); G. K. Green and Luis W. Alvarez, Phys. Rev. **55**, 417 (1939); R. D. Fowler and R. W. Dodson, Phys. Rev. **55**, 418 (1939); R. B. Roberts, R. C. Meyer and L. R. Hafstad, Phys. Rev. **55**, 417 (1939); W. Jentschke and F. Prankl, Naturwiss. **27**, 134 (1939); H. L. Anderson, E. T. Booth, J. R. Dunning, E. Fermi, G. N. Glasoe and F. G. Slack, Phys. Rev. **55**, 511 (1939).

[4] F. Joliot, Comptes rendus **208**, 341 (1939); L. Meitner and O. R. Frisch, Nature **143**, 471 (1939); H. L. Anderson, E. T. Booth, J. R. Dunning, E. Fermi, G. N. Glasoe and F. G. Slack, Phys. Rev. **55**, 511 (1939).

the second consists in the disintegration of this compound nucleus or its transition to a less excited state by the emission of radiation. For a heavy nucleus the disintegrative processes of the compound system which compete with the emission of radiation are the escape of a neutron and, according to the new discovery, the fission of the nucleus. While the first process demands the concentration on one particle at the nuclear surface of a large part of the excitation energy of the compound system which was initially distributed much as is thermal energy in a body of many degrees of freedom, the second process requires the transformation of a part of this energy into potential energy of a deformation of the nucleus sufficient to lead to division.[5]

Such a competition between the fission process and the neutron escape and capture processes seems in fact to be exhibited in a striking manner by the way in which the cross section for fission of thorium and uranium varies with the energy of the impinging neutrons. The remarkable difference observed by Meitner, Hahn, and Strassmann between the effects in these two elements seems also readily explained on such lines by the presence in uranium of several stable isotopes, a considerable part of the fission phenomena being reasonably attributable to the rare isotope U^{235} which, for a given neutron energy, will lead to a compound nucleus of higher excitation energy and smaller stability than that formed from the abundant uranium isotope.[6]

In the present article there is developed a more detailed treatment of the mechanism of the fission process and accompanying effects, based on the comparison between the nucleus and a liquid drop. The critical deformation energy is brought into connection with the potential energy of the drop in a state of unstable equilibrium, and is estimated in its dependence on nuclear charge and mass. Exactly how the excitation energy originally given to the nucleus is gradually exchanged among the various degrees of freedom and leads eventually to a critical deformation proves to be a question which needs not be discussed in order to determine the fission probability. In fact, simple statistical considerations lead to an approximate expression for the fission reaction rate which depends only on the critical energy of deformation and the properties of nuclear energy level distributions. The general theory presented appears to fit together well with the observations and to give a satisfactory description of the fission phenomenon.

For a first orientation as well as for the later considerations, we estimate quantitatively in Section I by means of the available evidence the energy which can be released by the division of a heavy nucleus in various ways, and in particular examine not only the energy released in the fission process itself, but also the energy required for subsequent neutron escape from the fragments and the energy available for beta-ray emission from these fragments.

In Section II the problem of the nuclear deformation is studied more closely from the point of view of the comparison between the nucleus and a liquid droplet in order to make an estimate of the energy required for different nuclei to realize the critical deformation necessary for fission.

In Section III the statistical mechanics of the fission process is considered in more detail, and an approximate estimate made of the fission probability. This is compared with the probability of radiation and of neutron escape. A discussion is then given on the basis of the theory for the variation with energy of the fission cross section.

In Section IV the preceding considerations are applied to an analysis of the observations of the cross sections for the fission of uranium and thorium by neutrons of various velocities. In particular it is shown how the comparison with the theory developed in Section III leads to values for the critical energies of fission for thorium and the various isotopes of uranium which are in good accord with the considerations of Section II.

In Section V the problem of the statistical distribution in size of the nuclear fragments arising from fission is considered, and also the questions of the excitation of these fragments and the origin of the secondary neutrons.

Finally, we consider in Section VI the fission effects to be expected for other elements than thorium and uranium at sufficiently high neutron velocities as well as the effect to be anticipated in

[5] N. Bohr, Nature **143**, 330 (1939).
[6] N. Bohr, Phys. Rev. **55**, 418 (1939).

thorium and uranium under deuteron and proton impact and radiative excitation.

I. Energy Released by Nuclear Division

The total energy released by the division of a nucleus into smaller parts is given by

$$\Delta E=(M_0-\Sigma M_i)c^2, \qquad (1)$$

where M_0 and M_i are the masses of the original and product nuclei at rest and unexcited. We have available no observations on the masses of nuclei with the abnormal charge to mass ratio formed for example by the division of such a heavy nucleus as uranium into two nearly equal parts. The difference between the mass of such a fragment and the corresponding stable nucleus of the same mass number may, however, if we look apart for the moment from fluctuations in energy due to odd-even alternations and the finer details of nuclear binding, be reasonably assumed, according to an argument of Gamow, to be representable in the form

$$M(Z,A)-M(Z_A,A)=\tfrac{1}{2}B_A(Z-Z_A)^2, \qquad (2)$$

where Z is the charge number of the fragment and Z_A is a quantity which in general will not be an integer. For the mass numbers $A=100$ to 140 this quantity Z_A is given by the dotted line in Fig. 8, and in a similar way it may be determined for lighter and heavier mass numbers.

B_A is a quantity which cannot as yet be determined directly from experiment but may be estimated in the following manner. Thus we may assume that the energies of nuclei with a given mass A will vary with the charge Z approximately according to the formula

$$M(Z,A)=C_A+\tfrac{1}{2}B_A'(Z-\tfrac{1}{2}A)^2 +(Z-\tfrac{1}{2}A)(M_p-M_n)+3Z^2e^2/5r_0A^{\frac{1}{3}}. \qquad (3)$$

Here the second term gives the comparative masses of the various isobars neglecting the influence of the difference M_p-M_n of the proton and neutron mass included in the third term and of the pure electrostatic energy given by the fourth term. In the latter term the usual assumption is made that the effective radius of the nucleus is equal to $r_0A^{\frac{1}{3}}$, with r_0 estimated as 1.48×10^{-13} from the theory of alpha-ray disintegration. Identifying the relative mass values given by expressions (2) and (3), we find

$$B_A'=(M_p-M_n+6Z_Ae^2/5r_0A^{\frac{1}{3}})/(\tfrac{1}{2}A-Z_A) \qquad (4)$$

and

$$B_A=B_A'+6e^2/5r_0A^{\frac{1}{3}} =(M_p-M_n+3A^{\frac{2}{3}}e^2/5r_0)/(\tfrac{1}{2}A-Z_A). \qquad (5)$$

The values of B_A obtained for various nuclei from this last relation are listed in Table I.

On the basis just discussed, we shall be able to estimate the mass of the nucleus (Z, A) with the help of the packing fraction of the known nuclei. Thus we may write

$$M(Z,A)=A(1+f_A) + \tfrac{1}{2}B_A(Z-Z_A)^2 \begin{cases} +0 & A \text{ odd} \\ -\tfrac{1}{2}\delta_A & A \text{ even, } Z \text{ even} \\ +\tfrac{1}{2}\delta_A & A \text{ even, } Z \text{ odd} \end{cases}, \qquad (6)$$

where f_A is to be taken as the average value of the packing fraction over a small region of atomic weights and the last term allows for the typical differences in binding energy among nuclei according to the odd and even character of their neutron and proton numbers. In using Dempster's measurements of packing fractions we must recognize that the average value of the second term in (6) is included in such measurements.[7] This correction, however, is, as may be read from Fig. 8, practically compensated by the influence of the third term, owing to the fact that the great majority of nuclei studied in the mass spectrograph are of even-even character.

From (6) we find the energy release involved in electron emission or absorption by a nucleus unstable with respect to a beta-ray

Table I. *Values of the quantities which appear in Eqs.* (6) *and* (7), *estimated for various values of the nuclear mass number A. Both* B_A and δ_A *are in Mev.*

A	Z_A	B_A	δ_A	A	Z_A	B_A	δ_A
50	23.0	$3._5$	$2._8$	150	62.5	1.2	$1._5$
60	27.5	$3._3$	$2._8$	160	65.4	1.1	$1._3$
70	31.2	$2._5$	$2._7$	170	69.1	1.1	$1._2$
80	35.0	$2._2$	$2._7$	180	72.9	1.0	$1._2$
90	39.4	$2._0$	$2._7$	190	76.4	1.0	$1._1$
100	44.0	$2._0$	$2._6$	200	80.0	0.9_5	$1._1$
110	47.7	1.7	$2._4$	210	83.5	0.9_2	$1._1$
120	50.8	1.5	$2._1$	220	87.0	0.8_8	$1._1$
130	53.9	1.3	$1._9$	230	$90._6$	0.8_6	$1._0$
140	58.0	1.2	$1._8$	240	$93._9$	0.8_3	$1._0$

[7] A. J. Dempster, Phys. Rev. **53**, 869 (1938).

transformation:

$$E_\beta = B_A\{|Z_A - Z| - \tfrac{1}{2}\} \begin{matrix} +0 \\ -\delta_A \\ +\delta_A \end{matrix} \left\{ \begin{matrix} A \text{ odd} \\ A \text{ even}, Z \text{ even} \\ A \text{ even}, Z \text{ odd} \end{matrix} \right\}. \quad (7)$$

This result gives us the possibility of estimating δ_A by an examination of the stability of isobars of even nuclei. In fact, if an even-even nucleus is stable or unstable, then δ_A is, respectively, greater or less than $B_A\{|Z_A - A| - \tfrac{1}{2}\}$. For nuclei of medium atomic weight this condition brackets δ_A very closely; for the region of very high mass numbers, on the other hand, we can estimate δ_A directly from the difference in energy release of the successive beta-ray transformations

$$UX_I \rightarrow (UX_{II}, UZ) \rightarrow U_{II},$$
$$MsTh_I \rightarrow MsTh_{II} \rightarrow RaTh, \ RaD \rightarrow RaE \rightarrow RaF.$$

The estimated values of δ_A are collected in Table I.

Applying the available measurements on nuclear masses supplemented by the above considerations, we obtain typical estimates as shown in Table II for the energy release on division of a nucleus into two approximately equal parts.[8]

Below mass number $A \sim 100$ nuclei are energetically stable with respect to division; above this limit energetic instability sets in with respect

TABLE II. *Estimates for the energy release on division of typical nuclei into two fragments are given in the third column. In the fourth is the estimated value of the total additional energy release associated with the subsequent beta-ray transformations. Energies are in Mev.*

ORIGINAL	TWO PRODUCTS	DIVISION	SUBSEQUENT
$_{28}Ni^{61}$	$_{14}Si^{30, \, 31}$	-11	2
$_{50}Sn^{117}$	$_{25}Mn^{58, \, 59}$	10	12
$_{68}Er^{167}$	$_{34}Se^{83, \, 84}$	94	13
$_{82}Pb^{206}$	$_{41}Nb^{103, \, 103}$	120	32
$_{92}U^{239}$	$_{46}Pd^{119, \, 120}$	200	31

to division into two nearly equal fragments, essentially because the decrease in electrostatic

[8] Even if there is no question of actual fission processes by which nuclei break up into more than two comparable parts, it may be of interest to point out that such divisions in many cases would be accompanied by the release of energy. Thus nuclei of mass number greater than $A = 110$ are unstable with respect to division into three nearly equal parts. For uranium the corresponding total energy liberation will be ~ 210 Mev, and thus is even somewhat greater than the release on division into two parts. The energy evolution on division of U^{239} into four comparable parts will, however, be about 150 Mev, and already division into as many as 15 comparable parts will be endothermic.

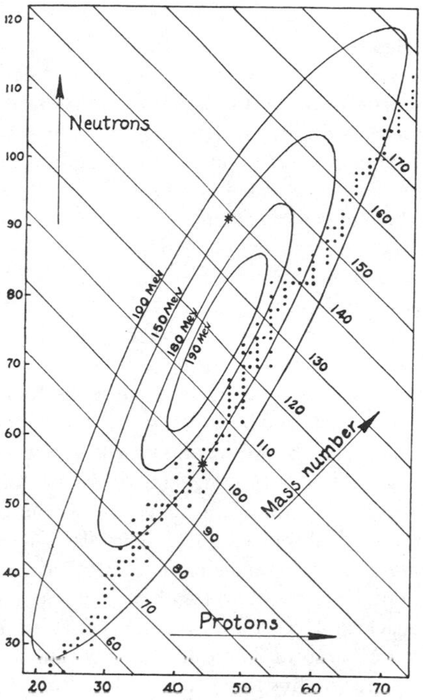

FIG. 1. The difference in energy between the nucleus $_{92}U^{239}$ in its normal state and the possible fragment nuclei $_{44}Ru^{100}$ and $_{48}Cd^{139}$ (indicated by the crosses in the figure) is estimated to be 150 Mev as shown by the corresponding contour line. In a similar way the estimated energy release for division of U^{239} into other possible fragments can be read from the figure. The region in the chart associated with the greatest energy release is seen to be at a distance from the region of the stable nuclei (dots in the figure) corresponding to the emission of from three to five beta-rays.

energy associated with the separation overcompensates the desaturation of short range forces consequent on the greater exposed nuclear surface. The energy evolved on division of the nucleus U^{239} into two fragments of any given charge and mass numbers is shown in Fig. 1. It is seen that there is a large range of atomic masses for which the energy liberated reaches nearly the maximum attainable value 200 Mev; but that for a given size of one fragment there is only a small range of charge numbers which correspond to an energy release at all near the maximum value. Thus the fragments formed by division of uranium in the *energetically* most favorable way lie in a narrow band in Fig. 1, separated from the region of the stable nuclei by an amount which corresponds to the change in nuclear charge

associated with the emission of three to six beta-particles.

The amount of energy released in the beta-ray transformations following the creation of the fragment nuclei may be estimated from Eq. (7), using the constants in Table I. Approximate values obtained in this way for the energy liberation in typical chains of beta-disintegrations are shown on the arrows in Fig. 8.

The magnitude of the energy available for beta-ray emission from typical fragment nuclei does not stand in conflict with the stability of these nuclei with respect to spontaneous neutron emission, as one sees at once from the fact that the energy change associated with an increase of the nuclear charge by one unit is given by the *difference* between binding energy of a proton and of a neutron, plus the neutron-proton mass difference. A direct estimate from Eq. (6) of the binding energy of a neutron in typical nuclear fragments lying in the band of greatest energy release (Fig. 1) gives the results summarized in the last column of Table III. The comparison of the figures in this table shows that the neutron binding is in certain cases considerably smaller than the energy which can be released by beta-ray transformation. This fact offers a reasonable explanation as we shall see in Section V for the delayed neutron emission accompanying the fission process.

II. Nuclear Stability with Respect to Deformations

According to the liquid drop model of atomic nuclei, the excitation energy of a nucleus must be

TABLE III. *Estimated values of energy release in beta-ray transformations and energy of neutron binding in final nucleus, in typical cases; also estimates of the neutron binding in the dividing nucleus. Values in Mev.*

Beta-Transition		Release	Binding
$_{40}Zr^{99}$	$_{41}Nb^{99}$	6.3	8.2
$_{41}Nb^{100}$	$_{42}Mo^{100}$	7.8	8.6
$_{46}Pd^{125}$	$_{47}Ag^{125}$	7.8	6.7
$_{47}Ag^{125}$	$_{48}Cd^{125}$	6.5	5.0
$_{49}In^{130}$	$_{50}Sn^{130}$	7.6	7.1
$_{52}Te^{140}$	$_{53}I^{140}$	5.0	3.5
$_{53}I^{140}$	$_{54}Xe^{140}$	7.4	5.9
Compound Nucleus			
	$_{92}U^{235}$		5.4
	$_{92}U^{236}$		6.4
	$_{92}U^{239}$		5.2
	$_{90}Th^{233}$		5.2
	$_{91}Pa^{232}$		6.4

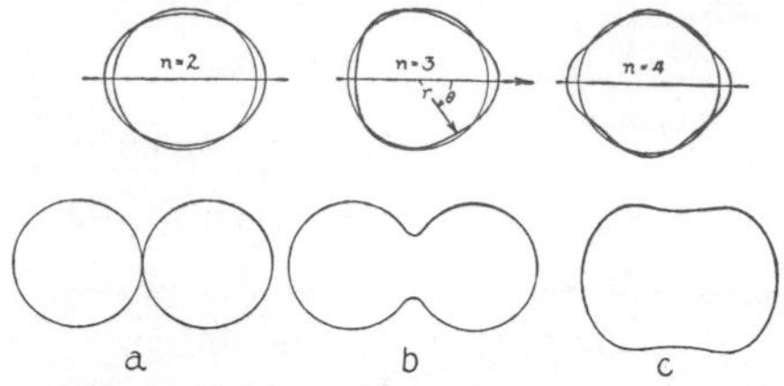

FIG. 2. Small deformations of a liquid drop of the type $\delta r(\theta) = \alpha_n P_n(\cos\theta)$ (upper portion of the figure) lead to characteristic oscillations of the fluid about the spherical form of stable equilibrium, even when the fluid has a uniform electrical charge. If the charge reaches the critical value $(10 \times \text{surface tension} \times \text{volume})^{\frac{1}{2}}$, however, the spherical form becomes unstable with respect to even infinitesimal deformations of the type $n=2$. For a slightly smaller charge, on the other hand, a finite deformation (c) will be required to lead to a configuration of *unstable equilibrium*, and with smaller and smaller charge densities the critical form gradually goes over (c, b, a) into that of two uncharged spheres an infinitesimal distance from each other (a).

expected to give rise to modes of motion of the nuclear matter similar to the oscillations of a fluid sphere under the influence of surface tension.[9] For heavy nuclei the high nuclear charge will, however, give rise to an effect which will to a large extent counteract the restoring force due to the short range attractions responsible for the surface tension of nuclear matter. This effect, the importance of which for the fission phenomenon was stressed by Frisch and Meitner, will be more closely considered in this section, where we shall investigate the stability of a nucleus for small deformations of various types[10] as well as for such large deformations that division may actually be expected to occur.

Consider a small arbitrary deformation of the liquid drop with which we compare the nucleus such that the distance from the center to an arbitrary point on the surface with colatitude θ is changed (see Fig. 2) from its original value R

[9] N. Bohr, Nature **137**, 344 and 351 (1936); N. Bohr and F. Kalckar, Kgl. Danske Vid. Selskab., Math. Phys. Medd. 14, No. 10 (1937).

[10] After the formulae given below were derived, expressions for the potential energy associated with spheroidal deformations of nuclei were published by E. Feenberg (Phys. Rev. **55**, 504 (1939)) and F. Weizsäcker (Naturwiss. **27**, 133 (1939)). Further, Professor Frenkel in Leningrad has kindly sent us in manuscript a copy of a more comprehensive paper on various aspects of the fission problem, to appear in the U.S.S.R. "Annales Physicae," which contains a deduction of Eq. (9) below for nuclear stability against arbitrary small deformations, as well as some remarks, similar to those made below (Eq. (14)) about the shape of a drop corresponding to unstable equilibrium. A short abstract of this paper has since appeared in Phys. Rev. **55**, 987 (1939).

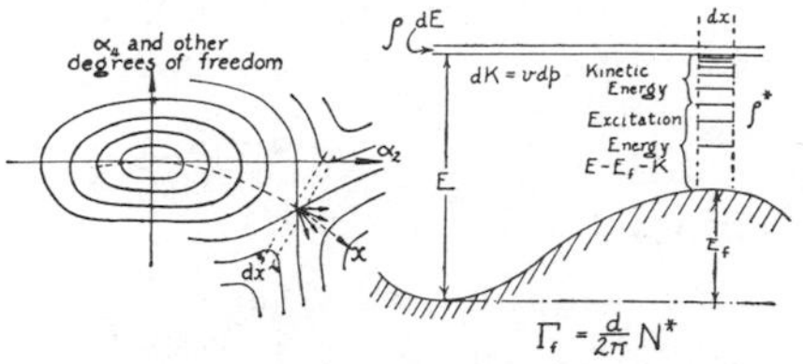

FIG. 3. The potential energy associated with any arbitrary deformation of the nuclear form may be plotted as a function of the parameters which specify the deformation, thus giving a contour surface which is represented schematically in the left-hand portion of the figure. The pass or saddle point corresponds to the critical deformation of unstable equilibrium. To the extent to which we may use classical terms, the course of the fission process may be symbolized by a ball lying in the hollow at the origin of coordinates (spherical form) which receives an impulse (neutron capture) which sets it to executing a complicated Lissajous figure of oscillation about equilibrium. If its energy is sufficient, it will in the course of time happen to move in the proper direction to pass over the saddle point (after which fission will occur), unless it loses its energy (radiation or neutron re-emission). At the right is a cross section taken through the fission barrier, illustrating the calculation in the text of the probability per unit time of fission occurring.

to the value

$$r(\theta)=R[1+\alpha_0+\alpha_2P_2(\cos\theta)+\alpha_3P_3(\cos\theta)+\cdots], \quad (8)$$

where the α_n are small quantities. Then a straightforward calculation shows that the surface energy plus the electrostatic energy of the comparison drop has increased to the value

$$\begin{aligned}E_{S+E}=&4\pi(r_0A^{\frac{1}{3}})^2O[1+2\alpha_2^2/5+5\alpha_3^2/7+\cdots\\&+(n-1)(n+2)\alpha_n^2/2(2n+1)+\cdots]\\&+3(Ze)^2/5r_0A^{\frac{1}{3}}[1-\alpha_2^2/5-10\alpha_3^2/49-\cdots\\&-5(n-1)\alpha_n^2/(2n+1)^2-\cdots], \quad (9)\end{aligned}$$

where we have assumed that the drop is composed of an incompressible fluid of volume $(4\pi/3)R^3=(4\pi/3)r_0^3A$, uniformly electrified to a charge Ze, and possessing a surface tension O. Examination of the coefficient of α_2^2 in the above expression for the distortion energy, namely,

$$4\pi r_0^2OA^{\frac{2}{3}}(2/5)\{1-(Z^2/A)\times[e^2/10(4\pi/3)r_0^3O]\} \quad (10)$$

makes it clear that with increasing value of the ratio Z^2/A we come finally to a limiting value

$$(Z^2/A)_{\text{limiting}}=10(4\pi/3)r_0^3O/e^2, \quad (11)$$

beyond which the nucleus is no longer stable with respect to deformations of the simplest type. The actual value of the numerical factors can be calculated with the help of the semi-empirical formula given by Bethe for the respective contributions to nuclear binding energies due to electrostatic and long range forces, the influence of the latter being divided into volume and surface effects. A revision of the constants in Bethe's formula has been carried through by Feenberg[11] in such a way as to obtain the best agreement with the mass defects of Dempster; he finds

$$r_0\doteqdot 1.4\times10^{-13}\text{ cm}, \quad 4\pi r_0^2O\doteqdot 14\text{ Mev}. \quad (12)$$

From these values a limit for the ratio Z^2/A is obtained which is 17 percent greater than the ratio $(92)^2/238$ characterizing U^{238}. Thus we can conclude that nuclei such as those of uranium and thorium are indeed near the limit of stability set by the exact compensation of the effects of electrostatic and short range forces. On the other hand, we cannot rely on the precise value of the limit given by these semi-empirical and indirect determinations of the ratio of surface energy to electrostatic energy, and we shall investigate below a method of obtaining the ratio in question from a study of the fission phenomenon itself.

Although nuclei for which the quantity Z^2/A is slightly less than the limiting value (11) are stable with respect to small arbitrary deformations, a larger deformation will give the long range repulsions more advantage over the short range attractions responsible for the surface tension, and it will therefore be possible for the nucleus, when suitably deformed, to divide spontaneously. Particularly important will be that critical deformation for which the nucleus is just on the verge of division. The drop will then possess a shape corresponding to unstable equilibrium: the work required to produce any infinitesimal displacement from this equilibrium configuration vanishes in the first order. To examine this point in more detail, let us consider the surface obtained by plotting the potential energy of an arbitrary distortion as a function of the parameters which specify its form and magnitude. Then we have to recognize the fact that the

[11] E. Feenberg, Phys. Rev. 55, 504 (1939).

potential barrier hindering division is to be compared with a pass or saddle point leading between two potential valleys on this surface. The energy relations are shown schematically in Fig. 3, where of course we are able to represent only two of the great number of parameters which are required to describe the shape of the system. The deformation parameters corresponding to the saddle point give us the critical form of the drop, and the potential energy required for this distortion we will term the critical energy for fission, E_f. If we consider a continuous change in the shape of the drop, leading from the original sphere to two spheres of half the size at infinite separation, then the critical energy in which we are interested is the lowest value which we can at all obtain, by suitable choice of this sequence of shapes, for the energy required to lead from the one configuration to the other.

Simple dimensional arguments show that the critical deformation energy for the droplet corresponding to a nucleus of given charge and mass number can be written as the product of the surface energy by a dimensionless function of the charge mass ratio:

$$E_f=4\pi r_0^2 O A^{2/3} f\{(Z^2/A)/(Z^2/A)_{\text{limiting}}\}. \quad (13)$$

We can determine E_f if we know the shape of the nucleus in the critical state; this will be given by solution of the well-known equation for the form of a surface in equilibrium under the action of a surface tension O and volume forces described by a potential φ:

$$\kappa O+\varphi=\text{constant}, \quad (14)$$

where κ is the total normal curvature of the surface. Because of the great mathematical difficulties of treating large deformations, we are however able to calculate the critical surface and the dimensionless function f in (13) only for certain special values of the argument, as follows: (1) if the volume potential in (14) vanishes altogether, we see from (14) that the surface of unstable equilibrium has constant curvature; we have in fact to deal with a division of the fluid into spheres. Thus, when there are no electrostatic forces at all to aid the fission, the critical energy for division into two equal fragments will just equal the total work done against surface tension in the separation process, i.e.,

$$E_f=2\cdot 4\pi r_0^2 O(A/2)^{2/3}-4\pi r_0^2 O A^{2/3}. \quad (15)$$

From this it follows that

$$f(0)=2^{1/3}-1=0.260. \quad (16)$$

(2) If the charge on the droplet is not zero, but is still very small, the critical shape will differ little from that of two spheres in contact. There will in fact exist only a narrow neck of fluid connecting the two portions of the figure, the radius of which, r_n, will be such as to bring about equilibrium; to a first approximation

$$2\pi r_n O=(Ze/2)^2/(2r_0(A/2)^{1/3})^2 \quad (17)$$

or

$$r_n/r_0 A^{1/3}=0.66\left(\frac{Z^2}{A}\right)\Big/\left(\frac{Z^2}{A}\right)_{\text{limiting}}. \quad (18)$$

To calculate the critical energy to the first order in Z^2/A, we can omit the influence of the neck as producing only a second-order change in the energy. Thus we need only compare the sum of surface and electrostatic energy for the original nucleus with the corresponding energy for two spherical nuclei of half the size in contact with each other. We find

$$\begin{aligned}E_f=&\,2\cdot 4\pi r_0^2 O(A/2)^{2/3}-4\pi r_0^2 O A^{2/3}\\&+2\cdot 3(Ze/2)^2/5r_0(A/2)^{1/3}\\&+(Ze/2)^2/2r_0(A/2)^{1/3}-3(Ze)^2/5r_0A^{1/3}, \quad (19)\end{aligned}$$

from which

$$E_f/4\pi r_0^2 O A^{2/3}\equiv f(x)=0.260-0.215x, \quad (20)$$

provided

$$x=\left(\frac{Z^2}{A}\right)\Big/\left(\frac{Z^2}{A}\right)_{\text{limiting}}=(\text{charge})^2/\text{surface tension}\times\text{volume}\times 10 \quad (21)$$

is a small quantity. (3) In the case of greatest actual interest, when Z^2/A is very close to the critical value, only a small deformation from a spherical form will be required to reach the critical state. According to Eq. (9), the potential energy required for an infinitesimal distortion will increase as the square of the amplitude, and

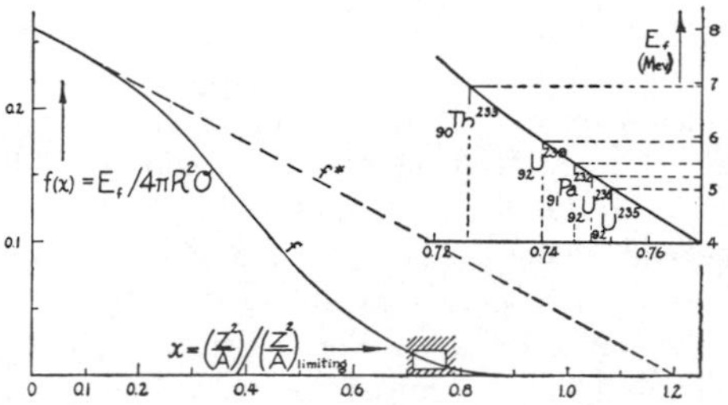

FIG. 4. The energy E_f required to produce a critical deformation leading to fission is divided by the surface energy $4\pi R^2O$ to obtain a dimensionless function of the quantity $x=(\text{charge})^2/(10\times\text{volume}\times\text{surface tension})$. The behavior of the function $f(x)$ is calculated in the text for $x=0$ and $x=1$, and a smooth curve is drawn here to connect these values. The curve $f^*(x)$ determines for comparison the energy required to deform the nucleus into two spheres in contact with each other. Over the cross-hatched region of the curve of interest for the heaviest nuclei the surface energy changes but little. Taking for it a value of 530 Mev, we obtain the energy scale in the upper part of the figure. In Section IV we estimate from the observations a value $E_f\sim 6$ Mev for U^{239}. Using the figure we thus find $(Z^2/A)_{\text{limiting}}=47.8$ and can estimate the fission barriers for other nuclei, as shown.

will moreover have the smallest possible value for a displacement of the form $P_2(\cos\theta)$. To find the deformation for which the potential energy has reached a maximum and is about to decrease, we have to carry out a more accurate calculation. We obtain for the distortion energy, accurate to the fourth order in α_2, the expression

$$\begin{aligned}\Delta E_{S+E}=&4\pi r_0^2OA^{2/3}[2\alpha_2^2/5+116\alpha_2^3/105\\&+101\alpha_2^4/35+2\alpha_2^2\alpha_4/35+\alpha_4^2]\\&-3(Ze)^2/5r_0A^{1/3}[\alpha_2^2/5+64\alpha_2^3/105\\&+58\alpha_2^4/35+8\alpha_2^2\alpha_4/35+5\alpha_4^2/27],\end{aligned} \quad (22)$$

in which it will be noted that we have had to include the terms in α_4^2 because of the coupling which sets in between the second and fourth modes of motion for appreciable amplitudes. Thus, on minimizing the potential energy with respect to α_4, we find

$$\alpha_4=-(243/595)\alpha_2^2 \quad (23)$$

in accordance with the fact that as the critical form becomes more elongated with decreasing Z^2/A, it must also develop a concavity about its equatorial belt such as to lead continuously with variation of the nuclear charge to the dumbbell shaped figure discussed in the preceding paragraph.

With the help of (23) we obtain the deformation energy as a function of α_2 alone. By a straightforward calculation we then find its maximum value as a function of α_2, thus determining the energy required to produce a distortion on the verge of leading to fission:

$$\begin{aligned}E_f/4\pi r_0^2OA^{2/3}=f(x)=&98(1-x)^3/135\\&-11368(1-x)^4/34425+\cdots\end{aligned} \quad (24)$$

for values of Z^2/A near the instability limit.

Interpolating in a reasonable way between the two limiting values which we have obtained for the critical energy for fission, we obtain the curve of Fig. 4 for f as a function of the ratio of the square of the charge number of the nucleus to its mass number. The upper part of the figure shows the interesting portion of the curve in enlargement and with a scale of energy values at the right based on the surface tension estimate of Eq. (12) and a nuclear mass of $A=235$. The slight variation of the factor $4\pi r_0^2OA^{2/3}$ among the various thorium and uranium isotopes may be neglected in comparison with the changes of the factor $f(x)$.

In Section IV we estimate from the observations that the critical fission energy for U^{239} is not far from 6 Mev. According to Fig. 4, this corresponds to a value of $x=0.74$, from which we conclude that $(Z^2/A)_{\text{limiting}}=(92)^2/239\times 0.74=47.8$. This result enables us to estimate the critical energies for other isotopes, as indicated in the figure. It is seen that protactinium would be particularly interesting as a subject for fission experiments.

As a by product, we are also able from Eq. (12) to compute the nuclear radius in terms of the surface energy of the nucleus; assuming Feenberg's value of 14 Mev for $4\pi r_0^2O$, we obtain $r_0=1.47\times 10^{-13}$ cm, which gives a satisfactory and quite independent check on Feenberg's determination of the nuclear radius from the packing fraction curve.

So far the considerations are purely classical, and any actual state of motion must of course be described in terms of quantum-mechanical concepts. The possibility of applying classical pictures to a certain extent will depend on the smallness of the ratio between the zero point amplitudes for oscillations of the type discussed above and the nuclear radius. A simple calculation gives for the square of the ratio in question the result

$$\langle\alpha_n^2\rangle_{\text{Av; zero point}}=A^{-7/6}\times\{(\hbar^2/12M_pr_0^2)/4\pi r_0^2O\}^{\frac{1}{2}}n^{\frac{1}{2}}(2n+1)^{\frac{3}{2}}\times\{(n-1)(n+2)(2n+1)-20(n-1)x\}^{-\frac{1}{2}}. \quad (25)$$

Since $\{(\hbar^2/12M_pr_0^2)/4\pi r_0^2O\}^{\frac{1}{2}}\doteqdot\frac{1}{3}$, this ratio is indeed a small quantity, and it follows that deformations of magnitudes comparable with nuclear dimensions can be described approximately classically by suitable wave packets built up from quantum states. In particular we may describe the critical deformations which lead to fission in an approximately classical way. This follows from a comparison of the critical energy $E_f\sim 6$ Mev required, as we shall see in Section IV, to account for the observations on uranium, with the zero point energy

$$\tfrac{1}{2}\hbar\omega_2=A^{-\frac{1}{2}}\{4\pi r_0^2O\cdot 2(1-x)\hbar^2/3M_pr_0^2\}^{\frac{1}{2}}\sim 0.4\text{ Mev} \quad (26)$$

of the simplest mode of capillary oscillation, from which it is apparent that the amplitude in question is considerably larger than the zero point disturbance:

$$\langle\alpha_2^2\rangle_{\text{Av}}/\langle\alpha_2^2\rangle_{\text{Av; zero point}}\approx E_f/\tfrac{1}{2}\hbar\omega_2\sim 15. \quad (27)$$

The drop with which we compare the nucleus will also in the critical state be capable of executing small oscillations about the shape of unstable equilibrium. If we study the distribution in frequency of these characteristic oscillations, we must expect for high frequencies to find a spectrum qualitatively not very different from that of the normal modes of oscillation about the form of stable equilibrium. The oscillations in question will be represented symbolically in Fig. 3 by motion of the representative point of the system in configuration space normal to the direction leading to fission. The distribution of the available energy of the system between such modes of motion and the mode of motion leading to fission will be determining for the probability of fission if the system is near the critical state. The statistical mechanics of this problem is considered in Section III. Here we would only like to point out that the fission process is from a practical point of view a nearly irreversible process. In fact if we imagine the fragment nuclei resulting from a fission to be reflected without loss of energy and to run directly towards each other, the electrostatic repulsion between the two nuclei will ordinarily prevent them from coming into contact. Thus, relative to the original nucleus, the energy of two spherical nuclei of half the size is given by Eq. (19) and corresponds to the values $f^*(x)$ shown by the dashed line in Fig. 4. To compare this with the energy required for the original fission process (smooth curve for $f(x)$ in the figure), we note that the surface energy $4\pi r_0^2OA^{\frac{2}{3}}$ is for the heaviest nuclei of the order of 500 Mev. We thus have to deal with a difference of $\sim 0.05\times 500$ Mev $=25$ Mev between the energy available when a heavy nucleus is just able to undergo fission, and the energy required to bring into contact two spherical fragments. There will of course be appreciable tidal forces exerted when the two fragments are brought together, and a simple estimate shows that this will lower the energy discrepancy just mentioned by something of the order of 10 Mev, which is not enough to alter our conclusions. That there is no paradox involved, however, follows from the fact that the fission process actually takes place for a configuration in which the sum of surface and electrostatic energy has a considerably smaller value than that corresponding to two rigid spheres in contact, or even two tidally distorted globes; namely, by arranging that in the division process the surface surrounding the original nucleus shall not tear until the mutual electrostatic energy of the two nascent nuclei has been brought down to a value essentially smaller than that corresponding to separated spheres, then there will be available enough electrostatic energy to provide the work required to tear the surface, which will of course have increased in total value to something more than that appropriate to two spheres. Thus it is clear that the two fragments formed by the division process will possess internal energy of excitation. Consequently, if we wish to reverse the fission process, we must take care that the fragments come together again sufficiently distorted, and indeed with the distortions so oriented, that contact can be made between projections on the two surfaces and the surface tension start drawing them together while the electrostatic repulsion between the effective electrical centers of gravity of the two parts is still not excessive. The probability that two atomic nuclei in any actual encounter will be suitably excited and possess the proper phase relations so that union will be possible to form a compound system will be extremely small. Such union processes, converse to fission, can be expected to occur for unexcited nuclei only when we have available much more kinetic energy than is released in the fission processes with which we are concerned.

The above considerations on the fission process, based on a comparison between the properties of a nucleus and those of a liquid drop, should be supplemented by remarking that the distortion which leads to fission, although associated with a greater effective mass and lower quantum frequency, and hence more nearly approaching the possibilities of a classical description than any of the higher order oscillation frequencies of the nucleus, will still be characterized by certain specific quantum-mechanical properties. Thus there will be an essential ambiguity in the definition of the critical fission energy of the order of magnitude of the zero point energy, $\hbar\omega_2/2$, which however as we have seen above is only a relatively small quantity. More important from the point of view of nuclear stability will be the possibility of quantum-mechanical tunnel effects, which will make it possible for a nucleus to divide even in its ground state by passage through a portion of configuration space where classically the kinetic energy is negative.

An accurate estimate for the stability of a heavy nucleus against fission in its ground state will, of course, involve a very complicated mathematical problem. In natural extension of the well-known theory of α-decay, we should in principle determine the probability per unit time of a fission process, λ_f, by the formula

$$\lambda_f(=\Gamma_f/\hbar)=5(\omega_f/2\pi)$$

$$\times\exp -2\int_{P_1}^{P_2}\{2(V-E)\sum_i m_i(dx_i/d\alpha)^2\}^{\frac{1}{2}}d\alpha/\hbar. \quad (28)$$

The factor 5 represents the degree of degeneracy of the oscillation leading to instability. The quantum of energy characterizing this vibration is, according to (26), $\hbar\omega\sim 0.8$ Mev. The integral in the exponent leads in the case of a single particle to the Gamow penetration factor. Similarly, in the present problem, the integral is extended in configuration space from the point P_1 of stable equilibrium over the fission saddle point S (as indicated by the dotted line in Fig. 3) and down on a path of steepest descent to the point P_2 where the classical value of the kinetic energy, $E-V$, is again zero. Along this path we may write the coordinate x_i of each elementary particle m_i in terms of a certain parameter α. Since the integral is invariant with respect to how the parameter is chosen, we may for convenience take α to represent the distance between the centers of gravity of the nascent nuclei. To make an accurate calculation on the basis of the liquid-drop model for the integral in (28) would be quite complicated, and we shall therefore estimate the result by assuming each elementary particle to move a distance $\frac{1}{2}\alpha$ in a straight line either to the right or the left according as it is associated with the one or the other nascent nucleus. Moreover, we shall take $V-E$ to be of the order of the fission energy E_f. Thus we obtain for the exponent in (28) approximately

$$(2ME_f)^{\frac{1}{2}}\alpha/\hbar. \quad (29)$$

With $M=239\times 1.66\times 10^{-24}$, $E_f\sim 6$ Mev$=10^{-5}$ erg, and the distance of separation intermediate between the diameter of the nucleus and its radius, say of the order $\sim 1.3\times 10^{-12}$ cm, we thus find a mean lifetime against fission in the ground state equal to

$$1/\lambda_f\sim 10^{-21}\exp\left[(2\times 4\times 10^{-22}\times 10^{-5})^{\frac{1}{2}}1.3\times 10^{-12}/10^{-27}\right]\sim 10^{30}\text{ sec.}\sim 10^{22}\text{ years.} \quad (30)$$

It will be seen that the lifetime thus estimated is not only enormously large compared with the time interval of the order 10^{-15} sec. involved in the actual fission processes initiated by neutron impacts, but that this is even large compared with the lifetime of uranium and thorium for α-ray decay. This remarkable stability of heavy nuclei against fission is as seen due to the large masses involved, a point which was already indicated in the cited article of Meitner and Frisch, where just the essential characteristics of the fission effect were stressed.

III. Break-up of the Compound System as a Monomolecular Reaction

To determine the fission probability, we consider a microcanonical ensemble of nuclei, all having excitation energies between E and $E+dE$. The number of nuclei will be chosen to be exactly equal to the number $\rho(E)dE$ of levels in this energy interval, so that there is one nucleus in each state. The number of nuclei which divide per unit time will then be $\rho(E)dE\Gamma_f/\hbar$, according to our definition of Γ_f. This number will be equal to the number of nuclei in the transition state which pass outward over the fission barrier per unit time.[11a] In a unit distance measured in the direction of fission there will be $(dp/h)\rho^*(E-E_f-K)dE$ quantum states of the microcanonical ensemble for which the momentum and kinetic energy associated with the fission distortion have values in the intervals dp and $dK=vdp$, respectively. Here ρ^* is the density of those levels of the compound nucleus in the transition state which arise from excitation of all degrees of freedom other than the fission itself. At the initial time we have one nucleus in each of the quantum states in question, and consequently the number of fissions per unit time will be

$$dE\int v(dp/h)\rho^*(E-E_f-K)=dEN^*/h, \quad (31)$$

where N^* is the number of levels in the transition state available with the given excitation. Comparing with our original expression for this number, we have

$$\Gamma_f=N^*/2\pi\rho(E)=(d/2\pi)N^* \quad (32)$$

for the fission width expressed in terms of the level density or the level spacing d of the compound nucleus.

The derivation just given for the level width will only be valid if N^* is sufficiently large compared to unity; that is, if the fission width is comparable with or greater than the level spacing. This corresponds to the conditions under which a correspondence principle treatment of the fission distortion becomes possible. On the other hand, when the excitation exceeds by only a little the critical energy, or falls below E_f, specific quantum-mechanical tunnel effects will begin to become of importance. The fission probability will of course fall off very rapidly with decreasing excitation energy at this point, the mathematical expression for the reaction rate eventually going over into the penetration formula of Eq. (28); this, as we have seen above, gives a negligible fission probability for uranium.

The probability of neutron re-emission, so important in limiting the fission yield for high excitation energies, has been estimated from statistical arguments by various authors, especially Weisskopf.[12] The result can be derived in a very simple form by considering the microcanonical ensemble introduced above. Only a few changes are necessary with respect to the reasoning used for the fission process. The transition state will be a spherical shell of unit thickness just outside the nuclear surface $4\pi R^2$; the critical energy is the neutron binding energy, E_n; and the density ρ^{**} of excitation levels in the transition state is given by the spectrum of the residual nucleus. The number of quantum states in the microcanonical ensemble which lie in the transition region and for which the neutron momentum lies in the range p to $p+dp$ and in the solid angle $d\Omega$ will be

$$(4\pi R^2\cdot p^2dpd\Omega/h^3)\rho^*(E-E_n-K)dE. \quad (33)$$

We multiply this by the normal velocity $v\cos\theta=(dK/dp)\cos\theta$ and integrate, obtaining

$$dE(4\pi R^2\cdot 2\pi m/h^3)\int\rho^*(E-E_n-K)KdK \quad (34)$$

for the number of neutron emission processes occurring per unit time. This is to be identified with $\rho(E)dE(\Gamma_n/\hbar)$. Therefore we have for the probability of neutron emission, expressed in energy units, the result

$$\Gamma_n=(1/2\pi\rho)(2mR^2/\hbar^2)\int\rho^{**}(E-E_n-K)KdK$$
$$=(d/2\pi)(A^{\frac{2}{3}}/K')\sum_i K_i \quad (35)$$

in complete analogy to the expression

$$\Gamma_f=(d/2\pi)\sum_i 1 \quad (36)$$

[11a] For a general discussion of the ideas involved in the concept of a transition state, reference is made to an article by E. Wigner, Trans. Faraday Soc. **34**, part 1, 29 (1938).

[12] V. Weisskopf, Phys. Rev. **52**, 295 (1937).

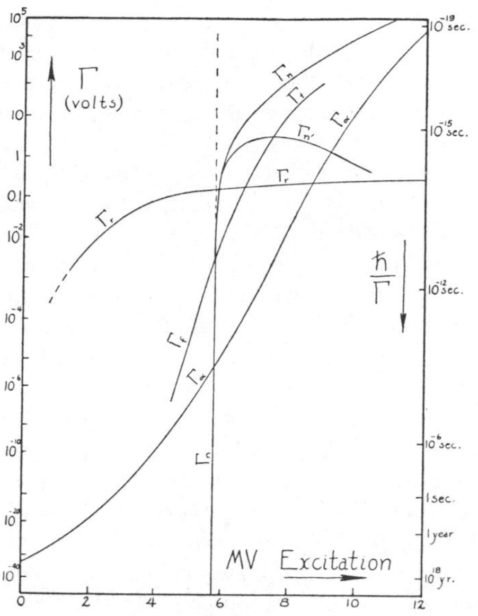

FIG. 5. Schematic diagram of the partial transition probabilities (multiplied by $\hbar$ and expressed in energy units) and their reciprocals (dimensions of a mean lifetime) for various excitation energies of a typical heavy nucleus. Γ_r, Γ_f, and Γ_α refer to radiation, fission, and alpha-particle emission, while $\Gamma_{n'}$ and Γ_n determine, respectively, the probability of a neutron emission leaving the residual nucleus in its ground state or in any state. The latter quantities are of course zero if the excitation is less than the neutron binding, which is taken here to be about 6 Mev.

for the fission width. Just as the summation in the latter equation goes over all those levels of the nucleus in the transition state which are available with the given excitation, so the sum in the former is taken over all available states of the residual nucleus, K_i denoting the corresponding kinetic energy $E-E_n-E_i$ which will be left for the neutron. K' represents, except for a factor, the zero point kinetic energy of an elementary particle in the nucleus; it is given by $A^{2/3}\hbar^2/2mR^2$ and will be 9.3 Mev if the nuclear radius is $A^{1/3}1.48\times10^{-13}$ cm.

No specification was made as to the angular momentum of the nucleus in the derivation of (35) and (36). Thus the expressions in question give us averages of the level widths over states of the compound system corresponding to many different values of the rotational quantum number J, while actually capture of a neutron of one- or two-Mev energy by a normal nucleus will give rise only to a restricted range of values of J. This point is of little importance in general, as the widths will not depend much on J, and therefore in the following considerations we shall apply the above estimates of Γ_f and Γ_n as they stand. In particular, d will represent the average spacing of levels of a given angular momentum. If, however, we wish to determine the partial width $\Gamma_{n'}$ giving the probability that the compound nucleus will break up leaving the residual nucleus in its ground state and giving the neutron its full kinetic energy, we shall not be justified in simply selecting out the corresponding term in the sum in (35) and identifying it with $\Gamma_{n'}$. In fact, a more detailed calculation along the above lines, specifying the angular momentum of the microcanonical ensemble as well as its energy, leads to the expression

$$\Sigma(2J+1)\Gamma_{n'}{}^{J} = (2s+1)(2i+1)(d/2\pi)(R^2/ƛ^2) \qquad (37)$$

for the partial neutron width, where the sum goes over those values of J which are realized when a nucleus of spin i is bombarded by a neutron of the given energy possessing spin $s=\frac{1}{2}$.

The smallness of the neutron mass in comparison with the reduced mass of two separating nascent nuclei will mean that we shall have in the former case to go to excitation energies much higher relative to the barrier than in the latter case before the condition is fulfilled for the application of the transition state method. In fact, only when the kinetic energy of the emerging particle is considerably greater than 1 Mev does the reduced wave-length $ƛ=\lambda/2\pi$ of the neutron become essentially smaller than the nuclear radius, allowing the use of the concepts of velocity and direction of the neutron emerging from the nuclear surface.

The absolute yield of the various processes initiated by neutron bombardment will depend upon the probability of absorption of the neutron to form a compound nucleus; this will be proportional to the converse probability $\Gamma_{n'}/\hbar$ of a neutron emission process which leaves the residual neutron emission process which leaves the residual nucleus in its ground state. $\Gamma_{n'}$ will vary as the neutron velocity itself for low neutron energies; according to the available information about nuclei of medium atomic weight, the width in volts is approximately 10^{-3} times the square root of the neutron energy in volts.[13] As the neutron energy increases from thermal values to 100 kev, we have to expect then an increase of $\Gamma_{n'}$ from something of the order of 10^{-4} ev to 0.1 or 1 ev. For high neutron energies we can use Eq. (37), according to which $\Gamma_{n'}$ will increase as the neutron energy itself, except as compensated by the decrease in level spacing as higher excitations are attained. As an order of magnitude estimate, we can take the level spacing in U to decrease from 100 kev for the lowest levels to 20 ev at 6 Mev (capture of thermal neutrons) to $\frac{1}{5}$ ev for $2\frac{1}{2}$-Mev neutrons. With $d=\frac{1}{5}$ ev we obtain $\Gamma_{n'}=(1/2\pi\times 5)(239^{\frac{2}{3}}/10)2\frac{1}{2}\doteq\frac{1}{2}$ ev for neutrons from the D+D reaction. The partial neutron width will not exceed for any energy a value of this order of magnitude, since the decrease in level spacing will be the dominating factor at higher energies.

The compound nucleus once formed, the outcome of the competition between the possibilities of fission, neutron emission, and radiation, will be determined by the relative magnitudes of Γ_f, Γ_n, and the corresponding radiation width Γ_r. From our knowledge of nuclei comparable with thorium and uranium we can conclude that the radiation width Γ_r will not exceed something of the order of 1 ev, and moreover that it will be nearly constant for the range of excitation energies which results from neutron absorption (see Fig. 5). The fission width will be extremely small for excitation energies below the critical energy E_f, but above this point Γ_f will become appreciable, soon exceeding the radiation width and rising almost exponentially for higher energies. Therefore, if the critical energy E_f required for fission is comparable with or greater than the excitation consequent on neutron capture, we have to expect that radiation will be more likely than fission; but if the barrier height is somewhat lower than the value of the neutron binding, and in any case if we irradiate with sufficiently energetic neutrons, radiative capture will always be less probable than division. As the speed of the bombarding neutrons is increased, we shall not expect an indefinite rise in the fission yield, however, for the output will be governed by the competition in the compound system between the possibilities of fission and of neutron emission. The width Γ_n which gives the probability of the latter process will for energies less than something of the order of 100 kev be equal to $\Gamma_{n'}$, the partial width for emissions leaving the residual nucleus in the ground state, since excitation of the product nucleus will be energetically impossible. For higher neutron energies, however, the number of available levels in the residual nucleus will rise rapidly, and Γ_n will be much larger than $\Gamma_{n'}$, increasing almost exponentially with energy.

In the energy region where the levels of the compound nucleus are well separated, the cross sections governing the yield of the various processes considered above can be obtained by direct application of the dispersion theory of Breit and Wigner.[14] In the case of resonance, where the energy E of the incident neutron is close to a special value E_0 characterizing an isolated level of the compound system, we shall have

$$\sigma_f=\pi\lambda\!\!\!^{-2}\frac{2J+1}{(2s+1)(2i+1)}\frac{\Gamma_{n'}\Gamma_f}{(E-E_0)^2+(\Gamma/2)^2} \tag{38}$$

and

$$\sigma_r=\pi\lambda\!\!\!^{-2}\frac{2J+1}{(2s+1)(2i+1)}\frac{\Gamma_{n'}\Gamma_r}{(E-E_0)^2+(\Gamma/2)^2}. \tag{39}$$

for the fission and radiation cross sections. Here $\lambda\!\!\!^{-}=\hbar/p=\hbar/(2mE)^{\frac{1}{2}}$ is the neutron wave-length divided by 2π, i and J are the rotational quantum numbers of the original and the compound nucleus, $s=\frac{1}{2}$, and $\Gamma=\Gamma_n+\Gamma_r+\Gamma_f$ is the total width of the resonance level at half-maximum.

In the energy region where the compound nucleus has many levels whose spacing, d, is comparable with or smaller than the total width, the dispersion theory cannot be directly applied due to the phase relations between the contributions of the different levels. A closer discussion[15] shows, however, that in cases like fission and radiative capture, the cross section will be obtained by summing many terms of the form (38) or (39). If the neutron wave-length is large compared with nuclear dimensions, only those states of the compound nucleus will contribute to the

[13] H. A. Bethe, Rev. Mod. Phys. **9**, 150 (1937).

[14] G. Breit and E. Wigner, Phys. Rev. **49**, 519 (1936). Cf. also H. Bethe and G. Placzek, Phys. Rev. **51**, 450 (1937).

[15] N. Bohr, R. Peierls and G. Placzek, Nature (in press).

sum which can be realized by capture of a neutron of zero angular momentum, and we shall obtain

$$\sigma_f = \pi\lambda^2\Gamma_{n'}(\Gamma_f/\Gamma)(2\pi/d)\times\begin{cases}1 \text{ if } i=0\\ \frac{1}{2} \text{ if } i>0\end{cases}, \quad (40)$$

$$\sigma_r = \pi\lambda^2\Gamma_{n'}(\Gamma_r/\Gamma)(2\pi/d)\times\begin{cases}1 \text{ if } i=0\\ \frac{1}{2} \text{ if } i>0\end{cases}. \quad (41)$$

On the other hand, if λ becomes essentially smaller than R, the nuclear radius (case of neutron energy over a million volts), the summation will give

$$\sigma_f = \frac{\pi\lambda^2\sum(2J+1)\Gamma_{n'}}{(2s+1)(2i+1)}(\Gamma_f/\Gamma)(2\pi/d)$$

$$= \pi R^2\Gamma_f/\Gamma, \quad (42)$$

$$\sigma_r = \pi R^2\Gamma_r/\Gamma. \quad (43)$$

The simple form of the result, which follows by use of the equation (37) derived above for $\Gamma_{n'}$, is of course an immediate consequence of the fact that the cross section for any given process for fast neutrons is given by the projected area of the nucleus times the ratio of the probability per unit time that the compound system react in the given way to the total probability of all reactions. Of course for extremely high bombarding energies it will no longer be possible to draw any simple distinction between neutron emission and fission; evaporation will go on simultaneously with the division process itself; and in general we shall have to expect then the production of numerous fragments of widely assorted sizes as the final result of the reaction.

IV. Discussion of the Observations

A. The resonance capture process

Meitner, Hahn, and Strassmann[16] observed that neutrons of some volts energy produced in uranium a beta-ray activity of 23 min. half-life whose chemistry is that of uranium itself. Moreover, neutrons of such energy gave no noticeable yield of the complex of periods which is produced in uranium by irradiation with either thermal or fast neutrons, and which is now known to arise from the beta-instability of the fragments arising from fission processes. The origin of the activity in question therefore had to be attributed to the ordinary type of radiative capture observed in other nuclei; like such processes it has a resonance character. The effective energy E_0 of the resonance level or levels was determined by comparing the absorption in boron of the neutrons producing the activity and of neutrons of thermal energy:

$$E_0 = (\pi kT/4)[\mu_{\text{thermal}}(B)/\mu_{\text{res}}(B)]^2$$

$$= 25\pm 10 \text{ ev}. \quad (44)$$

The absorption coefficient in uranium itself for the activating neutrons was found to be 3 cm^2/g, corresponding to an effective cross section of $3\ \text{cm}^2/\text{g}\times 238\times 1.66\times 10^{-24}\ \text{g} = 1.2\times 10^{-21}\ \text{cm}^2$. If we attribute the absorption to a single resonance level with no appreciable Doppler broadening, the cross section at exact resonance will be twice this amount, or $2.4\times 10^{-21}\ \text{cm}^2$; if on the other hand the true width Γ should be small compared with the Doppler broadening

$$\Delta = 2(E_0kT/238)^{\frac{1}{2}} = 0.12 \text{ ev},$$

we should have for the true cross section at exact resonance $2.7\times 10^{-21}\Delta/\Gamma$, which would be even greater.[17] If the activity is actually due to several comparable resonance levels, we will clearly obtain the same result for the cross section of each at exact resonance.

According to Nier[18] the abundances of U^{235} and U^{234} relative to U^{238} are 1/139 and 1/17,000; therefore, if the resonance absorption is due to either of the latter, the cross section at resonance will have to be at least $139\times 2.4\times 10^{-21}\ \text{cm}^2$ or $3.3\times 10^{-19}\ \text{cm}^2$. However, as Meitner, Hahn and Strassmann pointed out, this is excluded (cf. Eq. (39)) because it would be greater in order of magnitude than the square of the neutron wavelength. In fact, $\pi\lambda^2$ is only $25\times 10^{-21}\ \text{cm}^2$ for 25-volt neutrons. Therefore we have to attribute the capture to $U^{238}\rightarrow U^{239}$, a process in which the spin changes from $i=0$ to $J=\frac{1}{2}$. We apply the

[16] L. Meitner, O. Hahn and F. Strassmann, Zeits. f. Physik **106**, 249 (1937).

[17] We are using the treatment of Doppler broadening given by H. Bethe and G. Placzek, Phys. Rev. **51**, 450 (1937).

[18] A. O. Nier, Phys. Rev. **55**, 150 (1939).

resonance formula (39) and obtain

$$25\times 10^{-21}\times 4\Gamma_{n'}\Gamma_r/\Gamma^2 = 2.7\times 10^{-21}(\Delta/\Gamma) \text{ or } 2.4\times 10^{-21} \quad (45)$$

according as the level width $\Gamma=\Gamma_{n'}+\Gamma_r$ is or is not small compared with the Doppler broadening. In any case, we know[13] from experience with other nuclei for comparable neutron energies that $\Gamma_{n'}\ll\Gamma_r$; this condition makes the solution of (45) unique. We obtain $\Gamma_{n'}=\Gamma_r/40$ if the total width is greater than $\Delta=0.12$ ev; and if the total width is smaller than Δ we find $\Gamma_{n'}=0.003$ ev. Thus in neither case is the neutron width less than 0.003 ev. Comparison with observations on elements of medium atomic weight would lead us to expect a neutron width of $0.001\times(25)^{\frac{1}{2}}=0.005$ ev; and undoubtedly $\Gamma_{n'}$ can be no greater than this for uranium, in view of the small level spacing, or equivalently, in view of the small probability that enough energy be concentrated on a single particle in such a big nucleus to enable it to escape. We therefore conclude that $\Gamma_{n'}$ for 25-volt neutrons is approximately 0.003 ev.

Our result implies that the radiation width for the U^{239} resonance level cannot exceed $\sim$0.12 ev; it may be less, but not much less, first, because values as great as a volt or more have been observed for Γ_r in nuclei of medium atomic weight, and second, because values of a millivolt or more are observed in the transitions between individual levels of the radioactive elements, and for the excitation with which we are concerned the number of available lower levels is great and the corresponding radiation frequencies are higher.[13] A reasonable estimate of Γ_r would be 0.1 ev; of course direct measurement of the activation yield due to neutrons continuously distributed in energy near the resonance level would give a definite value for the radiation width.

The above considerations on the capture of neutrons to form U^{239} are expressed for simplicity as if there were a single resonance level, but the results are altered only slightly if several levels give absorption. However, the contribution of the resonance effect to the radiative capture cross section for *thermal* neutrons does depend essentially on the number of levels as well as their strength. On this basis Anderson and Fermi have been able to show that the radiative capture of slow neutrons cannot be due to the tail at low energies of only a single level.[19] In fact, if it were, we should have for the cross section from (39)

$$\sigma_r(\text{thermal})=\pi\lambda\!\!\!^{-}_{th}{}^2\Gamma_{n'}(\text{thermal})\Gamma_r/E_0^2; \quad (46)$$

since $\Gamma_{n'}$ is proportional to neutron velocity, we should obtain at the effective thermal energy $\pi kT/4=0.028$ ev.

$$\sigma_r(\text{thermal})\sim 23\times 10^{-18} \times 0.003(0.028/25)^{\frac{1}{2}}0.1/(25)^2 \sim 0.4\times 10^{-24}\ \text{cm}^2. \quad (47)$$

Anderson and Fermi however obtain for this cross section by direct measurement 1.2×10^{-24} cm^2.

The conclusion that the resonance absorption at the effective energy of 25 ev is actually due to more than one level gives the possibility of an order of magnitude estimate of the spacing between energy levels in U^{239} if for simplicity we assume random phase relations between their individual contributions. Taking into consideration the factor between the observations and the result (47) of the one level formula, and recalling that levels below thermal energies as well as above contribute to the absorption, we arrive at a level spacing of the order of $d=20$ ev as a reasonable figure at the excitation in question.

B. Fission produced by thermal neutrons

According to Meitner, Hahn and Strassmann[20] and other observers, irradiation of uranium by thermal neutrons actually gives a large number of radioactive periods which arise from fission fragments. By direct measurement the fission cross section for thermal neutrons is found to be between 2 and 3×10^{-24} cm^2 (averaged over the actual mixture of isotopes), that is, about twice the cross section for radiative capture. No appreciable part of this effect can come from the isotope U^{239}, however, because the observations on the $\sim$25-volt resonance capture of neutrons by this nucleus gave only the 23-minute activity; the inability of Meitner, Hahn, and Strassmann to find for neutrons of this energy any appreciable yield of the complex of periods

[19] H. L. Anderson and E. Fermi, Phys. Rev. **55**, 1106 (1939).

[20] L. Meitner, O. Hahn and F. Strassmann, Zeits. f. Physik **106**, 249 (1937).

now known to follow fission indicates that for slow neutrons in general the fission probability for this nucleus is certainly no greater than 1/10 of the radiation probability. Consequently, from comparison of (38) and (39), the fission cross section for this isotope cannot exceed something of the order $\sigma_f(\text{thermal}) = (1/10)\sigma_r(\text{thermal}) = 0.1\times 10^{-24}$ cm^2. From reasoning of this nature, as was pointed out in an earlier paper by Bohr, we have to attribute practically all of the fission observed with thermal neutrons to one of the rarer isotopes of uranium.[21] If we assign it to the compound nucleus U^{235}, we shall have 17,000 $\times 2.5\times 10^{-24}$ or 4×10^{-20} cm^2 for σ_f(thermal); if we attribute the division to U^{236}, σ_f will be between 3 and 4×10^{-22} cm^2.

We have to expect that the radiation width and the neutron width for slow neutrons will differ in no essential way between the various uranium isotopes. Therefore we will assume $\Gamma_{n'}(\text{thermal}) = 0.003(0.028/25)^{\frac{1}{2}} = 10^{-4}$ ev. The fission width, however, depends strongly on the barrier height; this is in turn a sensitive function of nuclear charge and mass numbers, as indicated in Fig. 4, and decreases strongly with decreasing isotopic weight. Thus it is reasonable that one of the lighter isotopes should be responsible for the fission.

Let us investigate first the possibility that the division produced by thermal neutrons is due to the compound nucleus U^{235}. If the level spacing d for this nucleus is essentially greater than the level width, the cross section will be due principally to one level ($J=\frac{1}{2}$ arising from $i=0$), and we shall have from

$$\sigma_f = \pi\lambda\!\!\!^-{}^2 \frac{2J+1}{(2s+1)(2i+1)} \frac{\Gamma_{n'}\Gamma_f}{(E-E_0)^2+(\Gamma/2)^2} \quad (38)$$

the equation

$$\Gamma_f/[E_0^2+\Gamma^2/4] = 4\times 10^{-20}/23 \times 10^{-18}\times 10^{-4} = 17(\text{ev})^{-1}. \quad (48)$$

Since $\Gamma > \Gamma_f$, this condition can be put as an inequality,

$$E_0^2 < (\Gamma/4)(4/17) - \Gamma) \quad (49)$$

from which it follows first, that $\Gamma \leq 4/17$ ev, and second, that $|E_0| < 2/17$ ev. Thus the level would have to be very narrow and very close to thermal energies. But in this case the fission cross section would have to fall off very rapidly with increasing neutron energy; since $\lambda\!\!\!^- \propto 1/v$, $E\propto v^2$, $\Gamma_{n'}\propto v$, we should have according to (38) $\sigma_f \propto 1/v^5$ for neutron energies greater than about half a volt. This behavior is quite inconsistent with the finding of the Columbia group that the fission cross section for cadmium resonance neutrons ($\sim$0.15 ev) and for the neutrons absorbed in boron (mean energy of several volts) stand to each other inversely in the ratio of the corresponding neutron velocities $(1/v)$.[22] Therefore, if the fission is to be attributed to U^{235}, we must assume that the level width is greater than the level spacing (many levels effective); but as the level spacing itself will certainly exceed the radiative width, we will then have a situation in which the total width will be essentially equal to Γ_f. Consequently we can write the cross section (40) for overlapping levels in the form

$$\sigma_f = \pi\lambda\!\!\!^-{}^2\Gamma_{n'}2\pi/d. \quad (50)$$

From this we find a level spacing

$$d = 23\times 10^{-18}\times 10^{-4}\times 2\pi/4\times 10^{-20} = 0.4 \text{ ev}$$

which is unreasonably small: According to the estimates of Table III, the nuclear excitations consequent on the capture of slow neutrons to form U^{235} and U^{239} are approximately 5.4 Mev and 5.2 Mev, respectively; moreover, the two nuclei have the same odd-even properties and should therefore possess similar level distributions. From the difference ΔE between the excitation energies in the two cases we can therefore obtain the ratio of the corresponding level spacings from the expression $\exp(\Delta E/T)$. Here T is the nuclear temperature, a low estimate for which is 0.5 Mev, giving a factor of $\exp 0.6 = 2$. From our conclusion in IV-A that the order of magnitude of the level spacing in U^{239} is 20 ev, we would expect then in U^{235} a spacing of the order of 10 ev. Therefore the result of Eq. (51) makes it seem quite unlikely that the fission observed for the thermal neutrons can be due to the rarest uranium isotope; we consequently attribute it almost entirely to the reaction $U^{235}+n_{th}\rightarrow U^{236}\rightarrow$fission.

[21] N. Bohr, Phys. Rev. **55**, 418 (1939).

[22] Anderson, Booth, Dunning, Fermi, Glasoe and Slack, reference 4.

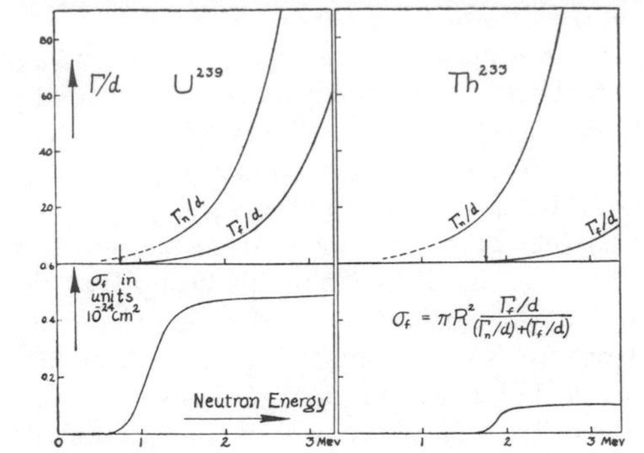

FIG. 6. Γ_n/d and Γ_f/d are the ratios of the neutron emission and fission probabilities (taken per unit of time and multiplied by $\hbar$) to the average level spacing in the compound nucleus at the given excitation. These ratios will vary with energy in nearly the same way for all heavy nuclei, except that the entire fission curve must be shifted to the left or right according as the critical fission energy E_f is less than or greater than the neutron binding E_n. The cross section for fission produced by fast neutrons depends on the ratio of the values in the two curves, and is given on the left for $E_f-E_n=(\frac{3}{4})$ Mev and on the right for $E_f-E_n=1\frac{3}{4}$ Mev, corresponding closely to the cases of U^{239} and Th^{233}, respectively.

We have two possibilities to account for the cross section $\sigma_f(\text{thermal})\sim 3.5\times 10^{-22}$ presented by the isotope U^{235} for formation of the compound nucleus U^{236}, according as the level width is smaller than or comparable with the level spacing. In the first case we shall have to attribute most of the fission to an isolated level, and by the reasoning which was employed previously, we conclude that for this level

$$\Gamma_f/[E_0^2+\Gamma^2/4] = [(2s+1)(2i+1)/(2J+1)]0.15(\text{ev})^{-1}=R. \quad (52)$$

If the spin of U^{235} is $\frac{3}{2}$ or greater, the right-hand side of (52) will be approximately 0.30 $(\text{ev})^{-1}$; but if i is as low as $\frac{1}{2}$, the right side will be either 0.6 or 0.2 $(\text{ev})^{-1}$. The resulting upper limits on the resonance energy and level width may be summarized as follows:

	$i\geq\frac{3}{2}$	$i=\frac{1}{2}, J=0$	$i=\frac{1}{2}, J=1$	
$\Gamma<4/R=$	13	7	20 ev	(53)
$\lvert E_0\rvert<1/R=$	3	1.7	5 ev.	

On the other hand, the indications[22] for low neutron energies of a $1/v$ variation of fission cross section with velocity lead us as in the discussion of the rarer uranium isotope to the conclusion that either E_0 or $\Gamma/2$ or both are greater than several electron volts. This allows us to obtain from (52) a lower limit also to Γ_f:

$$\Gamma_f=R[E_0^2+\Gamma^2/4]>10 \text{ to } 400 \text{ ev}. \quad (54)$$

In the present case, the various conditions are not inconsistent with each other, and it is therefore possible to attribute the fission to the effect of a single resonance level.

We can go further, however, by estimating the level spacing for the compound nucleus U^{236}. According to the values of Table III, the excitation following the neutron capture is considerably greater than in the case U^{239}, and we should therefore expect a rather smaller level spacing than the value ~ 20 ev estimated in the latter case. On the other hand, it is known that for similar energies the level density is lower in even even than odd even nuclei. Thus the level spacing in U^{236} may still be as great as 20 ev, but it is undoubtedly no greater. From (54) we conclude then that we have probably to do with a case of overlapping resonance levels rather than a single absorption line, although the latter possibility is not entirely excluded by the observations available.

In the case of overlapping levels we shall have from Eq. (40)

$$\sigma_f=(\pi\lambda\!\!\!^{-2}/2)\Gamma_{n'}(2\pi/d) \quad (55)$$

or consequently a level spacing

$$d=(23\times 10^{-18}/2)\times 10^{-4} \times 2\pi/3.5\times 10^{-22}=20 \text{ ev}; \quad (56)$$

and as we are attributing to the levels an unresolved structure, the fission width must be at least 10 ev. These values for level spacing and fission width give a reasonable account of the fission produced by slow neutrons.

C. Fission by fast neutrons

The discussion on the basis of theory of the fission produced by fast neutrons is simplified first by the fact that the probability of radiation can be neglected in comparison with the probabilities of fission and neutron escape and second by the circumstance that the neutron wavelength $/2\pi$ is small in comparison with the nuclear radius ($R\sim 9\times 10^{-13}$ cm) and we are in the region of continuous level distribution. Thus the fission cross section will be given by

$$\sigma_f=\pi R^2\Gamma_f/\Gamma\sim 2.4\times 10^{-24}\Gamma_f/(\Gamma_f+\Gamma_n), \quad (57)$$

or, in terms of the ratio of widths to level spacing,

$$\sigma_f \sim 2.4\times 10^{-24}(\Gamma_f/d)/[(\Gamma_f/d)+(\Gamma_n/d)]. \quad (58)$$

According to the results of Section III,

$$\Gamma_n/d=(1/2\pi)(A^{\frac{2}{3}}/10 \text{ Mev})\sum_i K_i \quad (59)$$

and

$$\Gamma_f/d=(1/2\pi)N^*. \quad (60)$$

In using Eq. (58) it is therefore seen that we do not have to know the level spacing d of the compound nucleus, but only that of the residual nucleus (Eq. (59)) and the number N^* of available levels of the dividing nucleus in the transition state (Eq. 60).

Considered as a function of energy, the ratio of fission width to level spacing will be extremely small for excitations less than the critical fission energy; with increase of the excitation above this value Eq. (60) will quickly become valid, and we shall have to anticipate a rapid rise in the ratio in question. If the spacing of levels in the transition state can be compared with that of the lower states of an ordinary heavy nucleus ($\sim$50 to 100 kev) we shall expect a value of $N^*=10$ to 20 for an energy 1 Mev above the fission barrier; but in any case the value of Γ_f/d will rise almost linearly with the available energy over a range of the order of a million volts, when the rise will become noticeably more rapid owing to the decrease to be expected at such excitations in the level spacing of the nucleus in the transition state. The associated behavior of Γ_f/d is illustrated in curves in Fig. 6. It should be remarked that the specific quantum-mechanical effects which set in at and below the critical fission energy may even show their influence to a certain extent above this energy and produce slight oscillations in the beginning of the Γ_f/d curve, allowing possibly a direct determination of N^*. How the ratio Γ_n/d will vary with energy is more accurately predictable than the ratio just considered. Denoting by K the neutron energy, we have for the number of levels which can be excited in the residual (=original) nucleus a figure of from $K/0.05$ Mev to $K/0.1$ Mev, and for the average kinetic energy of the inelastically scattered neutron $\sim K/2$, so that the sum K_i in (59) is easily evaluated, giving us, if we express K in Mev,

$$\Gamma_n/d \sim 3 \text{ to } 6 \text{ times } K^2. \quad (61)$$

This formula provides as a matter of fact however only a rough first orientation, since for energies below $K=1$ Mev it is not justified to apply the evaporation formula (a transition occurring until for slow neutrons Γ_n/d is proportional to velocity) and for energies above 1 Mev we have to take into account the gradual decrease which occurs in level spacing in the residual nucleus, and which has the effect of increasing the right-hand side of (61). An attempt has been made to estimate this increase in drawing Fig. 6.

The two ratios involved in the fast neutron fission cross section (58) will vary with energy in the same way for all the heaviest nuclei; the only difference from nucleus to nucleus will occur in the critical fission energy, which will have the effect of shifting one curve with respect to another as shown in the two portions of Fig. 6. Thus we can deduce the characteristic differences between nuclei to be expected in the variation with energy of the fast neutron cross section.

Meitner, Hahn, and Strassmann observed that fast neutrons as well as thermal ones produce in uranium the complex of activities which arise as a result of nuclear fission, and Ladenburg, Kanner, Barschall, and van Voorhis have made a direct measurement of the fission cross section for 2.5 Mev neutrons, obtaining 0.5×10^{-24} cm^2 ($\pm$25 percent).[23] Since the contribution to this cross section due to the U^{235} isotope cannot exceed $\pi R^2/139 \sim 0.02\times 10^{-24}$ cm^2, the effect must be attributed to the compound nucleus U^{239}. For this nucleus however as we have seen from the slow neutron observations the fission probability is negligible at low energies. Therefore we have to conclude that the variation with energy of the corresponding cross section resembles in its general features Fig. 6a. In this connection we have the further observation of Ladenburg *et al.* that the cross section changes little between 2 Mev and 3 Mev.[23] This points to a value of the critical fission energy for U^{239} definitely less

[23] R. Ladenburg, M. H. Kanner, H. H. Barschall and C. C. van Voorhis, Phys. Rev. **56**, 168 (1939).

than 2 Mev in excess of the neutron binding. Unpublished results of the Washington group[24] give $\sigma_f = 0.003 \times 10^{-24}$ at 0.6 Mev and 0.012×10^{-24} cm² at 1 Mev. With the Princeton observations[23] we have enough information to say that the critical energy for U^{239} is not far from $\frac{3}{4}$ Mev in excess of the neutron binding (~ 5.2 Mev from Table III):

$$E_f(U^{239}) \sim 6 \text{ Mev}. \qquad (62)$$

A second conclusion we can draw from the absolute cross section of Ladenburg *et al.* is that the ratio of (Γ_f/d) to (Γ_n/d) as indicated in the figure is substantially correct; this confirms our presumption that the energy level spacing in the transition state of the dividing nucleus is not different in order of magnitude from that of the low levels in the normal nucleus.

The fission cross section of Th^{232} for neutrons of 2 to 3 Mev energy has also been measured by the Princeton group; they find $\sigma_f = 0.1 \times 10^{-24}$ cm² in this energy range. On the basis of the considerations illustrated in Fig. 6 we are led in this case to a fission barrier $1\frac{3}{4}$ Mev greater than the neutron binding; hence, using Table III,

$$E_f(Th^{239}) \sim 7 \text{ Mev}. \qquad (63)$$

A check on the consistency of the values obtained for the fission barriers is furnished by the possibility pointed out in Section II and Fig. 4 of obtaining the critical energy for all nuclei once we know it for one nucleus. Taking $E_f(U^{239}) = 6$ Mev as standard, we obtain $E_f(Th^{232}) = 7$ Mev, in good accord with (63).

As in the preceding paragraph we deduce from Fig. 4 $E_f(U^{236}) = 5\frac{1}{4}$ Mev, $E_f(U^{235}) = 5$ Mev. Both values are *less* than the corresponding neutron binding energies estimated in Table III, $E_n(U^{236}) = 6.4$ Mev, $E_n(U^{235}) = 5.4$ Mev. From the values of $E_n - E_d$ we conclude along the lines of Fig. 6 that for thermal neutrons Γ_f/d is, respectively, ~ 5 and ~ 1 for the two isotopes. Thus it appears that in both cases the level distribution will be continuous. We can estimate the as yet entirely unmeasured fission cross section of the lightest uranium isotope for the thermal neutrons from

$$\sigma_f = \pi \lambda\!\!\!^{-2} \Gamma_{n'} 2\pi/d. \qquad (64)$$

[24] Reported by M. Tuve at the Princeton meeting of the American Physical Society, June 23, 1939.

d will not be much different from what it is for the similar compound nucleus U^{239}, say of the order of 20 ev. Thus

$$\sigma_f(\text{thermal}, U^{235}) \sim 23 \times 10^{-18} \times 10^{-4} \times 2\pi/20$$
$$\sim 500 \text{ to } 1000 \times 10^{-24} \text{ cm}^2, \qquad (65)$$

which is of course practically the same figure which holds for the next heaviest compound nucleus.

The various values estimated for fission barriers and fission and neutron widths are summarized in Fig. 7. The level spacing f for past neutrons has been estimated from its value for slow neutrons and the fact that nuclear level densities appear to increase, according to Weisskopf, approximately exponentially as $2(E/a)^{\frac{1}{2}}$, where a is a quantity related to the spacing of the lowest nuclear levels and roughly 0.1 Mev in magnitude.[25] The relative values of Γ_n, Γ_f and d for fast neutrons in Fig. 7, being obtained less indirectly, will be more reliable than their absolute values.

V. Neutrons, Delayed and Otherwise

Roberts, Meyer and Wang[26] have reported the emission of neutrons following a few seconds after the end of neutron bombardment of a

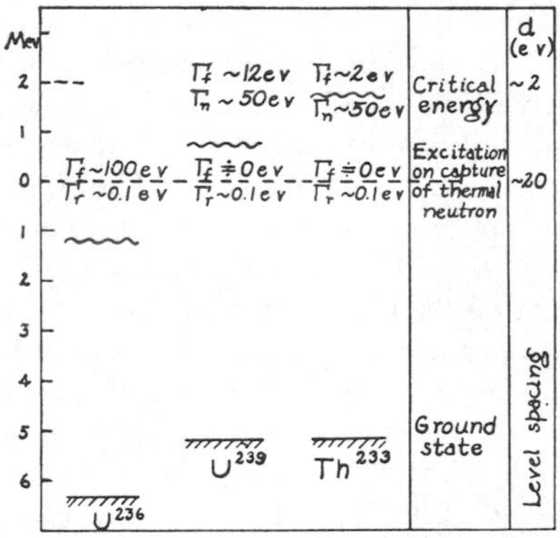

FIG. 7. Summary for comparative purposes of the estimated fission energies, neutron binding energies, level spacings, and neutron and fission widths for the three nuclei to which the observations refer. For fast neutrons the values of Γ_f, Γ_n, and d are less reliable than their ratios. The values in the top line refer to a neutron energy of 2 Mev in each case.

[25] V. Weisskopf, Phys. Rev. 52, 295 (1937).
[26] R. B. Roberts, R. C. Meyer and P. Wang, Phys. Rev. 55, 510 (1939).

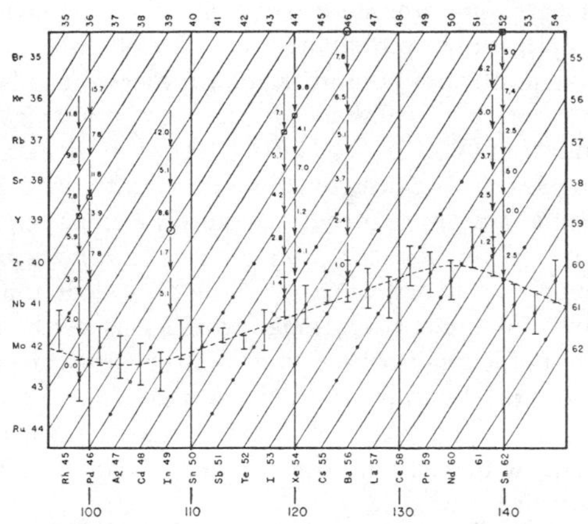

FIG. 8. Beta-decay of fission fragments leading to stable nuclei. Stable nuclei are represented by the small circles; thus the nucleus $_{50}Sn^{120}$ lies just under the arrow marked 4.1; the number indicates the estimated energy release in Mev (see Section I) in the beta-transformation of the preceding nucleus $_{49}In^{120}$. Characteristic differences are noted between nuclei of odd and even mass numbers in the energy of successive transformations, an aid in assigning activities to mass numbers. The dotted line has been drawn, as has been proposed by Gamow, in such a way as to lie within the indicated limits of nuclei of odd mass number; its use is described in Section I.

thorium or uranium target. Other observers have discovered the presence of additional neutrons following within an extremely short interval after the fission process.[27] We shall return later to the question as to the possible connection between the latter neutrons and the mechanism of the fission process. The delayed neutrons themselves are to be attributed however to a high nuclear excitation following beta-ray emission from a fission fragment, for the following reasons:

(1) The delayed neutrons are found only in association with nuclear fission, as is seen from the fact that the yields for both processes depend in the same way on the energy of the bombarding neutrons.

(2) They cannot, however, arise during the fission process itself, since the time required for the division is certainly less than 10^{-12} sec., according to the observations of Feather.[27a]

(3) Moreover, an excitation of a fission fragment in the course of the fission process to an energy sufficient for the subsequent evaporation of a neutron cannot be responsible for the *delayed* neutrons, since even by radiation alone such an excitation will disappear in a time of the order of 10^{-13} to 10^{-15} sec.

(4) The possibility that gamma-rays associated with the beta-ray transformations following fission might produce any appreciable number of photoneutrons in the source has been excluded by an experiment reported by Roberts, Hafstad, Meyer and Wang.[28]

(5) The energy release on beta-transformation is however in a number of cases sufficiently great to excite the product nucleus to a point where it can send out a neutron, as has been already pointed out in connection with the estimates in Table III. Typical values for the release are shown on the arrows in Fig. 8. The product nucleus will moreover have of the order of 10^4 to 10^5 levels to which beta-transformations can lead in this way, so that it will also be overwhelmingly probable that the product nucleus shall be highly excited.

We therefore conclude that the delayed emission of neutrons indeed arises as a result of nuclear excitation following the beta-decay of the nuclear fragments.

The actual probability of the occurrence of a nuclear excitation sufficient to make possible neutron emission will depend upon the comparative values of the matrix elements for the beta-ray transformation from the ground state of the original nucleus to the various excited states of the product nucleus. The simplest assumption we can make is that the matrix elements in question do not show any systematic variation with the energy of the final state. Then, according to the Fermi theory of beta-decay, the probability of a given beta-ray transition will be approximately proportional to the fifth power of the energy release.[29] If there are $\rho(E)dE$ excitation levels of the product nucleus in the range E to $E+dE$, it will follow from our assumptions that the probability of an excitation in the same energy interval will be given by

$$w(E)dE = \text{constant } (E_0-E)^5\rho(E)dE, \quad (66)$$

[27] H. L. Anderson, E. Fermi and H. B. Hanstein, Phys. Rev. **55**, 797 (1939); L. Szilard and W. H. Zinn, Phys. Rev. **55**, 799 (1939); H. von Halban, Jr., F. Joliot and L. Kowarski, Nature **143**, 680 (1939).

[27a] N. Feather, Nature **143**, 597 (1939).

[28] R. B. Roberts, L. R. Hafstad, R. C. Meyer and P. Wang, Phys. Rev. **55**, 664 (1939).

[29] L. W. Nordheim and F. L. Yost, Phys. Rev. **51**, 942 (1937).

where E_0 is the total available energy. According to (66) the probability $w(E)$ of a transition to the excited levels in a unit energy range at E reaches its maximum value for the energy $E=E_{max}$ given by

$$E_{max}=E_0-5/(d \ln \rho/dE)_{E_{max}}=E_0-5T, \quad (67)$$

where T is the temperature (in energy units) to which the product nucleus must be heated to have on the average the excitation energy E_{max}. Thus the most probable energy release on beta-transformation may be said to be five times the temperature of the product nucleus. According to our general information about the nuclei in question, an excitation of 4 Mev will correspond to a temperature of the order of 0.6 Mev. Therefore, on the basis of our assumptions, to realize an average excitation of 4 Mev by beta-transformation we shall require a total energy release of the order of $4+5\times0.6=7$ Mev.

The spacing of the lowest nuclear levels is of the order of 100 kev for elements of medium atomic weight, decreases to something of the order of 10 ev for excitations of the order of 8 Mev, and can, according to considerations of Weisskopf, be represented in terms of a nuclear level density varying approximately exponentially as the square root of the excitation energy.[23] Using such an expression for $\rho(E)$ in Eq. (66), we obtain the curve shown in Fig. 9 for the distribution function $w(E)$ giving the probability that an excitation E will result from the beta-decay of a

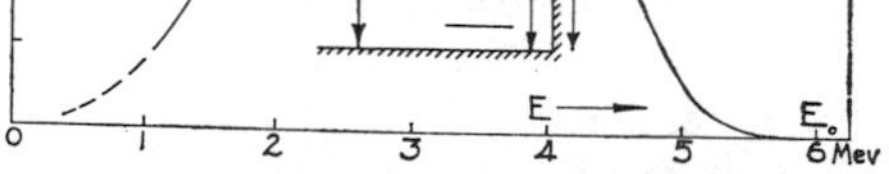

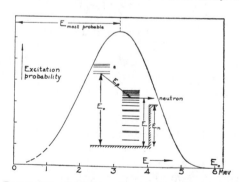

FIG. 9. The distribution in excitation of the product nuclei following beta-decay of fission fragments is estimated on the assumption of comparable matrix elements for the transformations to all excited levels. With sufficient available energy E_0 and a small enough neutron binding E_n it is seen that there will be an appreciable number of delayed neutrons. The quantity plotted is probability per unit range of excitation energy.

typical fission fragment. It is seen that there will be appreciable probability for neutron emission if the neutron binding is somewhat less than the total energy available for the beta-ray transformation. We can of course draw only general conclusions because of the uncertainty in our original assumption that the matrix elements for the various possible transitions show no systematic trend with energy. Still, it is clear that the above considerations provide us with a reasonable qualitative account of the observation of Booth, Dunning and Slack that there is a chance of the order of 1 in 60 that a nuclear fission will result in the delayed emission of a neutron.[30]

Another consequence of the high probability of transitions to excited levels will be to give a beta-ray spectrum which is the superposition of a very large number of elementary spectra. According to Bethe, Hoyle and Peierls, the observations on the beta-ray spectra of light elements point to the Fermi distribution in energy in the elementary spectra.[31] Adopting this result, and using the assumption of equal matrix elements discussed above, we obtain the curve of Fig. 10 for the qualitative type of intensity distribution to be expected for the electrons emitted in the beta-decay of a typical fission fragment. As is seen from the curve, we have to expect that the great majority of electrons will have energies much smaller in value than the actual transformation energy which is available. This is in accord with the failure of various observers to find any appreciable number of very high energy electrons following fission.[32]

The half-life for emission of a beta-ray of 8 Mev energy in an elementary transition will be something of the order of 1 to 1/10 sec., according to the empirical relation between lifetime and energy given by the first Sargent curve. Since we have to deal in the case of the nuclear fragments with transitions to 10^4 or 10^5 excited levels, we should therefore at first sight expect an extremely short lifetime with respect to electron emission. However, the existence of a

[30] E. T. Booth, J. R. Dunning and F. G. Slack, Phys. Rev. **55**, 876 (1939).

[31] H. A. Bethe, F. Hoyle and R. Peierls, Nature **143**, 200 (1939).

[32] H. H. Barschall, W. T. Harris, M. H. Kanner and L. A. Turner, Phys. Rev. **55**, 989 (1939).

sum rule for the matrix elements of the transitions in question has as a consequence that the individual matrix elements will actually be very much smaller than those involved in beta-ray transitions from which the Sargent curve is deduced. Consequently, there seems to be no difficulty in principle in understanding lifetimes of the order of seconds such as have been reported for typical beta-decay processes of the fission fragments.

In addition to the delayed neutrons discussed above there have been observed neutrons following within a very short time (within a time of the order of at most a second) after fission.[27] The corresponding yield has been reported as from two to three neutrons per fission.[33] To account for so many neutrons by the above considered mechanism of nuclear excitation following beta-ray transitions would require us to revise drastically the comparative estimates of beta-transformation energies and neutron binding made in Section I. As the estimates in question were based on indirect though simple arguments, it is in fact possible that they give misleading results. If however they are reasonably correct, we shall have to conclude that the neutrons arise either from the compound nucleus at the moment of fission or by evaporation from the fragments as a result of excitation imparted to them as they separate. In the latter case the time required for neutron emission will be 10^{-13} sec. or less (see Fig. 5). The time required to bring to rest a fragment with 100 Mev kinetic energy, on the other hand, will be at least the time required for a particle with average velocity 10^9 cm/sec. to traverse a distance of the order of 10^{-3} cm. Therefore the neutron will be evaporated before the fragment has lost much of its translational energy. The kinetic energy per particle in the fragment being about 1 Mev, a neutron evaporated in nearly the forward direction will thus have an energy which is certainly greater than 1 Mev, as has been emphasized by Szilard.[34] The observations so far published neither prove nor disprove the possibility of such an evaporation following fission.

[33] Anderson, Fermi and Hanstein, reference 27. Szilard and Zinn, reference 27. H. von Halban, Jr., F. Joliot and L. Kowarski, Nature **143** 680 (1939).

[34] Discussions, Washington meeting of American Physical Society, April 28, 1939.

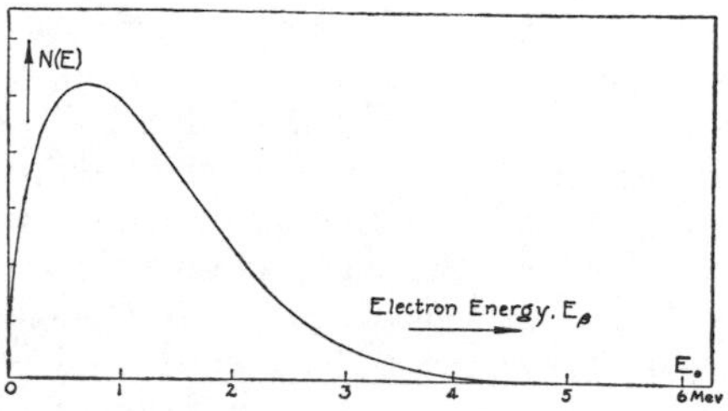

FIG. 10. The superposition of the beta-ray spectra corresponding to all the elementary transformations indicated in Fig. 9 gives a composite spectrum of a general type similar to that shown here, which is based on the assumption of comparable matrix elements and simple Fermi distributions for all transitions. The dependent variable is number of electrons per unit energy range.

We consider briefly the third possibility that the neutrons in question are produced during the fission process itself. In this connection attention may be called to observations on the manner in which a fluid mass of unstable form divides into two smaller masses of greater stability; it is found that tiny droplets are generally formed in the space where the original enveloping surface was torn apart. Although a detailed dynamical account of the division process will be even more complicated for a nucleus than for a fluid mass, the liquid drop model of the nucleus suggests that it is not unreasonable to expect at the moment of fission a production of neutrons from the nucleus analogous to the creation of the droplets from the fluid.

The statistical distribution in size of the fission fragments, like the possible production of neutrons at the moment of division, is essentially a problem of the dynamics of the fission process, rather than of the statistical mechanics of the critical state considered in Section II. Only after the deformation of the nucleus has exceeded the critical value, in fact, will there occur that rapid conversion of potential energy of distortion into energy of internal excitation and kinetic energy of separation which leads to the actual process of division.

For a classical liquid drop the course of the reaction in question will be completely determined by specifying the position and velocity in configuration space of the representative point of the system at the instant when it passes over the potential barrier in the direction of fission. If the energy of the original system is only infinitesimally greater than the critical energy, the representative point of the system must cross the barrier very near the saddle point and with a very small velocity. Still, the wide range of directions available for the velocity vector in this multidimensional space, as suggested schematically in Fig. 3, indicates that production of a considerable variety of fragment sizes may be expected even at energies very close to the threshold for the division process. When the excitation energy increases above the critical fission energy, however, it follows from the statistical arguments in Section III that the representative point of the system will in general pass over the fission barrier at some distance from the saddle point. With general displacements of the representative point along the ridge of the barrier away from the saddle point there are associated asymmetrical deformations from the critical form, and we therefore have to anticipate a somewhat larger difference in size of the fission fragments as more energy is made available to the nucleus in the transition state. Moreover, as an influence of the finer details of nuclear binding, it will also be expected that the relative probability of observing fission fragments of odd mass number will be less when we have to do with the division of a compound nucleus of even charge and mass than one with even charge and odd mass.[35]

VI. Fission Produced by Deuterons and Protons and by Irradiation

Regardless of what excitation process is used, it is clear that an appreciable yield of nuclear fissions will be obtained provided that the excitation energy is well above the critical energy for fission and that the probability of division of the compound nucleus is comparable with the probability of other processes leading to the break up of the system. Neutron escape being the most important process competing with fission, the latter condition will be satisfied if the fission energy does not much exceed the neutron binding, which is in fact the case, as we have seen, for the heaviest nuclei. Thus we have to expect for these nuclei that not only neutrons but also sufficiently energetic deuterons, protons, and gamma-rays will give rise to observable fission.

A. Fission produced by deuteron and proton bombardment

Oppenheimer and Phillips have pointed out that nuclei of high charge react with deuterons of not too great energy by a mechanism of polarization and dissociation of the neutron-proton binding in the field of the nucleus, the neutron being absorbed and the proton repulsed.[36] The excitation energy E of the newly formed nucleus is given by the kinetic energy E_d of the deuteron diminished by its dissociation energy I and the kinetic energy K of the lost proton, all increased by the binding energy E_n of the neutron in the product nucleus:

$$E = E_d - I - K + E_n. \tag{68}$$

The kinetic energy of the proton cannot exceed $E_d + E_n - I$, nor on the other hand will it fall below the potential energy which the proton will have in the Coulomb field at the greatest possible distance from the nucleus consistent with the deuteron reaction taking place with appreciable probability. This distance and the corresponding kinetic energy K_{min} have been calculated by Bethe.[37] For very low values of the bombarding energy E_D, he finds $K_{\text{min}} \sim 1$ Mev; when E_d rises to equality with the dissociation energy $I = 2.2$ Mev he obtains $K_{\text{min}} \sim E_d$; and even when the bombarding potential reaches a value corresponding to the height of the electrostatic barrier, K_{min} still continues to be of order E_d, although beyond this point increase of E_d produces no further rise in K_{min}. Since the barrier height for single charged particles will be of the order of 10 Mev for the heaviest nuclei, we can therefore assume $K_{\text{min}} \sim E_d$ for the ordinarily employed values of the deuteron bombarding energy. We conclude that the excitation energy of the product nucleus will have only a very small probability of exceeding the value

$$E_{\text{max}} \sim E_n - I. \tag{69}$$

Since this figure is considerably less than the

[35] S. Flugge and G. v. Droste also have raised the question of the possible influence of finer details of nuclear binding on the statistical distribution in size of the fission fragments, Zeits. f. physik. Chemie **B42** 274 (1939).

[36] R. Oppenheimer and M. Phillips, Phys. Rev. **48**, 500 (1935).

[37] H. A. Bethe, Phys. Rev. **53**, 39 (1938).

estimated values of the fission barriers in thorium and uranium, we have to expect that Oppenheimer-Phillips processes of the type discussed will be followed in general by radiation rather than fission, unless the kinetic energy of the deuteron is greater than 10 Mev.

We must still consider, particularly when the energy of the deuteron approaches 10 Mev, the possibility of processes in which the deuteron as a whole is captured, leading to the formation of a compound nucleus with excitation of the order of

$$E_d+2E_n-I\sim E_d+10 \text{ Mev.} \quad (70)$$

There will then ensue a competition between the possibilities of fission and neutron emission, the outcome of which will be determined by the comparative values of Γ_f and Γ_n (proton emission being negligible because of the height of the electrostatic barrier). The increase of charge associated with the deuteron capture will of course lower the critical energy of fission and increase the probability of fission relative to neutron evaporation compared to what its value would be for the original nucleus at the same excitation. If after the deuteron capture the evaporation of a neutron actually takes place, the fission barrier will again be decreased relative to the binding energy of a neutron. Since the kinetic energy of the evaporated neutron will be only of the order of thermal energies (≈ 1 Mev), the product nucleus has still an excitation of the order of E_d+3 Mev. Thus, if we are dealing with the capture of 6-Mev deuterons by uranium, we have a good possibility of obtaining fission at either one of two distinct stages of the ensuing nuclear reaction.

The cross section for fission in the double reaction just considered can be estimated by multiplying the corresponding fission cross section (42) for neutrons by a factor allowing for the effect of the electrostatic repulsion of the nucleus in hindering the capture of a deuteron:

$$\sigma_f\sim\pi R^2e^{-P}\{\Gamma_f(E')/\Gamma(E') + [\Gamma_n(E')/\Gamma(E')][\Gamma_f(E'')/\Gamma(E'')]\}. \quad (71)$$

Here P is the new Gamow penetration exponent for a deuteron of energy E and velocity v:[38]

$$P=(4Ze^2/\hbar v)\{\text{arc cos } x^{\frac{1}{2}}-x^{\frac{1}{2}}(1-x)^{\frac{1}{2}}\}, \quad (72)$$

[38] H. A. Bethe, Rev. Mod. Phys. 9, 163 (1937).

with $x=(ER/Ze^2)$. πR^2 is the projected area of the nucleus. E' is the excitation of the compound nucleus, and E'' the average excitation of the residual nucleus formed by neutron emission. For deuteron bombardment of U^{238} at 6 Mev we estimate a fission cross section of the order of

$$\pi(9\times10^{-13})^2 \exp(-12.9)\sim10^{-29} \text{ cm}^2 \quad (73)$$

if we make the reasonable assumption that the probability of fission following capture is of the order of magnitude unity. Observations are not yet available for comparison with our estimate.

Protons will be more efficient than deuterons for the same bombarding energy, since from (72) P will be smaller by the factor $2^{\frac{1}{2}}$ for the lighter particles. Thus for 6-Mev protons we estimate a cross section for production of fission in uranium of the order

$$\pi(9\times10^{-13})^2 \exp(-12.9/2^{\frac{1}{2}})(\Gamma_f/\Gamma)\sim10^{-28} \text{ cm}^2,$$

which should be observable.

B. Photo-fission

According to the dispersion theory of nuclear reactions, the cross section presented by a nucleus for fission by a gamma-ray of wavelength $2\pi\lambda\!\!\!^{-}$ and energy $E=\hbar\omega$ will be given by

$$\sigma_f=\pi\lambda\!\!\!^{-2}(2J+1)/2(2i+1)\frac{\Gamma_{r'}\Gamma_f}{(E-E_0)^2+(\Gamma/2)^2} \quad (74)$$

if we have to do with an isolated absorption line of natural frequency E_0/h. Here $\Gamma_{r'}/\hbar$ is the probability per unit time that the nucleus in the excited state will lose its entire excitation by emission of a single gamma-ray.

The situation of most interest, however, is that in which the excitation provided by the incident radiation is sufficient to carry the nucleus into the region of overlapping levels. On summing (74) over many levels, with average level spacing d, we obtain

$$\sigma_f=\pi\lambda\!\!\!^{-2}[(2J_{Av}+1)/2(2i+1)](2\pi/d)\Gamma_{r'}\Gamma_f/\Gamma. \quad (75)$$

Without entering into a detailed discussion of the orders of magnitude of the various quantities involved in (75), we can form an estimate of the cross section for photo-fission by comparison with the yields of photoneutrons reported by various observers. The ratio of the cross sections in question will be just Γ_f/Γ_n, so that

$$\sigma_f = (\Gamma_f/\Gamma_n)\sigma_n. \quad (76)$$

The observed values of σ_n for 12 to 17 Mev gamma-rays are $\sim 10^{-26}$ cm^2 for heavy elements.[39] In view of the comparative values of Γ_f and Γ_n arrived at in Section IV, it will therefore be reasonable to expect values of the order of 10^{-27} cm^2 for photo-fission of U^{238}, and 10^{-28} cm^2 for division of Th^{233}. Actually no radiative fission was found by Roberts, Meyer and Hafstad using the gamma-rays from 3 micro-amperes of 1-Mev protons bombarding either lithium or fluorine.[40] The former target gives the greater yield, about 7 quanta per 10^{10} protons, or 8×10^5 quanta/min. altogether. Under the most favorable circumstances, all these gamma-rays would have passed through that thickness, ~ 6 mg/cm^2, of a sheet of uranium from which the fission particles are able to emerge. Even then, adopting the cross section we have estimated, we should expect an effect of

$$8 \times 10^5 \times 10^{-27} \times 6 \times 10^{-3} \times 6.06 \times 10^{23}/238 \sim 1 \text{ count/80 min}; \quad (77)$$

which is too small to have been observed. Consequently, we have as yet no test of the estimated theoretical cross section.

Conclusion

The detailed account which we can give on the basis of the liquid drop model of the nucleus, not only for the possibility of fission, but also for the dependence of fission cross section on energy and the variation of the critical energy from nucleus to nucleus, appears to be verified in its major features by the comparison carried out above between the predictions and observations. In the present stage of nuclear theory we are not able to predict accurately such detailed quantities as the nuclear level density and the ratio in the nucleus between surface energy and electrostatic energy; but if one is content to make approximate estimates for them on the basis of the observations, as we have done above, then the other details fit together in a reasonable way to give a satisfactory picture of the mechanism of nuclear fission.

39 W. Bothe and W. Gentner, Zeits. f. Physik **112**, 45 (1939).

40 R. B. Roberts, R. C. Meyer and L. R. Hafstad, Phys. Rev. **55**, 417 (1939).

27

Reprinted from *Phys. Rev.* **55:**103 (1939)

ENERGY PRODUCTION IN STARS

H. A. Bethe
Cornell University

In several recent papers,[1-3] the present author has been quoted for investigations on the nuclear reactions responsible for the energy production in stars. As the publication of this work which was carried out last spring has been unduly delayed, it seems worth while to publish a short account of the principal results.

The most important source of stellar energy appears to be the reaction cycle:

$$\begin{aligned} &C^{12}+H^1=N^{13}\ (a),\ N^{13}=C^{13}+\epsilon^+\ (b)\\ &C^{13}+H^1=N^{14}\ (c)\\ &N^{14}+H^1=O^{15}\ (d),\ O^{15}=N^{15}+\epsilon^+\ (e)\\ &N^{15}+H^1=C^{12}+He^4\ (f). \end{aligned} \tag{1}$$

In this cycle, four protons are combined into one α-particle (plus two positrons which will be annihilated by two electrons). The carbon and nitrogen isotopes serve as catalysts for this combination. There are no alternative reactions between protons and the nuclei $C^{12}C^{13}N^{14}$; with N^{15}, there is the alternative process

$$N^{15}+H^1=O^{16},$$

but this radiative capture may be expected to be about 10,000 times less probable than the particle reaction (f). Thus practically no carbon and nitrogen will be consumed and the energy production will continue until all protons in the star are used up. At the present rate of energy production, the hydrogen content of the sun (35 percent by weight[4]) would suffice for 3.5×10^{10} years.

The reaction cycle (1) is preferred before all other nuclear reactions. Any element *lighter* than carbon, when reacting with protons, is destroyed permanently and will not be replaced. E.g., Be^9 would react in the following way:

$$\begin{aligned} &Be^9+H^1=Li^6+He^4\\ &Li^6+H^1=Be^7\\ &Be^7+\epsilon^-=Li^7\\ &Li^7+H^1=2He^4. \end{aligned}$$

Therefore, even if the star contained an appreciable amount of Li, Be or B when it was first formed, these elements would have been consumed in the early history of the star. This agrees with the extremely low abundance of these elements (if any) in the present stars. These considerations apply also to the heavy hydrogen isotopes H^2 and H^3.

The only abundant and very light elements are H^1 and He^4. Of these, He^4 will not react with protons at all because Li^5 is unstable, and the reaction between two protons, while possible, is rather slow[5] and will therefore be much less important[6] in ordinary stars than the cycle (1).

Elements heavier than nitrogen may be left out of consideration entirely because they will react more slowly with protons than carbon and nitrogen, even at temperatures much higher than those prevailing in stars. For the same reason, reactions between α-particles and other nuclei are of no importance.

To test the theory, we have calculated (Table I) the energy production in the sun for several nuclear reactions, making the following assumptions:

(1) The temperature at the center of the sun is 2×10^7 degrees. This value follows from the integration of the Eddington equations with any reasonable "star model."[4] The "point source model" with a convective core which is a very good approximation to reality gives 2.03×10^7 degrees.[7] The same calculation gives 50.2 for the density at the center of the sun. The central temperature is probably correct to within 10 percent.

(2) The concentration of hydrogen is assumed to be 35 percent by weight, that of the other reacting element 10 percent. In the reaction chain (1), the concentration of N^{14} was assumed to be 10 percent.

(3) The ratio of the average energy production to the production at the center was calculated[7] from the temperature-density dependence of the nuclear reaction and the temperature-density distribution in the star.

TABLE I. *Energy production in the sun for several nuclear reactions.*

Reaction	Average Energy Production ϵ(erg/g sec.)
$H^1+H^1=H^2+\epsilon^++f.$*	0.2
$H^2+H^1=He^3$	3×10^{16}
$Li^7+H^1=2He^4$	4×10^{14}
$B^{10}+H^1=C^{11}+f.$	3×10^5
$B^{11}+H^1=3He^4$	10^{10}
$N^{14}+H^1=O^{15}+f.$	3
$O^{16}+H^1=F^{17}+f.$	10^{-4}

* "$+f.$" means that the energy production in the reactions following the one listed, is included. E.g. the figure for the $N^{14}+H^1$ includes the complete chain (1).

It is evident from Table I that only the nitrogen reaction gives agreement with the observed energy production of 2 ergs/g sec. All the reactions with lighter elements would give energy productions which are too large by many orders of magnitude if they were abundant enough, whereas the next heavier element, O^{16}, already gives more than 10,000 times too small a value. In view of the extremely strong dependence on the atomic number, the agreement of the nitrogen-carbon cycle with observation is excellent.

The nitrogen-carbon reactions also explain correctly the dependence of mass on luminosity for main sequence stars. In this connection, the strong dependence of the reaction rate on temperature ($\sim T^{18}$) is important, because massive stars have much greater luminosities with only slightly higher central temperatures (e.g., Y Cygni has $T=3.2\times10^7$ and $\epsilon=1200$ ergs/g sec.).

With the assumed reaction chain, there will be no appreciable change in the abundance of elements heavier than helium during the evolution of the star but only a transmutation of hydrogen into helium. This result which is more general than the reaction chain (1) is in contrast to the commonly accepted "Aufbauhypothese."

A detailed account of these investigations will be published soon.

[1] C. F. v. Weizsaecker, Physik. Zeits. **39**, 639 (1938).
[2] J. Oppenheimer and R. Serber, Phys. Rev. **54**, 540 (1938).
[3] G. Gamow, Phys. Rev. (in print).
[4] B. Strömgren, Ergebn. d. Exak. Naturwiss. **16** (1937).
[5] H. Bethe and C. Critchfield, Phys. Rev. **54**, 248 (1938).
[6] Only for very cool stars (red dwarfs) the H+H reaction may become important.
[7] The author is indebted to Mr. Marshak for these calculations.

28

NUCLEAR RESONANCE ABSORPTION OF GAMMA RADIATION IN Ir^{191}

Rudolf L. Mössbauer

This article was translated expressly for this Benchmark volume by R. Bruce Lindsay, Brown University, from Naturwissenschaften ***45:****538–539 (1958), by permission of the publisher, Springer-Verlag, Heidelberg. The figures have been reproduced from the original article.*

The nuclear-resonance fluorescence of gamma radiation is an effect hard to detect under normal conditions, since in their emission and absorption, the quanta suffer such large recoil-energy losses that the emission and absorption lines are shifted with respect to each other to a considerable extent, and hence in general the resonance condition is masked. One possibility for the compensation of the recoil-energy losses of the quanta is by inducing relative motion of the radiation source and the absorber with the help of a very rapid centrifuge [1] Another possibility is provided by the increase in the thermal motion of the emitting and absorbing atoms.[2] In investigations of the latter kind with Ir^{191}, we had found[3] that contrary to classical expectation at low temperature, a strong increase in the nuclear resonance absorption takes place. With the help of a theory due to Lamb,[4] this result can be interpreted as follows: In the solid body, the recoil momentum does not always lead to a change in the state of oscillation of the crystal lattice but rather that for a part of the quantum transitions the crystal as a whole takes up the recoil momentum. Correspondingly the theory leads, for the emission and absorption spectra, over and beyond the broad distribution reflecting the thermal motion of the bound atoms in the crystal grating, to some extraordinarily strong lines with the natural line width. Because of the vanishing recoil-energy losses, the lines appear undisplaced in the position associated with the resonance energy (the excitation energy of the nuclear level under investigation).

We have now demonstrated the existence of these undisplaced resonance lines with the help of a "centrifuge" method involving velocities of only a few cm per sec. Fig. 1 shows the experimental arrangement and Fig. 2 the experimental results for the 129 keV transition in Ir^{191}.

The above provides a new method for the immediate measurement of the level widths of the lower excited nuclear states. In our case described here, the level width of the 129 keV level in Ir^{191} agrees within the limits of experimental error with that obtained earlier[3] by a less direct method, the value found being 6.5×10^{-6}eV, corresponding to a state lifetime of 1.0×10^{-10} sec.

Laboratory of Technical Physics of
the Technische Hochschule, Munich
and the Max Planck Institute for
Medical Research, Heidelberg. (Aug. 13, 1958)

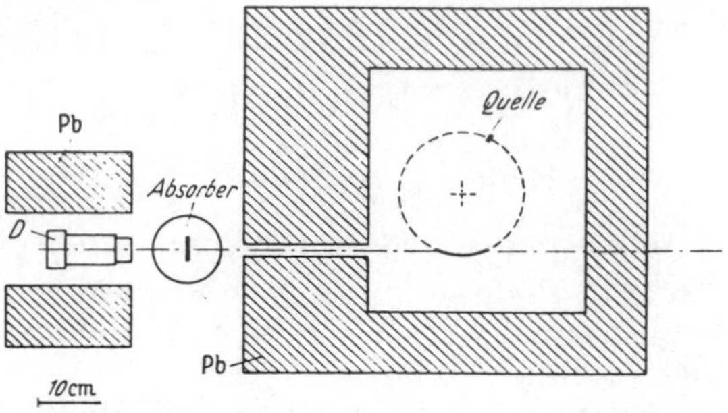

Figure 1. Experimental set-up. The detector D registers only those quanta emitted by the radiating source while passing point A in the circular motion of the source.

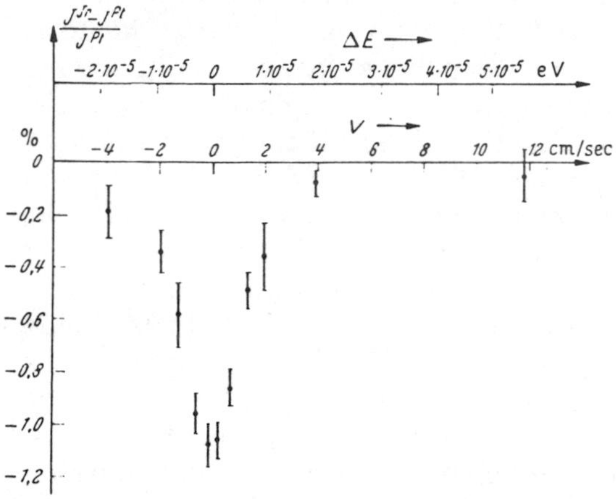

Figure 2. The difference in intensity ΔE between the gamma radiation (from the 129 keV transition in Ir^{191}) measured behind the resonance absorber and that measured behind a comparison platinum absorber, plotted as a function of the velocity v of the source relative to the absorber. The temperature of the source and absorber is kept at 88K. $\Delta E = (v/c)E_0$ deonotes the energy shift of the 129 keV gamma ray quantum.

REFERENCES

1. a. P. B. Moon, *Proc. Physical Society A 64,* 76, 1951.
   b. W. G. Davy and P. B. Moon, *Proc. Physical Society A* 66, 956, 1953.
2. a. K. G. Mahmfors; Ark. Fysik 6, 49, 1952.
   b. F. R. Metzger and W. B. Todd, *Physical Review, 95,* 853, 1954.
   c. F. R. Metzger, *Physical Review 97* 1258, 1955 and J. Franklin Institute *261,* 219, 1956.
   d. H. Schopper, *Zs. f. Physik, 144,* 476, 1956.
3. R. L. Mössbauer, *Zs. f. Physik, 151,* 124, 1958.
4. W. E. Lamb, Jr., *Physical Review, 55,* 190, 1939.

Part VI

HIGH-ENERGY ATOMIC DEVICES

Editor's Comments on Papers 29 Through 35

Interest in the collisions of charged particles began with the discovery of radioactivity and the emission of alpha and beta rays. The alpha particles were used as early atomic projectiles, but their velocity and energy depend on the substance from which they are emitted; hence they are not subject to the control of the experimenter. The desirability of being able to produce a stream of charged particles with controllable energy over a considerable range was soon realized, stimulating efforts to construct devices capable of accelerating such particles to any desired velocity. This section is devoted to reference to early attempts in this direction during the period covered by the present volume. The review is limited to the historical development of the earlier accelerators; no attempt is made to cover the more recent devices in the Gev range. These are adequately taken care of in the currently available literature. The bibliography at the end of the commentary may be consulted for further details.

The incentive to begin to think about high-energy particle accelerators seemed to strike several persons simultaneously in the late 1920s and early 1930s. From the standpoint of fundamental operation, the simplest of such devices was that invented by Robert J. Van De Graaff (1901-1967) in 1930. It continually provided charge to a metal sphere above ground by means of a rapidly ascending charged insulated belt to which charge was given at the bottom. The first machines developed in this manner were able to push particle energies up to 1.5 Mev. We reproduce here the first report on this device by the inventor (in 1931) as Paper 29. Later developments are mentioned in the bibliography at the end of the commentary.

More or less contemporaneously with Van De Graaff, Ernest O. Lawrence (1901-1958) was investigating an entirely different idea, the cyclotron. The basic principle of this device is the resonance observed in the motion of positively charged particles moving in circular orbits in the presence of a magnetic field perpendicular to the plane of the motion. A radio-frequency electric field is applied at certain points in the path so that the particle is given a further push every time it passes around the circle. This idea was first described in a short article in *Science* in 1930, reproduced here as Paper 30. A more detailed history of the development of the cyclotron and its various forms is given in articles published in *Physics Today* in 1959 by M. Stanley Livingston (1905-) and Edwin M. McMillan (1907-). These are reproduced here as Papers 31 and 32, respectively. While Lawrence was working on the cyclotron, John D. Cockcroft (1897-1967) and Ernest T. S. Walton (1903-) in England were pursuing a different approach: the invention of a particle accelerator based on a voltage multiplier connected with a vacuum tube provided with a proton source. Their first announcement of this device was a brief note in *Nature* in February 1932. Their first model was able to accelerate protons to 710 Kev. They later constructed more elaborate models providing much higher energies. The preliminary announcement is reproduced here as Paper 33.

The cyclotron and the Cockcroft-Walton accelerator proved useful primarily in accelerating protons or other positive ions. The acceleration of electrons covering high energy proved to be somewhat more difficult. The problem was first solved in 1941 by Donald W. Kerst (1911-) with the use of a variable magnetic field to provide a varying electric field on electrons moving in a circular orbit. The magnetic induction effect accelerates the electrons, and the field is so arranged as to keep the electrons in a circular orbit. In his first device, later called the betatron, Kerst achieved energies of the order of 350 Mev. His first article (1941) is reproduced here as Paper 34.

The early cyclotrons had difficulty keeping the charged particles in step with the electric field as they proceed around the ring. This trouble becomes particularly evident in the case of very high energies due to the relativistic change of mass under such conditions. A better method of synchronizing the field and the particle motion was devised by Edwin A. McMillan in 1945. This led to the development of the so-called synchrotron, with the achievement of much higher energies than with the use of the original cyclotron and betatron. The initial article by McMillan is reproduced here as Paper 35.

BIBLIOGRAPHY

Gouiran, R., 1967, *Particles and Accelerators,* McGraw-Hill, New York.

Richtmyer, F. K., E. H. Kennard, and J. M. Cooper, 1969, *Introduction to Modern Physics,* 3rd ed., McGraw-Hill, New York.

Wilson, R. R., and R. M. Littauer, 1960, *Accelerators, Machines of Nuclear Physics,* Anchor Books, Garden City, New York.

29

Reprinted from *Phys. Rev.* **38:**1919-1920 (1931)

A 1,500,000 VOLT ELECTROSTATIC GENERATOR

Robert J. Van De Graaff
National Research Fellow, Princeton University

The application of extremely high potentials to discharge tubes affords a powerful means for the investigation of the atomic nucleus and other fundamental problems. The electrostatic generator here described was developed to supply suitable potentials for such investigations. In recent preliminary trials, spark-gap measurements showed a potential of approximately 1,500,000 volts, the only apparent limit being brush discharge from the whole surface of the 24-inch spherical electrodes. The generator has the basic advantage of supplying a direct steady potential, thus eliminating certain difficulties inherent in the application of non-steady high potentials. The machine is simple, inexpensive, and portable. An ordinary lamp socket furnishes the only power needed. The apparatus is composed of two identical units, generating opposite potentials. The high potential electrode of each unit consists of a 24-inch hollow copper sphere mounted upon a 7 foot upright Pyrex rod. Each sphere is charged by a silk belt running between a pulley in its interior and a grounded motor driven pulley at the base of the rod. The ascending surface of the belt is charged near the lower pulley by a brush discharge, maintained by a 10,000 volt transformer kenotron set, and is subsequently discharged by points inside the sphere.

30

Reprinted from *Science* **72**:376–377 (1930)

ON THE PRODUCTION OF HIGH SPEED PROTONS

Ernest O. Lawrence and N. E. Edlefsen

Very little is known about nuclear properties of atoms because of the difficulties inherent in excitation of nuclear transitions in the laboratory. The study of the nucleus would be greatly facilitated by the development of a source of high speed protons having kinetic energies of about one million volt-electrons. The straightforward method of accelerating protons through the requisite difference of potential presents great difficulties associated with the high electric fields necessarily involved. Apart from obvious difficulties in obtaining such high potentials with proper insulation, there is the problem of development of a vacuum tube suitable for such voltages. A method for the acceleration of protons to high speeds which does not involve these difficulties is as follows. Semicircular hollow plates in a vacuum not unlike duants of an electrometer are placed in a uniform magnetic field which is normal to the plane of the plates. The diametral edges of the plates are crossed by a grid of wires so that inside each pair of plates there is an electric field free region. The two pairs of plates are joined to an inductance thereby serving as the condenser of a high frequency oscillatory circuit. Impressed oscillations then produce an alternating electric field in the space between the grids of the two pairs of plates which is perpendicular to the magnetic field. Thus during one half cycle the electric field accelerates protons into the region between one of the pairs of plates where they are bent around on a circular path by the magnetic field and eventually emerge again into the region between the grids. If now the time required for the passage along the semi-circular path inside the plates equals the half period of the oscillations, the protons will enter the region between the grids when the field has reversed direction and thereby will receive an additional acceleration. Passing into the interior of the other pair of plates the protons continue on a circular path of larger radius coming out between the grids where again the field has reversed and the protons are accelerated into the region of the first pair of plates, etc. Because the radii of the circular paths are proportional to the velocities of the protons the time required for traversal of a semicircular path is independent of the radius of the circle. Therefore once the protons are in synchronism with the oscillating field they continue indefinitely to be accelerated on passing through the region between the grids, and spiraling around on ever-widening circles gain more and more kinetic energy from the oscillating field. For example, oscillations of 10,000 volts and 20 meters wave-length impressed on plates of 10 cm radius in a magnetic field of 15,000 Gauss will yield protons having about one million volt-electrons of kinetic energy. The method is being developed in this laboratory, and preliminary experiments indicate that there are probably no serious difficulties in the way of obtaining protons having high enough speeds to be useful for studies of atomic nuclei.

31

Reprinted from *Phys. Today* **12:**18–23 (1959)

PART I

History of the CYCLOTRON

*By **M. Stanley Livingston***

On May 1, 1959, in memory of the late Ernest Orlando Lawrence, two invited lectures on the history of the cyclotron were presented as part of the American Physical Society's annual spring meeting in Washington, D. C. The present article is based on Prof. Livingston's talk on that occasion. The second speaker was E. M. McMillan, whose illustrated account also appears in this issue beginning on p. 24.

THE principle of the magnetic resonance accelerator, now known as the cyclotron, was proposed by Professor Ernest O. Lawrence of the University of California in 1930, in a short article in *Science* by Lawrence and N. E. Edlefsen.[1] It was suggested by the experiment of Wideröe [2] in 1928, in which ions of Na and K were accelerated to twice the applied voltage while traversing two tubular electrodes in line between which an oscillatory electric field was applied—an elementary linear accelerator. In 1953 Professor Lawrence described to the writer the origin of the idea, as he then remembered it.

The conception of the idea occurred in the library of the University of California in the early summer of 1929, when Lawrence was browsing through the current journals and read Wideröe's paper in the *Archiv für Elektrotechnik*. Lawrence speculated on possible variations of this resonance principle, including the use of a magnetic field to deflect particles in circular paths so they would return to the first electrode, and thus reuse the electric field in the gap. He discovered that the equations of motion predicted a constant period of revolution, so that particles could be accelerated indefinitely in resonance with an oscillatory electric field —the "cyclotron resonance" principle.

Lawrence seems to have discussed the idea with others during this early formative period. For example, Thomas H. Johnson has told the writer that Lawrence discussed it with himself and Jesse W. Beams during a conference at the Bartol Institute in Philadelphia during that summer, and that further details grew out of the discussion.

The first opportunity to test the idea came during the spring of 1930, when Lawrence asked Edlefsen, then a graduate student at Berkeley who had completed his thesis and was awaiting the June degree date, to set up an experimental system. Edlefsen used an existing small magnet in the laboratory and built a glass vacuum chamber with two hollow internal electrodes to which radiofrequency voltage could be applied, with an unshielded probe electrode at the periphery. The current to the probe varied with magnetic field, and a broad resonance peak was observed which was interpreted as due to the resonant acceleration of hydrogen ions.

However, Lawrence and Edlefsen had not in fact observed true cyclotron resonance; this came a little later. Nevertheless, this first paper was the initial announcement of a principle of acceleration which was soon found to be valid and which became the basis for all future cyclotron development.

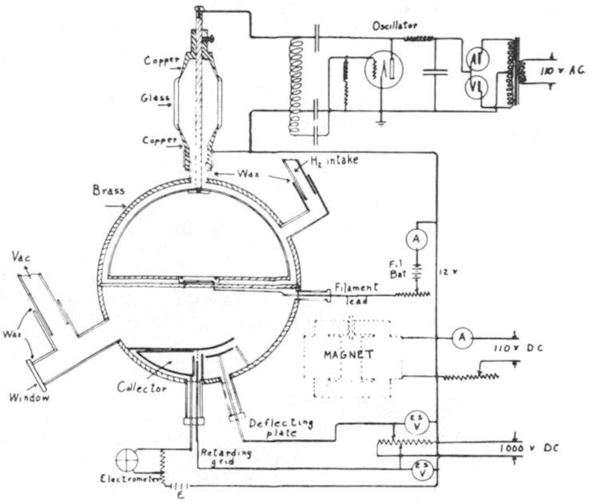

Fig. 1. Vacuum chamber of the first cyclotron. (PhD Thesis, M. S. Livingston, University of California, April 14, 1931)

M. Stanley Livingston, professor of physics at the Massachusetts Institute of Technology, is director of the Cambridge Electron Accelerator project at Harvard University, a program conducted under the joint auspices of Harvard and MIT.

Doctoral Thesis

IN the summer of 1930 Professor Lawrence suggested the problem of resonance acceleration to the author, then a graduate student at Berkeley, as an experimental research investigation. In my early efforts to confirm Edlefsen's results I found that the broad peak observed by him was probably due to single acceleration of N and O ions from the residual gas, which curved in the magnetic field and struck the unshielded electrode at the edge of the chamber.

It was my opportunity and responsibility to continue the study and to demonstrate true cyclotron resonance. A Doctoral Thesis [3] by the author dated April 14, 1931, reported the results of the study. It was not published but is on file at the University of California library. The electromagnet available was of 4-inch pole diameter. Fig. 1 is an illustration from this thesis, showing the arrangement of components which is still a basic feature of all cyclotrons. The vacuum chamber was made of brass and copper. Only one "D" was used, on this and several subsequent models; the need for a more efficient electrical circuit for the radiofrequency electrodes came later with the effort to increase energy. A vacuum tube oscillator provided up to 1000 volts on the electrode, at a frequency which could be varied by adjusting the number of turns in a resonant inductance. Hydrogen ions (H_2^+ and later H^+) were produced through ionization of hydrogen gas in the chamber, by electrons emitted from a tungsten-wire cathode at the center. Resonant ions which reached the edge of the chamber were observed in a shielded collector cup and had to traverse a deflecting electric field. Sharp peaks were observed in the collected current at the magnetic field for resonance with H_2^+ ions as shown in Fig. 2, a typical resonance curve taken from the thesis. Also present were 3/2 and 5/2 resonance peaks at proportionately lower magnetic fields, due to harmonic resonances of H_2^+ ions. By varying the frequency of the applied electric field, resonance was observed over a wide range of frequency and magnetic field, as shown in Fig. 3, proving conclusively the validity of the resonance principle.

The small magnet used in these resonance studies had a maximum field of 5200 gauss, for which resonance with H_2^+ ions occurred at 76 meters wavelength or 4.0 megacycles frequency. In this small chamber the final ion energy was 13 000 electron volts, obtained with the application of a minimum of 160 volts peak on the D. This corresponds to about 40 turns or 80 accelerations. A stronger magnet was borrowed for a short time, capable of producing 13 000 gauss, with which it was possible to extend the resonance curve and to produce hydrogen ions of 80 000 ev energy. This goal was reached on January 2, 1931.

The First 1-Mev Cyclotron

LAWRENCE moved promptly to exploit this breakthrough. In the spring of 1931 he applied for and was awarded a grant by the National Research Council (about $1000) for a machine which could give useful energies for nuclear research. The writer was appointed as an instructor at the University of California on completion of the doctorate in order to continue the research. During the summer and fall of 1931, the writer, under the supervision of Lawrence, designed and built a 9-inch diameter magnet and brought it into operation, first with H_2^+ ions of 0.5-Mev energy. Then the poles were enlarged to 11 inches and protons were accelerated to 1.2 Mev. This was the first time in scientific history that artificially accelerated ions of this energy had been produced. The beam intensity available at a target was about 0.01 microampere. The progress and results were reported in a series of three

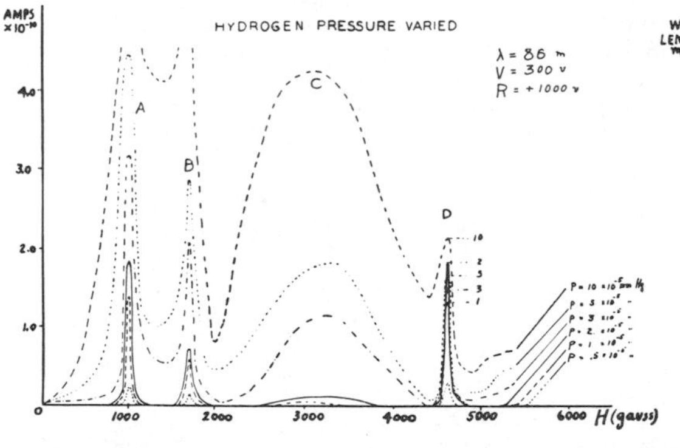

Fig. 2. Typical curves of current at the collector vs. magnetic field, showing resonant H_2^+ ions of 13 000 ev energy (peak D) and the variation of intensity with hydrogen gas pressure. (Thesis—Livingston)

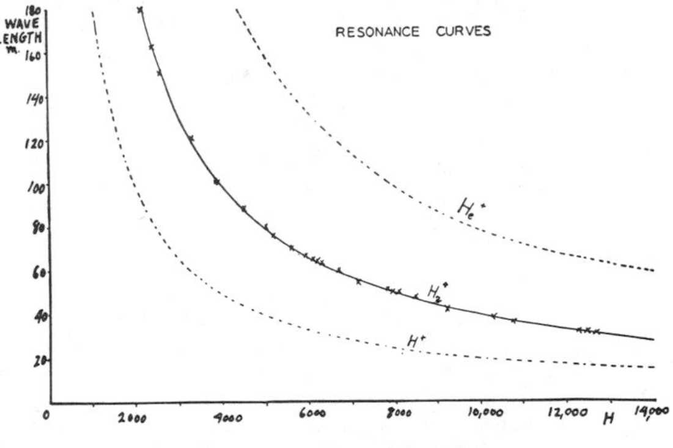

Fig. 3. Experimental values of cyclotron resonance for H_2^+ ions. (Thesis—Livingston)

[*Editor's Note:* Figures 4 and 5 appear at the end of this article.]

abstracts and papers by Lawrence and Livingston in *The Physical Review*.[4] Figs. 4 and 5 show the size and general arrangements of this first practical cyclotron.

Of course, Lawrence had other interests and other students in the laboratory. Milton White continued research with the first cyclotron. David Sloan developed a series of linear accelerators for heavy ions, limited by the radio power tubes and techniques available at that time, for Hg ions and later for Li ions. With Wesley Coates, Robert Thornton, and Bernard Kinsey, Sloan also invented and developed a resonance transformer using a radiofrequency coil in a vacuum chamber which developed 1 million volts. With Jack Livingood and Frank Exner he tried for a time to make this into an electron accelerator. I must again thank Dave Sloan for the many times that he assisted me in solving problems of the cyclotron oscillator.

The Race for High Voltage

TO understand the meaning of this achievement we must look at it from the perspective of the status of science throughout the world. When Rutherford demonstrated in 1919 that the nitrogen nucleus could be disintegrated by the naturally occurring alpha particles from radium and thorium, a new era was opened in physics. For the first time man was able to modify the structure of the atomic nucleus, but in submicroscopic quantities and only by borrowing the enormous energies (5 to 8 Mev) of radioactive matter. During the 1920's x-ray techniques were developed so machines could be built for 100 to 200 kilovolts. Development to still higher voltages was limited by corona discharge and insulation breakdown, and the multimillion volt range seemed out of reach.

Physicists recognized the potential value of artificial sources of accelerated particles. In a speech before the Royal Society in 1927 Rutherford expressed his hope that accelerators of sufficient energy to disintegrate nuclei could be built. Then in 1928 Gamow and also Condon and Gurney showed how the new wave mechanics, which was to be so successful in atomic science, could be used to describe the penetration of nuclear potential barriers by charged particles. Their theories made it seem probable that energies of 500 kilovolts or less would be sufficient to cause the disintegration of light nuclei. This more modest goal seemed feasible. Experimentation started around 1929 in several laboratories to develop the necessary accelerating devices.

This race for high voltage started on several fronts. Cockcroft and Walton in the Cavendish Laboratory of Cambridge University, urged on by Rutherford, chose to extend the known engineering techniques of the voltage-multiplier, which had already been successful in some x-ray installations. Van de Graaff chose the long-known phenomena of electrostatics and developed a new type of belt-charged static generator to obtain high voltages. Others explored the Tesla coil transformer with an oil-insulated high-voltage coil, or the "surge-generator" in which capacitors are charged in parallel and discharged in series, and still others used transformers stacked in cascade on insulated platforms.

The first to succeed were Cockcroft and Walton.[5] They reported the disintegration of lithium by protons of about 400 kilovolts energy, in 1932. I like to consider this as the first significant date in accelerator history and the practical start of experimental nuclear physics.

All the schemes and techniques described above have the same basic limitation in energy; the breakdown of dielectrics or gases sets a practical limit to the voltages which can be successfully used. This limit has been raised by improved technology, especially in the pressure-insulated electrostatic generator, but it still remains as a technological limit. The cyclotron avoids this voltage-breakdown limitation by the principle of resonance acceleration. It provides a method of obtaining high particle energies without the use of high voltage.

[*Editor's Note:* Figures 6 and 7 appear at the end of this article.]

The Cyclotron Splits its First Atoms

THE above digression into the story of the state of the art shows why the 1.2-Mev protons from the 11-inch Berkeley cyclotron were so important. This small and relatively inexpensive machine could split atoms! This was Lawrence's goal. This was why Lawrence literally danced with glee when, watching over my shoulder as I tuned the magnet through resonance, the galvanometer spot swung across the scale indicating that 1 000 000-volt ions were reaching the collector. The story quickly spread around the laboratory and we were busy all that day demonstrating million-volt protons to eager viewers.

We had barely confirmed our results and I was busy with revisions to increase beam intensity when we received the issue of the *Proceedings of the Royal Society* describing the results of Cockcroft and Walton in disintegrating lithium with protons of only 400 000 electron volts. We were unprepared at that time to observe disintegrations with adequate instruments. Lawrence sent an emergency call to his friend and former colleague, Donald Cooksey at Yale, who came out to Berkeley for the summer with Franz Kurie; they helped develop the necessary counters and instruments for disintegration measurements. Within a few months after hearing the news from Cambridge we were ready to try for ourselves. Targets of various elements were mounted on removable stems which could be swung into the beam of ions. The counters clicked, and we were observing disintegrations! These first early results were published on October 1, 1932, as confirmation of the work of Cockcroft and Walton, by Lawrence, Livingston, and White.[6]

The "27-inch" Cyclotron

LONG before I had completed the 11-inch machine as a working accelerator, Lawrence was planning the next step. His aims were ambitious, but supporting funds were small and slow in arriving. He was forced to use many economies and substitutes to reach his goals. He located a magnet core from an obsolete Poulsen arc magnet with a 45-inch core, which was donated by the Federal Telegraph Company. Two pole cores were used and machined to form the symmetrical, flat pole faces for a cyclotron. In the initial arrangement the pole faces were tapered to a 27½-inch diameter pole face; in later years this was expanded to 34 inches and still higher energies were obtained. The windings were layer-wound of strip copper and immersed in oil tanks for cooling. (The oil tanks leaked! We all wore paper hats when working between coils to keep oil out of our hair.) The magnet was installed in the "old radiation lab" in December 1931; this was an old frame warehouse building near the University of California Physics Building which was for years the center of cyclotron and other accelerator activities. Fig. 6 is a photograph of this magnet with the vacuum chamber rolled out for modifications.

Other dodges were necessary to meet the mounting bills for materials and parts. The Physics Department shops were kept filled with orders for machining. Willing graduate students worked with the mechanics installing the components. My appointment as instructor terminated, and for the following year Lawrence arranged for me an appointment as research assistant in which I not only continued development on the cyclotron but also supervised the design and installation of a 1-Mev resonance transformer x-ray installation of the Sloan design in the University Hospital in San Francisco.

The vacuum chamber for the 27-inch machine was a brass ring with many radial spouts, fitted with "lids" of iron plate on top and bottom which were extensions of the pole faces. This chamber is shown in Fig. 7. Sealing wax and a special soft mixture of beeswax and rosin were first used for vacuum seals, but were ultimately replaced by gasket seals. In the initial model only one insulated D-shaped electrode was used, facing a slotted bar at ground potential which was called a "dummy D". In the space behind the bar the collector could be mounted at any chosen radius. The beam was first observed at a small radius, and the magnet was "shimmed" and other adjustments made to give maximum beam intensity. Then the chamber was opened, the collector moved to a larger radius, and the tuning and shimming extended. Thus we learned, the hard way, of the necessity of a radially decreasing magnetic field for focusing. If our optimism persuaded us to install the collector at too large a radius, we made a "strategic retreat" to a smaller radius and recovered the beam. Eventually we reached a practical maximum radius of 10 inches and installed two symmetrical D's with which higher energies could be attained. Technical improvements and new gadgets were added day by day as we gained experience. The progress during this period of development from 1-Mev protons to 5-Mev deuterons was reported in *The Physical Review* by Livingston [7] in 1932 and by Lawrence and Livingston [8] in 1934.

I am indebted to Edwin M. McMillan for a brief chronological account of these early developments on the 27-inch cyclotron. (It seems that earlier laboratory notebooks were lost.) These records show, for example:

June 13, 1932. 16-cm radius, 28-meter wavelength, beam of 1.24-Mev H_2^+ ions.

August 20, 1932. 18-cm radius, 29 meters, 1.58-Mev H_2^+ ions.

August 24, 1932. Sylphon bellows put on filament for adjustment.

September 28, 1932. 25.4-cm radius, 25.8 meters, 2.6-Mev H_2^+ ions.

October 20, 1932. Installed two D's in tank, radius fixed at 10 in.

November 16, 1932. 4.8-Mev H_2^+ ions, ion current 10^{-9} amps.

December 2 5, 1932. Installed target chamber for studies of disintegrations with Geiger counter. Start of long series of experiments.

March 20, 1933. 5 Mev of H_2^+; 1.5 Mev of He^+; 2 Mev of $(HD)^+$. Deuterium ions acelerated for first time.

September 27, 1933. Observed neutrons from targets bombarded by D^+.

December 3, 1933. Automatic magnet current control circuit installed.

February 24, 1934. Observed induced radioactivity in C by deuteron bombardment. 3-Mev D^+ ions, beam current 0.1 microampere.

March 16, 1934. 1.6-Mev H^+ ions, beam current 0.8 microampere.

April–May, 1934. 5.0-Mev D^+ ions, beam current 0.3 microampere.

Those were busy and exciting times. Other young scientists joined the group, some to assist in the continuing development of the cyclotron and others to develop the instruments for research instrumentation. Malcolm Henderson came in 1933 and developed counting instruments and magnet control circuits, and also spent long hours repairing leaks and helping with the development of the cyclotron. Franz Kurie joined the team, and Jack Livingood and Dave Sloan continued with their linear accelerators and resonance transformers, but were always available to help with problems on the cyclotron. Edwin McMillan was a major thinker in the planning and design of research experiments. And we all had a fond regard for Commander Telesio Lucci, retired from the Italian Navy, who became our self-appointed laboratory assistant. As the experiments began to show results we depended heavily on Robert Oppenheimer for discussions and theoretical interpretation.

One of the exciting periods was our first use of deuterons in the cyclotron. Professor G. N. Lewis of the Chemistry Department had succeeded in concentrating "heavy water" with about 20% deuterium from battery acid residues, and we electrolyzed it to obtain gas for our ion source. Soon after we tuned in the first beam we observed alpha particles from a Li target with longer range and higher energy than any previously found in natural radioactivities—14.5-cm range, coming from the Li^6 (d,p) reaction. These results were reported in 1933 by Lewis, Livingston, and Lawrence,[9] and led to an extensive program of research in deuteron reactions. Neutrons were also observed, in much higher intensities when deuterons were used as bombarding particles, and were put to use in a variety of ways.

We had frustrations—repairing vacuum leaks in the wax seals of the chamber or "tank" was a continuing problem. The ion source filament was another weak point, and required continuous development. And sometimes Lawrence could be *very* enthusiastic. I recall working till midnight one night to replace a filament and to reseal the tank. The next morning I cautiously warmed up and tuned the cyclotron to a new beam intensity record. Lawrence was so pleased and excited when he came into the laboratory that morning that he jubilantly ran the filament current higher and higher, exclaiming each time at the new high beam intensity, until he pushed too high and burned out the filament!

We made mistakes too, due to inexperience in research and the general feeling of urgency in the laboratory. The neutron had been identified by Chadwick in 1932. By 1933 we were producing and observing neutrons from every target bombarded by deuterons.[10] They showed a striking similarity in energy, independent of the target, and each target also gave a proton group of constant energy. This led to the now forgotten mistake in which the neutron mass was calculated on the assumption that the deuteron was breaking up into a proton and a neutron in the nuclear field. The neutron mass was computed from the energy of the common proton group,[11] and was much lower than the value determined by Chadwick. Shortly afterward, Tuve, Hafstad, and Dahl in Washington, D. C., using the first electrostatic generator to be completed and used for research, showed that these protons and neutrons came from the $D(d,p)$ and $D(d,n)$ reactions,

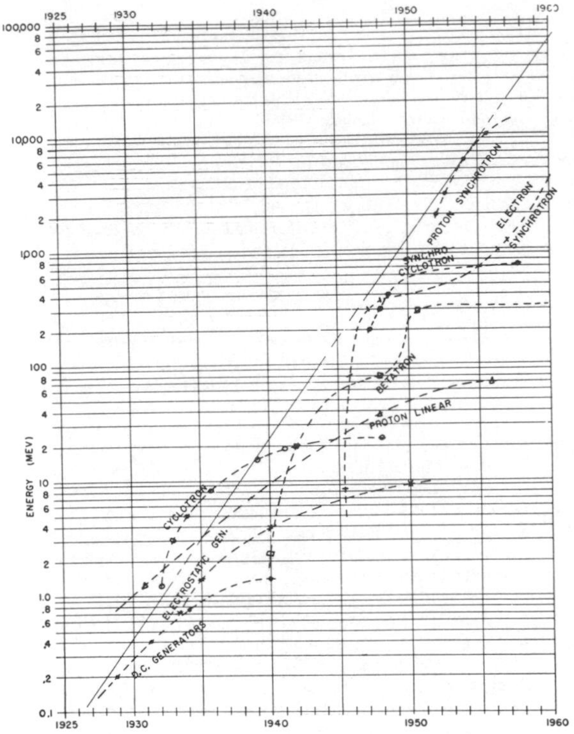

Fig. 8. Energies attained with accelerators as a function of time.

in which the target was deuterium gas deposited in all targets by the beam. We were chagrined, and vowed to be more careful in the future.

We also had many successful and exciting moments. I recall the day early in 1934 (February 24) when Lawrence came racing into the lab waving a copy of the *Comptes Rendus* and excitedly told us of the discovery of induced radioactivity by Curie and Joliot in Paris, using natural alpha particles on boron and other light elements. They predicted that the same activities could be produced by deuterons on other targets, such as carbon. Now it just so happened that we had a wheel of targets inside the cyclotron which could be turned into the beam by a greased joint, and a thin mica window on a re-entrant seal through which we had been observing the long-range alpha particles from deuteron bombardment. We also had a Geiger point counter and counting circuits at hand. We had been making 1-minute runs on alpha particles, with the counter switch connected to one terminal of a double-pole knife-switch used to turn the oscillator on and off. We quickly disconnected this counter switch, turned the target wheel to carbon, adjusted the counter circuits, and then bombarded the target for 5 minutes. When the oscillator switch was opened this time, the counter was turned on, and click-click--click---click----click. We were observing induced radioactivity within less than a half-hour after hearing of the Curie-Joliot results. This result was first reported by Henderson, Livingston, and Lawrence [12] in March, 1934.

I left the laboratory in July, 1934, to go to Cornell (and later to MIT) as the first missionary from the Lawrence cyclotron group. Edwin McMillan overlapped my term of apprenticeship by a few months, and stayed on to win the Nobel Prize and ultimately to succeed Professor Lawrence as director of the laboratory which he founded. McMillan can tell the rest of the story.

But it would be unfair to the spirit of Professor Lawrence if I failed to indicate some gleam of great things to come, some vision of the future. Recently I prepared a graph of the growth of particle energies obtained with accelerators with time, shown in Fig. 8. To keep this rapidly rising curve on the plot, the energies are plotted on a logarithmic scale. The curves show the growth of accelerator energy for each type of accelerator plotted at the dates when new voltage records were achieved. The cyclotron was the first resonance accelerator to be successful, and it led to the much more sophisticated synchronous accelerators which are still in the process of growth. The over-all envelope to the curve of log E vs time is almost linear, which means an exponential rise in energy, with a 10-fold increase occurring every 6 years and with a total increase in particle energy of over 10 000 since the days of the first practical accelerators. The end is not yet in sight. If you are tempted to extrapolate this curve to 1960, or even to 1970, then you are truly sensing the exponentially rising spirit of the Berkeley Radiation Laboratory in those early days, stimulated by our unique leader, Professor Lawrence.

References

1. E. O. Lawrence and N. E. Edlefsen, Science **72**, 376 (1930).
2. R. Wideröe, Arch. Elektrotech. **21**, 387 (1928).
3. M. S. Livingston, "The Production of High-Velocity Hydrogen Ions without the Use of High Voltages". PhD thesis, University of California, April 14, 1931.
4. E. O. Lawrence and M. S. Livingston, Phys. Rev. **37**, 1707 (1931); Phys. Rev. **38**, 136 (1931); Phys. Rev. **40**, 19 (1932).
5. Sir John Cockcroft and E. T. S. Walton, Proc. Roy. Soc. **136A**, 619 (1932); Proc. Roy. Soc. **137A**, 229 (1932).
6. E. O. Lawrence, M. S. Livingston, and M. G. White, Phys. Rev. **42**, 150 (1932).
7. M. S. Livingston, Phys. Rev. **42**, 441 (1932).
8. E. O. Lawrence and M. S. Livingston, Phys. Rev. **45**, 608 (1934).
9. G. N. Lewis, M. S. Livingston, and E. O. Lawrence, Phys. Rev. **44**, 55 (1933); E. O. Lawrence, M. S. Livingston, and G. N. Lewis, Phys. Rev. **44**, 56 (1933).
10. M. S. Livingston, M. C. Henderson, and E. O. Lawrence, Phys. Rev. **44**, 782 (1933); E. O. Lawrence and M. S. Livingston, Phys. Rev. **45**, 220 (1934).
11. M. S. Livingston, M. C. Henderson, and E. O. Lawrence, Phys. Rev. **44**, 781 (1933); G. N. Lewis, M. S. Livingston, M. C. Henderson, and E. O. Lawrence, Phys. Rev. **45**, 242 (1934); Phys. Rev. **45**, 497 (1934); M. C. Henderson, M. S. Livingston, and E. O. Lawrence, Phys. Rev. **46**, 38 (1934).
12. M. C. Henderson, M. S. Livingston, and E. O. Lawrence, Phys. Rev. **45**, 428 (1934); M. S. Livingston and E. M. McMillan, Phys. Rev. **46**, 437 (1934); M. S. Livingston, M. C. Henderson, and E. O. Lawrence, Proc. Natl. Acad. Sci. US **20**, 470 (1934); E. M. McMillan and M. S. Livingston, Phys. Rev. **47**, 452 (1935).

Fig. 4. 1.2 Mev H^+ cyclotron at the University of California.[4]

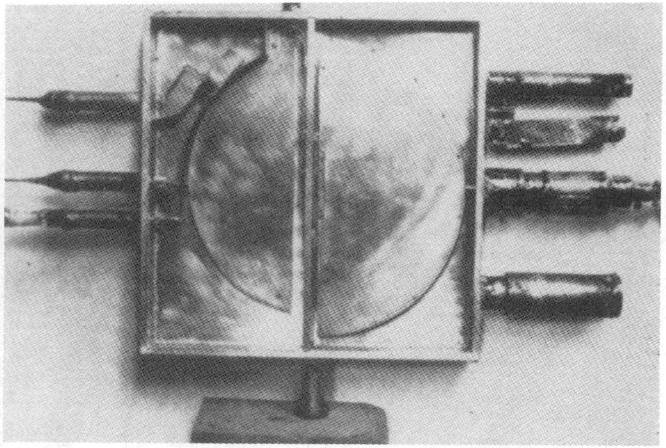

Fig. 5. Vacuum chamber for 1.2 Mev cyclotron with 11-inch pole faces.[4]

Fig. 6. The "27-inch" cyclotron which produced 5 Mev D^+ ions, with chamber rolled out.[7, 8]

Fig. 7. Vacuum chamber for the "27-inch" cyclotron.[7, 8]

32

Reprinted from *Phys. Today* **12**:24-34 (1959)

PART II

History of the CYCLOTRON

By ***Edwin M. McMillan***

Slide 1

As Dr. Livingston has told you, our activities overlapped by a few months, so that between us we can give a continuous story of cyclotron development as carried out at Berkeley under the guidance of Professor Lawrence. My start in his laboratory was in April of 1934, but I was around Berkeley before that working in Le Conte Hall on a molecular beam problem. Therefore, I have two kinds of early memories of the Radiation Laboratory at that time. One is as a place that I visited occasionally before I was working there; the other is as a place where I came to work, which I remember better, although it still seems like a very long time ago. The whole way of working was rather different from what it is in most laboratories today. We did practically everything ourselves. We had no professional engineers, so we had to design our own apparatus; we made sketches for the shop, and did much of our own machine work; we took all of our own data, did all our own calculations, and wrote all our own papers. Things are now quite different from that, because everybody does just his share and the operations have become much larger and more professional. While the modern method produces more results, perhaps this older way may have been more fun.

What I have done in preparing a paper to give here is to let it be based mainly on a set of lantern slides, because I think pictures are more interesting than words. I would like to run through these pictures and try to recall what they illustrate and the various incidents, some amusing, some otherwise, that go along with them.

I'm going to start with another picture of the 27″ cyclotron. This shows the machine as it looked in 1934

Nobel Laureate Edwin M. McMillan is director of the Lawrence Radiation Laboratory at the University of California at Berkeley, having succeeded to that post following the death of the Laboratory's original director, E. O. Lawrence, in 1958. The article is based on the second of two talks presented before the American Physical Society last May in memory of Prof. Lawrence.

Slide 2

when Stan and I were both there. (Slide 1.) Dr. Livingston is in the picture, and Professor Lawrence. The machine is the same as in the views shown by Stan, but here it is all assembled with the 27″ chamber in place. I have another view here of Professor Lawrence sitting at the control table, showing how one operated the machine. (Slide 2.) This was the major tool of nuclear research of that day and this was the control station. The switchboard in back had to do with magnet control, and the beam current was observed on the galvanometer scale.

As an illustration of the kind of experimental equipment one used, I have this drawing which was taken from a publication of about that period, early in 1935. (Slide 3.) This was an experiment to disintegrate aluminum with deuterons. You'll notice that in those days they were called deutons. The story was told that Ernest Rutherford objected to the name deuton; he didn't like the sound of it, but agreed that it would be all right if we put in his initials, E.R. (I don't think this story is really true, but at least the fact that it was told is true.) Well, these deutons came along inside the cyclotron vacuum chamber. This box is a cylinder soldered into the side of the brass wall of the cyclotron chamber. The beam that's inside passes through a thin target of aluminum foil. The secondary particles studied in this case were protons, making this an example of a *(d,p)* reaction. We didn't have that notation then, but that is what it would be called now. The secondary protons came out through a mica window, real old-fashioned mica, and into an ionization chamber counter and were counted. We measured the energy of these protons by simply sliding this counter back and forth inside of the tube, varying the range. We were measuring the range in air and plotting range curves in the way that one did in those days. This was considered a piece of research in physics; this was published, but nowadays, of course, nobody would think of doing a thing quite that way.

Now, let us go on to the development of the cyclotron itself. The two principal parameters of the cyclotron, as far as its use is concerned, are the energy of the particles and the intensity. With that older vacuum tank that we saw, the one that was in place

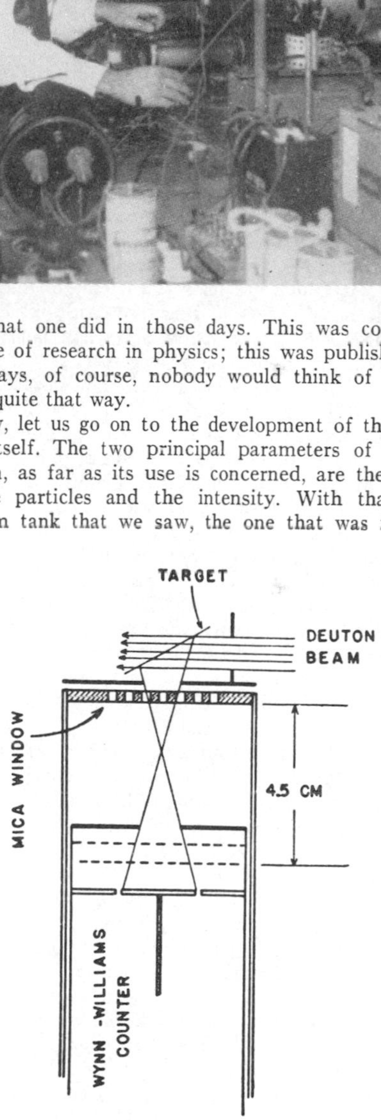

Slide 3: Arrangement of target, screens, and counter for bombarding in vacuum.

Slide 4

in Slide 1, the energy was up to about 3 Mev (this is the energy for deuterons). In 1936 a new chamber was built which is shown in the next slide. (Slide 4.) Comparing it with the chamber that Livingston showed, you'll see that there are many changes. For instance, the insulators for the two dees are made of Pyrex, with flanged ends which are clamped and bolted together rather than being waxed together, as the older ones were. The whole structure is more rugged, but there are still old-fashioned touches. You'll notice, coming into the center, a filament-type ion source that was still used then. Over in one corner you can see a glass liquid air trap, which was a very fragile and troublesome thing. People were always bumping into it and, of course, when it was bumped into, we'd have to pull the tank out, clean out the broken glass, and put the tank together all over again. With this new tank in place giving higher energies, up to 6 Mev for deuterons, and also larger currents, new types of experiments could be tried.

It was at about this time that an interest in biological work started in the laboratory, which has continued to the present. This was really started by John Lawrence, Ernest Lawrence's brother, who came out to the laboratory in 1935 to see what we were doing, and to see if there were any interest in the medical side. At this time biological experiments were started. I can recall the first time that a mouse was irradiated with neutrons. We put the mouse in a little cage and stuck him up on the side of the cyclotron tank and left him there for a while. Of course, nothing happened because there was not enough intensity. Then a serious attempt was made to see what neutrons did to mice. The first time this was done, it was done with an arrangement designed by Paul Aebersold in which the mouse could be put into the re-entrant tube shown in Slide 3, which was built into the cyclotron tank wall. In this way he could be close enough to the target to get some intensity. This mouse came out dead. This created a great impression at the time and I think perhaps was one reason why, in the Lawrence Radiation Laboratory, people have always been careful with radiation even though it was soon discovered that somebody had forgotten to turn on the air supply which was supposed to provide ventilation for this mouse so that he died of anoxia. Anyhow, it was a very dramatic thing at the time.

Also at about this same time the first radioactive tracer experiments on human beings were tried. The first one that I recall, and I think the first use anywhere of an artificially produced radioisotope in human beings, was an early experiment of Joseph Hamilton in which he measured the circulation time of the blood by a very primitive method. The experimental subject takes some radioactive sodium dissolved in water in the form of sodium chloride, drinks it, and then has a Geiger counter which he holds in his hand, so that when the radioactive sodium reaches the hand, it starts to register. His hand is in a lead box so that the stuff that's just in his body doesn't affect the counter by gamma rays. I brought along a picture of this setup. (Slide 5.) This drawing, I believe, was made by Dr. Hamilton's wife, who is an artist. It shows the hand in the box, you see this cutaway lead box, holding a Geiger counter; the beaker with the radio sodium isn't shown but you might have shown him in the act of drinking it. After he does this, within just a few seconds, you begin to get some registration. After a few minutes, you begin to get equilibrium, and from these observations you get the circulation time of the blood. This, of course, is a very simple beginning, just like the simple beginning in physics that I showed with the primitive experiment of a (d,p) reaction. There were also simple beginnings of therapeutic use, coming a little bit later, in which neutron radiation was used, for instance, in the treatment of cancer. These things have gone on and built up so that there's now a whole field of radio medicine which had its beginning back in that time.

Another highlight from 1936 was the first time that anyone tried to make artificially a naturally occurring radionuclide (of course, we didn't have the word nuclide

Slide 5

Slide 6

then, but that is what it would now be called). This, I think, was a fairly classical experiment because there were then some people who didn't quite believe that the artificial radioactive materials were on the same status as the naturally occurring ones. Jack Livingood put some bismuth in the deuteron beam of the cyclotron, with an energy of about 6 Mev. This is high enough that one does get an appreciable yield of the (d,p) reaction forming radium E, a bismuth isotope, which then decays into polonium. The periods and energies were identical to those of natural radium E and polonium, so everybody was happy. This was the first time that one had gotten up that far in the periodic table with a charged-particle disintegration experiment.

Another thing that we were trying to do then was to bring the beam out of the tank. It seemed that there might some day be a use for a beam extractor. And so these experiments, which were spoken of as snouting experiments—getting the beam out of a snout—were done. Of course, in that re-entrant tube I showed you in Slide 3 you could get the beam in air by putting a little window on one side and letting the beam travel about two inches across the diameter of that brass tube. It was in air but it wasn't really outside the tank, because it plunged back into the wall of the tube. To get the beam the rest of the way out, we had to increase the strength of the deflecting field and move the deflector plate out some, so as to get enough radial displacement that the beam would come out to the edge of the magnetic field. The next slide I'm going to show is the first time that a beam was brought outside the tank in this sense. I remember this occasion very well because when we first tried, the beam didn't quite clear the edge of the tank; it was coming almost tangentially and the thickness of the tank wall stopped it, so I spent about half a day with a file, curled up alongside the cyclotron, filing a groove in the thickness of the tank wall so that the beam could come out. This beam is shown in the next picture. (Slide 6.) There's a copper fitting, which is truly a snout, since it is a nose-shaped affair, which is fastened to the side of the tank, and the beam comes out through it, with the meter stick indicating the range. A little later, about two months after this, the beam was carried farther around—about a quarter of the way around the magnet. (Slide 7.) This shows where it came out of the window, way outside the cyclotron field. This, one might say, is the ancestor of modern beam extraction which has become a very sophisticated art in comparison to what it was in those days.

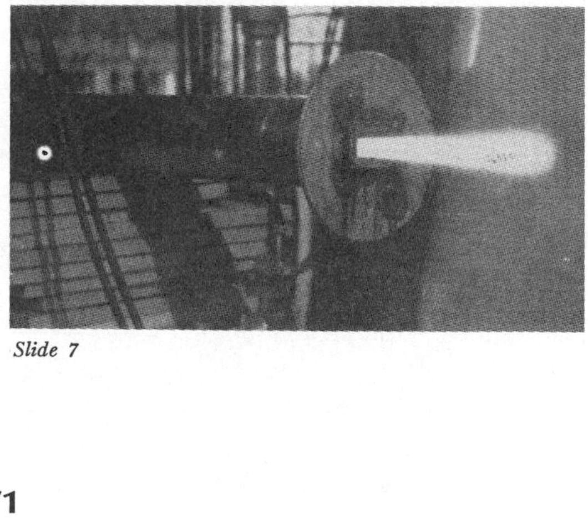

Slide 7

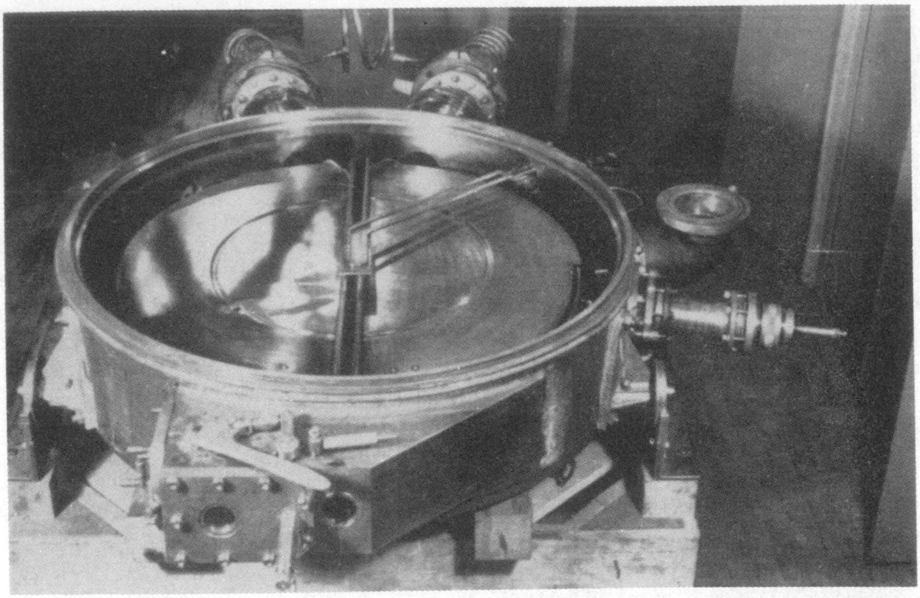

Slide 8

Everything up to now has been about the so-called 27-inch cyclotron. By the way, one thing I should apologize for at some point is my concentration on work at Berkeley. This is supposed to be the history of the cyclotron. But, in the first place, for some time this was the only place where there was a cyclotron, so that's where cyclotron history was being made. Secondly, this talk is in honor of Professor Lawrence, and that's where he was doing his work. Nevertheless, when we get to about 1936 or 1937, there did begin to be feedback of cyclotron lore from other parts of the world. At the end of 1936 there were about twenty other cyclotrons in the world; so the art had spread and things were coming back—improved ion sources, improved arrangements of radiofrequency systems, magnet control circuits, and all kinds of things. And from then on, of course, development of the cyclotron really became an international matter. Nevertheless, I shall continue to show pictures taken at Berkeley.

This is the 37-inch cyclotron, which used the same magnet as the 27-inch. (Slide 8.) All one had to do was to take out the old pole pieces, which had a reduced diameter, and put in larger diameter poles and the new tank shown on this slide. This was in late 1937 and begins to show signs of professionalism. You'll notice a gasket groove around the top, you'll notice nicely machined surfaces and things welded together, bolted together, and gasketed together, showing improved standards of design and construction. Still, you see a few old-fashioned touches; I think that the tank coil on the top side looks a bit primitive. We were still using a simple resonant circuit and two dees, plus an inductance forming the resonant circuit, which was loosely coupled to an oscillator. With this larger diameter and better designed tank, the deuteron energy was now up to 8 Mev. The energy was climbing; currents were getting up to 100 microamperes which were tremendous currents at that time. Experiments were beginning to get sophisticated. It was in 1938 that Dr. Alvarez first introduced the method of time of flight for neutrons. By keying the cyclotron beam and then having a gated detector, one could use the time of flight to measure the velocity and to select out given energy ranges. That was the birth of that method.

Also in this period the first artificial element, technetium, was discovered by Segrè and Perrier, using a piece of the cyclotron. As you know, where the beam emerges from the dee there is a deflecting plate, and just next to the deflecting plate the boundary of the dee is made of a thin sheet of metal which has to decide whether a given turn of the beam is inside the dee or outside. Because the front edge of this metal sheet gets a lot of bombardment it is always made of a refractory metal. In this case it was made of molybdenum, and when the old tank was dismantled and thrown away and the new tank went in (the one I just showed you), Segrè said he wanted the old molybdenum strip, so we gave it to him. He was then in Italy and, with the help of Perrier, was able to get a definite proof that it contained the new element technetium made by deuteron bombardment of the molybdenum. If it hadn't been for the fact that this particular spot —this particular item—in the anatomy of the cyclotron gets a lot of bombardment, this new discovery would have been considerably delayed.

Another thing that started in this period is that the theorists were getting interested in the cyclotron. Before, you see, it was an experimental art, and the people that worked on the cyclotron sort of knew what they were doing, but they weren't very sophisticated about it. They didn't stop to think much about how and why it worked; they knew that it worked and that was enough. But it was at this time that Bethe and Rose first pointed out the relativistic limit on cyclotron energies and, a little after that, that L. H. Thomas devised an answer to the relativistic limit. This answer turned out to be a little hard for the experimenters to understand, so it lay fallow for many years. Now, of course, everybody wants to build Thomas-type cyclotrons or FFAG machines (which are, in a sense, extreme examples of Thomas cyclotrons), so it is now a great thing; but it lay dormant for quite a while because nobody took it very seriously at first. Also, at that time in 1937, cyclotron energies were limited by other factors such as sizes, budgets, and things like that, and not by the relativistic effect, which was thought of before it became a practical limit.

Shortly after, in my history, comes the 60-inch cyclotron, which was the first really professionally designed cyclotron that was built in Berkeley. There were some elsewhere in the world, but this was the first in Berkeley. Before I get to that, as a sort of transition, I want to show a picture, taken around 1938, that illustrates several things. (Slide 9.) Now, let's see, what does this illustrate? First, it illustrates that people had started worrying about shielding against radiation around the cyclotron. Those were 5-gallon cans that were filled with water and simply stacked around and above the cyclotron to give shielding. As a matter of fact, the cans in this picture were originally on top of the cyclotron. They developed leaks, and the people that worked underneath would get tired of having water drip on them, and then they would take the leaky ones down and kick big dents in them so that nobody would be tempted to put them back.

The second thing that this slide illustrates is the type of building this work was done in, the Old Radiation Laboratory. I might inject a slightly sad touch, in that as I left Berkeley to come to this meeting, the last boards of the Old Radiation Laboratory were being battered down by a great big clam shell. We managed to save a few pieces as historical relics; otherwise it is all gone now. The third thing illustrated is that the man pictured here is Bill Brobeck, who was our first professional engineer hired at the Laboratory, showing the coming in of the more professional approach to the design and building of accelerators.

Now I will say a little about the 60-inch cyclotron, starting with a picture that was taken in 1938, showing

Slide 9

Slide 10 (Left to right and top to bottom): A. S. Langsdorf, S. J. Simmons, J. G. Hamilton, D. H. Sloan, J. R. Oppenheimer, W. M. Brobeck, R. Cornog, R. R. Wilson, E. Viez, J. J. Livingood, J. Backus, W. B. Mann, P. C. Aebersold, E. M. McMillan, E. M. Lyman, M. D. Kamen, D. C. Kalbfell, W. W. Salisbury, J. H. Lawrence, R. Serber, F. N. D. Kurie, R. T. Birge, E. O. Lawrence, D. Cooksey, A. H. Snell, L. W. Alvarez, P. H. Abelson.

Slide 11

the magnet, which had just been installed, and (approximately) the scientific staff of the Radiation Laboratory as of that time. (Slide 10.) You can see Professor Lawrence in the center, with Professor Birge, who was then chairman of the Physics Department, at his right, and Dr. Cooksey at his left. There are probably quite a few people here who can recognize themselves in that picture. It is always a little shocking to look at these old pictures and realize what time has done to us all!

This is the 60-inch cyclotron shortly after it was put together. (Slide 11.) A good many modifications in design were embodied in this machine and one of the most important ones is one of the things that fed back from outside; that is, the idea of getting away from glass insulators altogether, and having the dees plus their stems form a resonant system which is entirely inside the vacuum. The two tanks at the right hold the dee stems. This system has no insulators except in the lead-in for radiofrequency power. The power lead-ins come down the slanting copper cylinders at the right. The round tank on top of the magnetic yoke contains the deflector voltage supply, a rectified voltage supply under oil. And I think you can recognize the people in there: Don Cooksey, Dale Corson, Ernest Lawrence, Robert Thornton, John Backus, Winfield Salisbury, Luis Alvarez on the magnet coil, and myself on a dee-stem tank.

Now, just to show that physicists are not always serious, I have made a slide of the following pose: Laslett, Thornton, and Backus posing in the dee-stem tank of the 60-inch cyclotron before it was assembled. (Slide 12.) The next slide shows the control station of the 60 inch; now we have a real control desk, designed and not thrown together. (Slide 13.) At the desk are

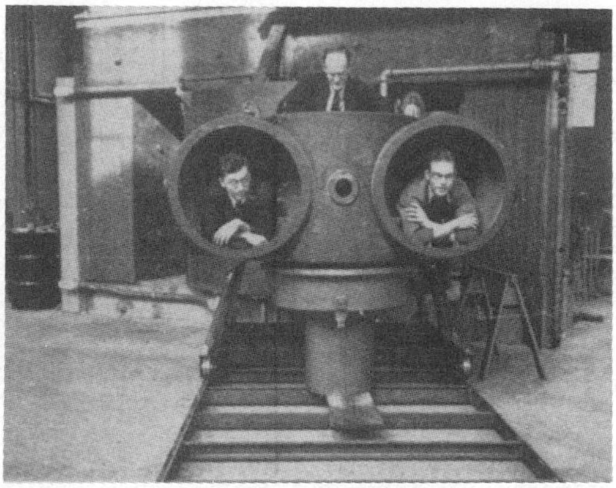

Slide 12

Professor Lawrence and his brother, John Lawrence, who initiated the medical work and is still continuing it at the Lawrence Radiation Laboratory.

We are now up to 1939. Fission has been discovered. I should point out that the old 37-inch cyclotron was still running, since the 60-inch had a new magnet and a new building, the Crocker Laboratory. So some of these things I mention now were done on the old 37-inch, which ran, with some interruptions, right up to the time when it was used for the first model test on the principle of the synchrocyclotron in 1946. But when fission was discovered, everybody in the Laboratory immediately jumped on the band wagon the way people do, and tried to think of an experiment having to do with fission. They did things with cloud chambers and counters and and made recoil experiments and various things of that kind.

Slide 13

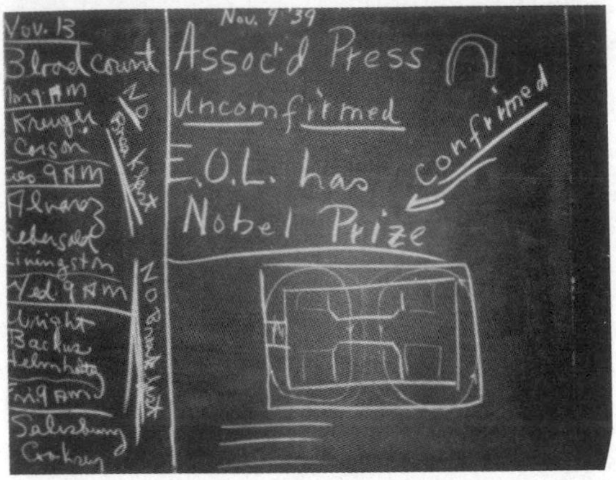

Slide 14

Slide 15

In 1940 came the first production of a transuranium element, which was done with the 60-inch cyclotron, although some of the experiments that led up to it had been done with the 37-inch. Carbon 14, which is perhaps the most important of all the tracer isotopes, came in this period. Kamen and Ruben finally pinned that down. Carbon 14 was something people had been trying to discover for a long time. I tried once myself but didn't quite get it. The mass 3 isotopes, hydrogen 3 and helium 3, were discovered then, helium 3 being found by an unusual use of a cyclotron. It was used as a mass spectrometer rather than as a cyclotron; that is, it was set for a resonance point for particles with charge 2 and mass 3, and when something came through at that resonance it had to be helium 3. This was done by Alvarez.

Perhaps the crowning event of that time was the award of the Nobel Prize to Professor Lawrence. Somebody, I think Cooksey, had the foresight to take a photograph of what appeared on the blackboard then. (Slide 14.) You see there is a two-stage announcement: first it says ASSOCIATED PRESS—UNCONFIRMED and then it says CONFIRMED with an arrow. The column down the left is a schedule of dates when people in the Laboratory received blood counts. I see Kruger, Corson, Alvarez, Aebersold, Livingston, Wright, Backus, Helmholz, Salisbury, and Cooksey. That's the other Livingston, Bob Livingston.

Now Ernest Lawrence was never a man who wanted to rest on achievement; he always wanted to go a step farther. I think it was this forward-looking spirit, and his ability to communicate it to others, that was his true greatness. So, even though the 60-inch cyclotron was a beautiful machine, was running fine, and was doing a great deal of important work, he had this dream of 100 million volts. I've looked at some of his old correspondence and it's always referred to as "100 million volts"; and he believed this could be achieved with the cyclotron. When he got the Nobel Prize, this helped things by focusing attention on this whole concept, and he set out on a campaign to see if he could raise the money to build a 100-million-volt cyclotron. Of course, in those days, money was essentially private money. There was no Manhattan District; there was no Atomic Energy Commission; and so he was trying to get this money by private funds.

In the course of this effort a good many things were written, plans and calculations were made, and one rather interesting picture was drawn which I will show you now. This was an artist's concept of a cyclotron for 100 million volts. (Slide 15.) This is what is now called the 184-inch cyclotron. You can see that this concept is rather different from the way the machine really looks. The magnet yoke is the same, but you see two tremendous tanks projecting on either side. Those were the dee-stem tanks; the beam was supposed to be deflected at one dee, make a complete turn inside, pass through a slit in one dee stem, and emerge as shown in the picture. But the important point this illustrates is that one was designing this as a conventional cyclotron, and one could easily estimate what dee voltages would be required to reach a given particle voltage, following the ideas of Rose and Bethe. We estimated that to reach 100 million electron volts for deuterons with this sort of design we would have wanted about

Slide 16

1.4 million volts between dees, or 700 000 volts to ground on each dee. We were planning to go ahead with floods of rf power to reach this voltage, and perhaps we would have, who knows?

The next picture shows a conference in the Old Radiation Laboratory, the building that has just been torn down, between Ernest Lawrence, Arthur Compton, Vannevar Bush, James Conant, Karl Compton, and Alfred Loomis. (Slide 16.) They were discussing ways of getting support for the project, and were obviously in a happy mood. Dr. Cooksey, who took the picture, tells me that someone had just told a joke, but the happiness may have had a deeper justification, for a few days later, on April 8, 1940, the Rockefeller Foundation decided to give 1.15 million dollars for the cyclotron. This grant, with help from the Regents of the University and others, made it possible for the project to go ahead.

But then the war came along and the whole effort of the Laboratory was diverted to other things. The magnet for this cyclotron was used for research on the electromagnetic isotope separation process, and it wasn't until quite a while later that it came back to use as a cyclotron. By that time other ideas had come out—the idea of the use of phase stability and frequency modulation—and so when the machine finally was built as a cyclotron, it didn't look like that picture on Slide 15 but looked like this one. (Slide 17.) Here

Slide 17

Slide 18

is what the 184-inch cyclotron looked like when it was first assembled. You can get some idea of the size, since there's a man there for scale. Of course, by now this is a synchrocyclotron. When I think of the history of the cyclotron in the sense of this talk, I think of it as the history of the fixed frequency cyclotron, so I won't say much more about this machine except that it does work. I'll show you a picture of about the way it looks today, encased in concrete blocks for shielding, which is a better solution to the shielding problem than 5-gallon cans of water. (Slide 18.) If you look hard, you can see a man in this picture, too.

I shall close this talk with an aerial view of the present establishment in Berkeley of the Lawrence Radiation Laboratory. (Slide 19.) In the foreground, in the circular building, is the Bevatron, which is of course a descendant of the cyclotron since it does use the magnetic resonance principle. A little farther back is another circular building which houses the 184-inch cyclotron, the machine I just showed you. The other buildings house other accelerators, research laboratories, shops, and all the things which make up the laboratory which really, one can say in all truth, is the outgrowth of the ideas and the faith and the strength of Professor Lawrence, in whose memory we have spoken today.

Slide 19

33

Reprinted from *Nature* **129:**242 (1932)

ARTIFICIAL PRODUCTION OF FAST PROTONS

J. D. Cockcroft
E. T. S. Walton

A HIGH potential laboratory has been developed at the Cavendish Laboratory for the study of the properties of high speed positive ions. The potential from a high voltage transformer is rectified and multiplied four times by a special arrangement of rectifiers and condensers, giving a working steady potential of 800 kilovolts. Currents of the order of a milliampere may be obtained at a potential constant to 1-2 per cent.

Protons from a discharge in hydrogen are directed down the axis of two glass cylinders 14 in. in diameter and 36 in. long, and accelerated by the steady potentials of the rectifier. They are then passed into an experimental chamber at atmospheric pressure through a mica window having a stopping power of about 1 mm. air equivalent. Luminescence of the air can easily be observed.

The ranges of the protons in air and hydrogen have been measured using a fluorescent screen as a detector. The range in air at S.T.P. of a proton having a velocity of 10^9 cm./sec. is found to be 8·2 mm., whilst the corresponding range for hydrogen is 3·2 cm. The observed ranges support the general conclusions of Blackett on the relative ranges of protons and α-particles, although the absolute values of the ranges are lower for both gases. The ranges and stopping power will be measured more accurately by an ionisation method.

The maximum energy of the protons produced up to the present has been 710 kilovolts with a velocity of $1 \cdot 16 \times 10^9$ cm./sec. and a corresponding range in air of 13·5 mm. at S.T.P. We do not anticipate any difficulty in working up to 800 kilovolts with our present apparatus.

Cavendish Laboratory,
Cambridge, Feb. 2.

34

Reprinted from *Phys. Rev.* **60:**47-53 (1941)

The Acceleration of Electrons by Magnetic Induction

D. W. Kerst*
University of Illinois, Urbana, Illinois

In the past the acceleration of electrons to very high voltage has required the generation of the full voltage and the application of that voltage to an accelerating tube containing the electron beam. No convenient method for repeated acceleration through a small potential difference has been available for electrons, although the method has been highly successful in the cyclotron for the heavier positive ions at velocities much less than the velocity of light.

Several investigators[1-4] have considered the possibility of using the electric field associated with a time-varying magnetic field as an accelerating force. This is a very attractive possibility because the magnetic field can be used to cause a circular or spiral orbit for the electron while the magnetic flux within the orbit increases and causes a tangential electric field along the orbit. The energy gained by the electron in one revolution is about equal to the instantaneous voltage induced in one turn of a wire placed at the position of the orbit. Since the electron can make many revolutions in a short time, it can gain much energy. The comparatively small momentum of a high energy electron requires correspondingly small values of Hr for high energy orbits. For example, the energy of an electron when $v \sim c$ is $KE = 3\times10^{-4}Hr - 0.51$ million electron volt. Thus with $H = 3000$ oersteds and $r = 5$ cm, the energy of the electron would be about 4 Mev, and the orbit could be held between the poles of a small magnet.

Because of the experimental experiences of previous investigators[1-3] with this method of acceleration, a rather detailed study of the focusing to be expected was made, and it is presented in the paper immediately following this one. With the results of this theoretical investigation to guide the design, it was possible to make an induction accelerator which produced x-rays of 2.3 Mev.[5,6] Briefly, in the focusing theory it is shown that:

1. The electrons have a stable orbit, "equilibrium orbit," where

$$\phi_0 = 2\pi r_0^2 H_0. \tag{1}$$

ϕ_0 is the flux within the orbit at r_0, and H_0 is the magnetic field at r_0. Both ϕ_0 and H_0 are increased during the acceleration process. This flux condition holds for all velocities of the electrons, and it shows that if a maximum flux density of 10,000 gauss is allowed in the iron then 5000 oersteds is the maximum magnetic field which can be used at the orbit.

2. In the plane of their orbits the electrons oscillate about their instantaneous circles, circles for which $p = eHr/c$, with an increasing frequency

$$\omega_r = \Omega(1-n)^{\frac{1}{2}}, \tag{2}$$

where Ω is the angular velocity of the electron in its orbit, and ω_r is 2π times the radial focusing frequency. The number n is determined by the

---

* On leave at the General Electric Company Research Laboratory.

1 G. Breit and M. A. Tuve, Carnegie Institution Year Book (1927-28) No. 27, p. 209.

2 R. Wideröe, Arch. f. Electrotechnik **21**, 400 (1928).

3 E. T. S. Walton, Proc. Camb. Phil. Soc. **25**, 469-81 (1929).

4 W. W. Jassinsky, Arch. f. Electrotechnik **30**, 500 (1936).

5 D. W. Kerst, Phys. Rev. **58**, 841 (1940).

6 D. W. Kerst, Phys. Rev. **59**, 110 (1941).

radial dependence of the magnetic field, which we take to be of the form $H \sim 1/r^n$. For radial focusing n must be less than unity.

3. Axial oscillations, oscillations perpendicular to the plane of the orbit, have

$$\omega_A = \Omega n^{\frac{1}{2}}. \quad (3)$$

For axial stability n must be greater than zero. If the beam is to be smaller in an axial direction than it is in a radial direction then $n > \frac{1}{2}$.

4. Decrease of the amplitude of both axial and radial vibration occurs because of the increase of the restoring force with increasing magnetic field. At nonrelativistic velocities the damping is

$$da/a = -dE/4E, \quad (4)$$

where dE/E is the fractional increase of the kinetic energy of an electron and da/a is the fractional decrease in amplitude of the oscillation about the instantaneous circle. This holds for both axial and radial oscillations.

5. Instantaneous circles not coincident with the equilibrium orbit shrink or expand toward coincidence:

$$dx/x = -dE/2E, \quad (5)$$

where x is the displacement of the instantaneous circle from the equilibrium orbit and dx is the shift of the circle toward the equilibrium orbit while the electron's energy increases by the fraction dE/E.

Because of the shrinking or expansion of the instantaneous circle of a displaced electron toward the equilibrium orbit and the decrease of the amplitude of oscillation of an electron about its instantaneous circle, it was expected that a cathode or an electron injector placed outside of the equilibrium orbit could shoot in electrons which would miss the injector on successive revolutions around the magnet. Furthermore, since an instantaneous circle for electrons with a constant injection energy exists within the acceleration chamber for a small but finite interval of time while the magnetic field is increasing, a finite amount of charge should be captured in orbits not striking the walls. With constant potential on an injector the electrons first would hit the outer wall of the chamber before the magnetic field had grown large enough to give the electrons a radius of curvature less than the radius of the wall. Then as the field increased it would reach values which give instantaneous circles within the vacuum tube. Eventually the field would be so large that the electrons coming out of the injector strike the walls or spiral around in small circles.

From Eq. (2) it can be seen that the spreading rays from the injector will form injector images $\pi(1-n)^{-\frac{1}{2}}$ radian apart, since if each ray of the beam oscillates about the instantaneous circle, it also oscillates about the central ray from the injector.

Equations (4) and (5) indicate that for large da or dx it is desirable to use a small injection voltage E, and a high voltage per turn dE. A shift da in two revolutions or dx in one revolution of about 1 mm when the displacement of the injector from the equilibrium orbit is $x = a = 1.5$ cm would require $dE/E \simeq \frac{1}{8}$. It is not possible to decrease the injection voltage E indefinitely since scattering increases when this is done, and it may be that the beam would be lost by scattering through an angle greater than that which magnetic focusing can handle. Conservative estimates on the scattering out of a cone of 7° half-angle, which is about the focusing limit, showed that at 10^{-6} mm of Hg air pressure in the tube, a maximum $Hr = 15{,}000$ gauss cm, and $f = 600$ cycles/sec. for the frequency of oscillation of the magnetic field, the injection voltage must exceed 100 volts for less than a 20-percent loss of beam. This then requires that dE or the voltage per turn at the equilibrium orbit must be at least 12 volts at the time of injection. In the magnet which was built this condition was easily satisfied, for 25 volts per revolution could be attained. There is no trouble with energy loss by the electrons in their long paths through the residual gas in the vacuum tube. Wideröe[2] was aware of the flux requirement (1), but apparently he did not realize that the voltage gain per revolution could not be too small compared with the injection voltage, or that it would be necessary to inject the electrons nearly tangent to the equilibrium orbit.

After the injected electrons have orbits approximately coincident with the equilibrium orbit, they should remain close to r_0. A simple method was used to shift the equilibrium orbit so that the electrons struck a thin tungsten target.

A portion of the flux through the center of the orbit passed through disks cemented to the center of the pole faces and made of material which saturated easily. As saturation progressed, the electrons had to move inward to a smaller radius of curvature in order that $\phi = 2\pi r^2 H$. Eventually the electrons grazed past the target producing thin target x-radiation.

Space Charge Effects

It is difficult to estimate the effect of space charge within the beam on the formation of images, since there are two focusing oscillations, axial and radial, in general having different frequencies. However, an estimate can be made of the upper limit of current which the magnetic forces can hold.

At injection the magnetic forces are small so that the estimate should be made with the magnetic field at this time. The electrons at the edge of the beam, which is assumed to be a cylinder of radius Δ, determined by the distance between the equilibrium orbit and the injector, will experience a repulsive force of

$$F_e = (cQe)/(10\pi r_0 \Delta) \quad (6)$$

dynes. Q is the number of coulombs in the orbit, and r_0 is the radius of the orbit.

For radial unbalanced magnetic forces we get

$$F_m = e^2 H^2 (1-n)\Delta / mc^2 \quad (7)$$

dynes acting on the electrons toward r_0. If $F_e = F_m$ at the edge of the beam, then this balance will hold within the beam for all other distances δ smaller than Δ, because the Q of (6) is proportional to δ^2. Thus

$$I = 10\pi f e r_0 \Delta^2 H^2 (1-n)/mc^3. \quad (8)$$

Or in terms of the injection voltage E,

$$I = \pi \Delta^2 f E (1-n)/(15 r_0 c). \quad (9)$$

I is the target current in amperes, f the frequency of oscillation of the magnetic field, and n gives the dependence of the magnetic field on r. Since the value of E which can be used depends upon the voltage per turn dE, which is also dependent on the peak magnetic field reached or the output energy, and also upon the orbit shift da which is required to miss the injector, we get

$$I = 2\pi^3 \Delta^3 f^2 (KE + \tfrac{1}{2})(1-n)/(45cda \times 10^4), \quad (10)$$

where KE is the final energy of the electrons in Mev. Inserting into (10) the constants of the machine which was built, we have $\Delta = 1.5$ cm, $f = 600$ cycles/sec., $KE = 2.3$ Mev, and $da \simeq 0.1$ cm. The target current should be about 0.03 microampere. This is fairly close to the lower limit of current as estimated from the x-ray output of the accelerator. This lower limit was found by using the thick target yield of Van Atta and Northrup[7] in the direction of the electron beam for comparison with the x-ray yield from the induction accelerator. It is found that the target current in the induction accelerator must be greater than 0.02 microampere. It is only possible to find the lower limit of this current since the output is thin target radiation.

A curious behavior which was immediately noticed when the apparatus began to work was that injection voltages much greater than the largest voltage which would allow the orbit to miss the injector could be used. In fact the yield increased greatly as this voltage was raised. This seems understandable on the basis of space charge forces spreading the beam away from the central ray of the injector.

At relativistic energies space charge forces are completely balanced by magnetic self-focusing of the beam, for the electric force on a stray electron at a distance Δ from the beam center is

$$e\mathcal{E} = 2\sigma e/\Delta, \quad (11)$$

where σ is the linear charge density in e.s.u./cm. The magnetic attraction due to the main current in the beam is

$$evH/c = (v/c)^2 2\sigma e/\Delta. \quad (12)$$

Thus it is evident that when $v \to c$, the magnetic pull of the beam for a stray electron just equals the electrostatic repulsion. Or, from the point of view of an observer on the electron, the spacing of the fixed number of electrons around the orbit will increase, since as $v \to c$ his yardstick becomes a smaller fraction of the circumference of the orbit.

[7] L. C. Van Atta and D. L. Northrup, Am. J. Roentgen. Rad. Ther. **41**, 633 (1939).

The Magnet

A photograph of the apparatus is shown in Fig. 1.

For the production of the rapidly changing magnetic field, with the proper dependence on the radius, finely laminated iron pole faces were made from 0.003-inch silicon steel sheets. The return magnetic circuit made from the same material was interleaved for mechanical strength; and the roughly circular central pole pieces were

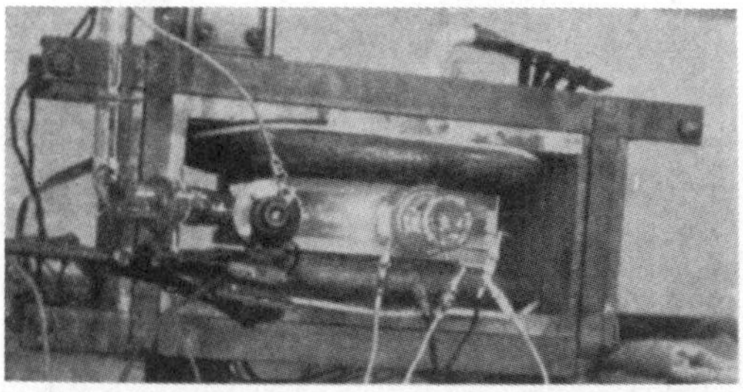

Fig. 1. The induction accelerator. The glass doughnut is between pole faces which are held apart by eccentric wedges.

formed by stacking with different widths of laminations as shown in Fig. 2. Each pole piece was capped by a disk of radially arranged laminations so that perfect circular symmetry was achieved at the pole surfaces (*B*, Fig. 2). The whole pole face was held together by a thick Transite or asbestos board ring about its perimeter, with cement of water glass and flint dust filling the cracks between the laminations and hardened by baking. The pole caps were held against the pole pieces by eccentric wedges between the Transite rings.

To supply the flux within the orbit which was necessary to hold the electrons out at the equilibrium orbit, two disks about two inches in diameter, each made of pressed iron dust of permeability about eight, were cemented onto the flat center portions of the pole faces. The thickness of these disks was chosen so that the equilibrium orbit was formed at about 7.5 centimeters radius, and the final adjustment of the position of the equilibrium orbit was made by painting a mixture of water glass and iron filings, or the turnings from the compressed iron dust disks, onto the surface of the iron dust disks so that the reluctance of the gap was correct. Since the iron particles in these disks are separated, the flux density through them is greater than the average flux density through the disks. This causes the saturation which shrinks the equilibrium orbit down to the target.

Because of the large leakage flux the main coil of the magnet had to be highly subdivided. Two hundred strands of No. 20 double enameled wire were made into a cable approximately 10 yards long which was twisted about ten times and which had its inside wires interchanged with its outside wires several times. This was formed into a 10-turn coil with about $\frac{1}{32}$ inch of insulation between turns. Two such 10-turn coils were made and fitted with large lugs of 1-inch copper tubing. The coils were wrapped with cotton tape, dipped in Bakelite varnish and baked. Figure 3 shows the circuit with the coils connected to a total of eighty 5-microfarad Pyranol condensers which were rated at 660 volts a.c. Energy was supplied to this resonating circuit by a 2-turn primary of stranded enameled wire around the pole pieces. The r.m.s. electromotive force in this primary

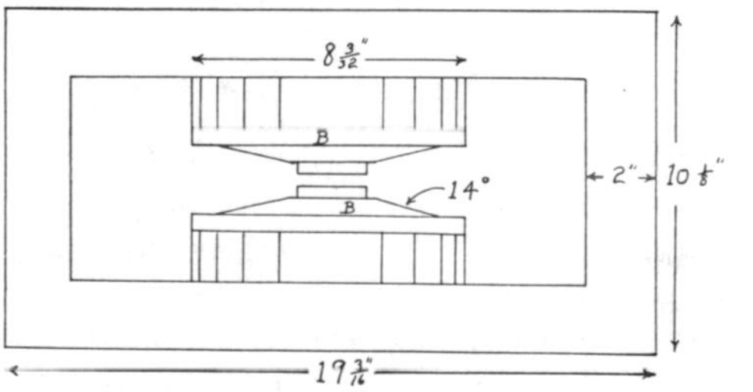

Fig. 2. Dimensions of the magnet. Parts *B* are the pole caps made with all laminations placed in a radial direction. The iron dust disks supply the central flux. Outside of the 14° conical surface a flat rim on the pole face tends to prevent too rapid decrease of the field at the edge of the gap.

could be run up above 100 volts, but 60 to 80 volts were usually used in operation since this was sufficient to saturate the powdered iron and to collapse the orbit. Power was supplied by a 4-kilowatt 600-cycle alternator driven by a direct-current motor with an adjustable speed.

To determine how the magnetic field decreased with r a small search coil was arranged so that it could be held between the poles at different radial distances from the center. The e.m.f. from this search coil was bucked against the e.m.f. from a voltage divider connected across a 1-turn coil about the leg of the magnet, equality of e.m.f. being determined by no deflection of the beam of an oscillograph. Since the wave form of the voltage on the search coil differed slightly from the wave form from the coil around the core, balance could be obtained only at one phase, and the phase of interest was that of zero magnetic field. The use of the oscillograph as a null instrument in this way proved to be sufficiently accurate. The pole faces gave a field following the $1/r^{\frac{2}{3}}$ law between a radius of 4.5 centimeters and 9.25 centimeters when the separation between the flat central portions of the pole faces was 2.8 centimeters. At $r=10$ cm the law was $1/r$.

To determine the position of the equilibrium orbit use was made of a concentric system of seven 1-turn coils set in grooves in a Bakelite disk which fit snugly around one of the iron dust disks. Relative values for the e.m.f.'s induced in these coils were determined by the same null method which was used for determining the shape of the field. These data could be used in several ways to find r_0. The simplest method is to find the radius of minimum electric field, since this will be r_0.

The Acceleration Chamber

It was desirable to have the volume available for electron orbits as large as possible. This required a doughnut-shaped glass vessel with walls parallel to the conical pole faces. The outer diameter was 20 cm, and the center of the doughnut was 7 cm in diameter. The glass work for this vessel required great skill.[8] A 20-cm spherical bulb with a wall thickness of 2.5 mm was flattened and dished conically on both sides so that it would fit between the magnet poles. The central 7-cm hole of the doughnut was formed by pushing the centers of the dished faces together and picking out the glass.

The inside of this bulb was coated with a thin conducting layer by chemical silvering. This coating is necessary to prevent stray charges from building up potentials on the walls. Contact is made to the silver coating which is then grounded outside of the tube. Resistances of the silver coat between 20 ohms and 300 ohms between test probes put in the side arms were used.

[8] I am indebted to the Vacuum Tube Department of the General Electric X-Ray Corporation for producing a correctly shaped tube.

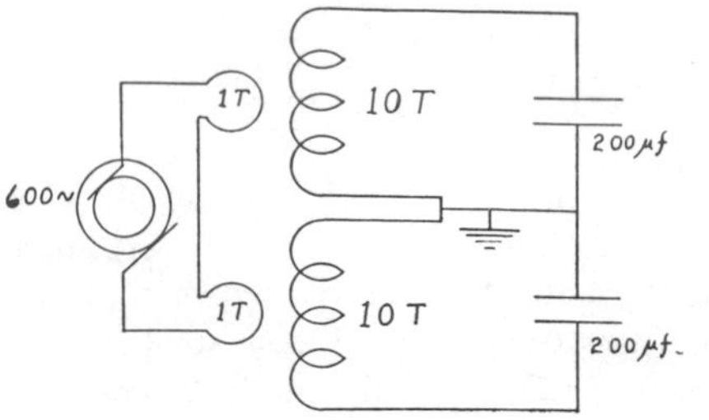

Fig. 3. The resonant circuit which energizes the magnet. Losses are supplied by a 4-kilowatt 600-cycle-per-second generator.

Both target and injector were mounted on the glass pinch seal which was waxed to a flare on the doughnut. The target was a piece of 0.015-inch tungsten sheet folded to present an edge of large radius of curvature to the electron orbit which shrinks inward toward it. Thin tungsten is used so that eddy currents will not be excessive. The injector was made of thin molybdenum sheets and a cylindrically spiraled tungsten wire for the filament. The whole assembly is shown in Fig. 4.

Fortunately the presence of this much metal near the orbit does not seem to disturb the local magnetic field too much. However, the orbits are destroyed if a piece of metal is brought up to the side of the glass doughnut while the machine is running. This was observed when a block of beryllium was put near the tube for experiments with photo-disintegration. The x-ray yield disappeared when the beryllium was too close. The original injector was arranged to send a beam of electrons in both directions around the magnet so that both directions of field could be used. Thus two beams of x-rays oppositely directed were produced.

Operation

Although the focusing theory shows that for the present design of the accelerator electrons should be injected at approximately 200 volts, it was found that the yield increased with voltage and that it was still increasing at 600 volts. Likewise the negative voltage on the small focusing plates G beside the filament gave an

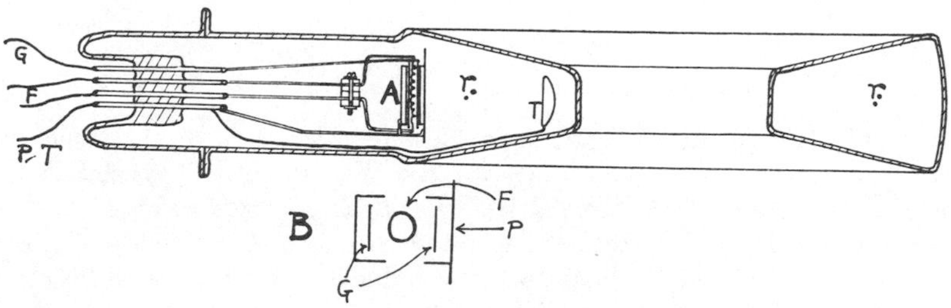

FIG. 4. Cross section of the doughnut-shaped acceleration chamber. The equilibrium orbit is at r_0, T is the tungsten target, A is the injector, B is a top view of the injector. A ribbon beam of electrons from the filament F is shot out through slots in the positive plates P. G are negative focusing electrodes.

increasing yield as it was raised to 200 volts. The beam from the injector was allowed to pour into the vacuum tube continuously.

Initially the yield obtained was approximately equivalent to 10 millicuries of radium when measured in a direction from the target at right angles to the electron beam as it strikes the target, and the yield was approximately ten times greater in the direction of the electron beam. Later the yield in the direction of the beam was increased to the equivalent of one gram of radium.

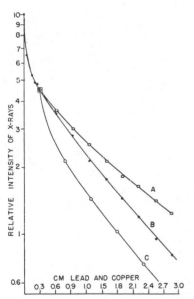

FIG. 5. The relative intensity of x-rays from the induction accelerator as a function of absorber thickness. A is absorption in copper. The absorption coefficient is 0.454 cm^{-1} which gives a monochromatic equivalent of 1.35 Mev for the x-rays. B is the curve for lead. The coefficient is 0.62 cm^{-1}, which corresponds to 1.4 Mev monochromatic equivalent. C is the lead absorption curve of Van Atta and Northrup for x-rays produced by 2.0-Mev electrons striking a thick target.

Figure 5 shows an absorption curve of the radiation as it passes through lead, and the comparison with two-million-volt x-rays from the M. I. T. electrostatic machine[7] shows that the energy reached in the induction accelerator is approximately 2.0 million electron volts and that the radiation produced is thin target radiation. With other iron dust disks cemented on the pole faces, absorption coefficients indicating 2.3-Mev x-rays have been obtained. By means of a collimated Geiger-Müller counter the source of the x-rays was shown to be the target. A large lead block with a small hole in it let an abundant amount of x-rays through to the Geiger-Müller tube only when it was pointed at the target and not when it was pointed at the injector or other portions of the vacuum tube. By connecting the vertical deflection plates of an oscillograph to the Geiger-Müller counter and the horizontal deflection plates to a 1-turn coil around one leg of the magnet, the phase of the magnetic field at which electrons struck the target could be determined. As was expected it was found that the greater the excitation of the dynamo which energizes the magnet, the earlier in the cycle the iron dust saturated and brought the beam in to the target. Lowering the primary voltage postponed sufficient saturation in the iron dust until the peak of the magnetic field was reached. The Geiger-Müller counter then gave x-ray pulses at the center of the oscillograph screen. This indicated a path length of about sixty miles from injector to target. If the primary voltage was lowered beyond this point, the yield disappeared, for the electrons were not drawn in to the target but were slowed down by the decreasing magnetic field. Fortunately the operation of the accelerator is not sensitive to the alignment of the pole faces. No difference in the output can be detected when the pole faces are placed off axis as far as a thirty-second of an inch. It is also surprising that vacuum requirements are not as severe as was expected. No rigorous outgassing is necessary and the apparatus has been run with a vacuum as poor as 10^{-5} mm Hg. The tube can be opened for changes and operated three-quarters of an hour after sealing shut.

At present, low flux densities have been used at the orbit. When these are increased, it should be possible to go to 5 million volts even with this small model. One of the promising possibilities for the induction accelerator as a research tool is that the electrons from the beam can come out through the glass walls of the doughnut after they strike the target. They should be fairly homogeneous in energy provided that the target has a high atomic number. The great increase in bremsstrahlung production with rising electron energy in addition to the concentration of this radiation in a cone of solid angle mc^2/E about the original electron direction gives the induction accelerator the possibility of providing an intense source of x-radiation for nuclear investigations. Since there is no evident limit on the energy which can be reached by induction acceleration, it may soon be possible to produce some small scale cosmic-ray phenomena in the laboratory.

I am indebted to Professor H. M. Mott-Smith and Professor R. Serber for many discussions of the theoretical aspects of this problem and to Mr. R. P. Jones for assistance in the construction of the magnet.

35

Reprinted from *Phys. Rev.* **68:**143–144 (1945)

The Synchrotron—A Proposed High Energy Particle Accelerator

Edwin M. McMillan
University of California, Berkeley, California
September, 5, 1945

ONE of the most successful methods for accelerating charged particles to very high energies involves the repeated application of an oscillating electric field, as in the cyclotron. If a very large number of individual accelerations is required, there may be difficulty in keeping the particles in step with the electric field. In the case of the cyclotron this difficulty appears when the relativistic mass change causes an appreciable variation in the angular velocity of the particles.

The device proposed here makes use of a "phase stability" possessed by certain orbits in a cyclotron. Consider, for example, a particle whose energy is such that its angular velocity is just right to match the frequency of the electric field. This will be called the equilibrium energy. Suppose further that the particle crosses the accelerating gaps just as the electric field passes through zero, changing in such a sense that an earlier arrival of the particle would result in an acceleration. This orbit is obviously stationary. To show that it is stable, suppose that a displacement in phase is made such that the particle arrives at the gaps too early. It is then accelerated; the increase in energy causes a decrease in angular velocity, which makes the time of arrival tend to become later. A similar argument shows that a change of energy from the equilibrium value tends to correct itself. These displaced orbits will continue to oscillate, with both phase and energy varying about their equilibrium values.

In order to accelerate the particles it is now necessary to change the value of the equilibrium energy, which can be done by varying either the magnetic field or the frequency. While the equilibrium energy is changing, the phase of the motion will shift ahead just enough to provide the necessary accelerating force; the similarity of this behavior to that of a synchronous motor suggested the name of the device.

The equations describing the phase and energy variations have been derived by taking into account time variation of both magnetic field and frequency, acceleration by the "betatron effect" (rate of change of flux), variation of the latter with orbit radius during the oscillations, and energy losses by ionization or radiation. It was assumed that the period of the phase oscillations is long compared to the period of orbital motion. The charge was taken to be one electronic charge. Equation (1) defines the equilibrium energy; (2) gives the instantaneous energy in terms of the equilibrium value and the phase variation, and (3) is the "equation of motion" for the phase. Equation (4) determines the radius of the orbit.

$$E_0=(300cH)/(2\pi f), \tag{1}$$

$$E=E_0[1-(d\phi)/(d\theta)], \tag{2}$$

$$2\pi\frac{d}{d\theta}\left(E_0\frac{d\phi}{d\theta}\right)+V\sin\phi = \left[\frac{1}{f}\frac{dE_0}{dt}-\frac{300}{c}\frac{dF_0}{dt}+L\right]+\left[\frac{E_0}{f^2}\frac{df}{dt}\right]\frac{d\phi}{d\theta}, \tag{3}$$

$$R=(E^2-E_r^2)^{\frac{1}{2}}/300H. \tag{4}$$

The symbols are:

E = total energy of particle (kinetic plus rest energy),
E_0 = equilibrium value of E,
E_r = rest energy,
V = energy gain per turn from electric field, at most favorable phase for acceleration,
L = loss of energy per turn from ionization and radiation,
H = magnetic field at orbit,
F_0 = magnetic flux through equilibrium orbit,
ϕ = phase of particle (angular position with respect to gap when electric field = 0),
θ = angular displacement of particle,
f = frequency of electric field,
c = light velocity,
R = radius of orbit.

(Energies are in electron volts, magnetic quantities in e.m.u., angles in radians, other quantities in c.g.s. units.)

Equation (3) is seen to be identical with the equation of motion of a pendulum of unrestricted amplitude, the terms on the right representing a constant torque and a damping force. The phase variation is, therefore, oscillatory so long as the amplitude is not too great, the allowable amplitude being $\pm\pi$ when the first bracket on the right is zero, and vanishing when that bracket is equal to V. According to the adiabatic theorem, the amplitude will diminish as the inverse fourth root of E_0, since E_0 occupies the role of a slowly varying mass in the first term of the equation; if the frequency is diminished, the last term on the right furnishes additional damping.

The application of the method will depend on the type of particles to be accelerated, since the initial energy will in any case be near the rest energy. In the case of electrons, E_0 will vary during the acceleration by a large factor. It is not practical at present to vary the frequency by such a large factor, so one would choose to vary H, which has the additional advantage that the orbit approaches a constant radius. In the case of heavy particles E_0 will vary much less; for example, in the acceleration of protons to 300 Mev it changes by 30 percent. Thus it may be practical to vary the frequency for heavy particle acceleration.

A possible design for a 300 Mev electron accelerator is outlined below:

peak H = 10,000 gauss,
final radius of orbit = 100 cm,
frequency = 48 megacycles/sec.,
injection energy = 300 kv,
initial radius of orbit = 78 cm.

Since the radius expands 22 cm during the acceleration, the magnetic field needs to cover only a ring of this width, with of course some additional width to shape the field properly. The field should decrease with radius slightly in order to give radial and axial stability to the orbits. The total magnetic flux is about $\frac{1}{5}$ of what would be needed to satisfy the betatron flux condition for the same final energy.

The voltage needed on the accelerating electrodes depends on the rate of change of the magnetic field. If the magnet is excited at 60 cycles, the peak value of $(1/f)(dE_0/dt)$ is 2300 volts. (The betatron term containing dF_0/dt is about $\frac{1}{5}$ of this and will be neglected.) If we let V = 10,000 volts, the greatest phase shift will be 13°. The number of turns per phase oscillation will vary from 22 to 440 during the acceleration. The relative variation of E_0 during one period of the phase oscillation will be 6.3 percent at the time of injection, and will then diminish. Therefore, the assumptions of slow variation during a period used in deriving the equations are valid. The energy loss by radiation is discussed in the letter following this, and is shown not to be serious in the above case.

The application to heavy particles will not be discussed in detail, but it seems probable that the best method will be the variation of frequency. Since this variation does not have to be extremely rapid, it could be accomplished by means of motor-driven mechanical turning devices.

The synchrotron offers the possibility of reaching energies in the billion-volt range with either electrons or heavy particles; in the former case, it will accomplish this end at a smaller cost in materials and power than the betatron; in the latter, it lacks the relativistic energy limit of the cyclotron.

Construction of a 300-Mev electron accelerator using the above principle at the Radiation Laboratory of the University of California at Berkeley is now being planned.

Part VII

ENERGY IN PLASMA, SOLID STATE, AND LIQUID PHYSICS

Editor's Comments on Papers 36 Through 46

An important aspect of the role of energy in atomic physics is the emission of electrons from hot bodies, generally known as thermionic emission. This is the basic phenomenon responsible for the operation of the thermionic vacuum tube.[1] Its history is a complicated affair, but the first definite discovery of thermionic emission was by Thomas A. Edison (1847–1931) in 1883. He did not follow up the scientific implications of his discovery. Detailed research on the phenomenon was carried out by the British physicist Owen W. Richardson (1879–1959) around 1915. He derived a fundamental equation for the thermionic current as a function of temperature and the properties of the emitter. Much further work was carried out during the period covered by the present volume. Much of this is well summarized in the review article by Saul Dushman (1883–1954) published in 1930. This article is reproduced in large part here as Paper 36.

Plasma physics is the study of the behavior of highly ionized gases. During the period 1925–1960 it has gained great importance in such fields as the state of matter in stars, in electronic vacuum tubes, magneto hydrodynamics, and the ultra-high-temperature nuclear-fusion process. The term *plasma* was first introduced in the sense used here by Irving Langmuir (1881–1957) in an article entitled "The Interaction of Electron and Positive Ion Space Charges in Cathode Sheaths" (June, 1929). This was based on the desire to understand better what goes on in electronic vacuum tubes, which were increasingly employed at that time for the production and amplification of electrical oscillations. A substantial part of Langmuir's article is reproduced here as Paper 37.

Another early concern with plasma physics was the development of magneto hydrodynamics by the Swedish physicist Hannes O. G. Alfven (1908–). So-called hydromagnetic waves are produced in a plasma exposed to a constant magnetic field, giving rise to a new branch of hydrodynamics. The seminal article by Alfven is reproduced here as Paper 38.

As a result of the introduction and development of quantum mechanics, great strides were made in solid-state physics in the period 1925–1960. This development was indeed foreshadowed by Sommerfeld's detailed calculations in the electron theory of metals based on the new quantum statistics (Paper 13). It was soon realized, however, that further progress in the physics of metals would depend on the investigation of the interaction of the electrons with the atoms of the solid. First steps in this direction were taken by Felix Bloch (1905–). They were materially extended by R. de L. Kronig (1904–) and William G. Penney (1909–) in a paper in which they replaced the interaction of the electrons with the lattice atoms by the quantum mechanical treatment of the motion of the electrons through a periodic potential field, in this way determining the electron energy eigenvalues. This seminal article is reproduced here as Paper 39. It was followed by a host of papers by many other investigators, which became increasingly elaborate. See the bibliography at the end of the commentary.

Semiconductors, materials with electrical conductivity intermediate between that of a metal and an insulator, were scientific curiosities for a long time save perhaps for the use of galena (zinc sulphide crystal) as a detector in early wireless telegraphy sets. Study of them has made enormous strides, however, in the last half-century, largely due to their increasing applications in current rectification and more recently in the transistor, which has largely replaced the thermionic vacuum tube in modern electronic devices. The associated literature is extensive. For a general review of the properties of semiconductors, we have chosen to include one by Frederick Seitz (1911-   ), and one by Robert J. Maurer (1913-   ). Seitz stresses in particular the chemical properties and lays emphasis on the role of free energy (*E-TS*, where *E* is the total energy of the system, *S* the entropy, and *T* the absolute temperature) in the behavior of semiconductors (it is a minimum for the stable state). Maurer stresses the role of impurities in semiconductor behavior; he also brings in the production of "holes" and "hole conduction." The articles of Seitz and Maurer are reproduced in part here as Papers 40 and 41, respectively.

Probably the most spectacular result of the study of semiconductors has been the invention of the transistor and its manifold elaborations. The first work on this device was performed by John Bardeen (1908-   ) and Walter H. Brattain (1902-   ) at the Bell Laboratories. They reported their fundamental discovery (1948) in a short paper that indicated how a block of semiconducting germanium appropriately introduced into an electrical circuit could serve effectively as a triode for the amplification, rectification, and production of electrical oscillations. The consequences of this invention in all branches of electronics have been enormous, particularly in the radio and television fields, as well as more recently in the construction of computers through the production of integrated circuits on single chips of a semiconductor like silicon. Most of the last-named developments have taken place since 1960, and hence no papers connected with them are reproduced here, although the bibliography at the end of this commentary makes reference to them. The seminal article by Bardeen and Brattain is reproduced here as Paper 42.

Another important development in solid-state physics in which energy plays a preeminent role is the discovery of nuclear magnetic resonance in 1945. Edward M. Purcell (1912-   ), Henry C. Torrey (1911-   ), and Robert V. Pound (1919-   ) were the first to detect this effect in a classical experiment in which they observed the resonance absorption of radio-frequency energy due to a transition between the energy levels corresponding to different orientations of nuclear spin when solid paraffin is placed in a constant magnetic field. This led to the extensive study of the fundamental properties of solids over a wide range. The seminal article by the three physicists is reproduced here as Paper 43.

The 1950s saw the introduction of another solid-state device (with applications to liquids and gases as well). This was the *maser* (acronym for *m*icrowave *a*mplification by *s*timulated *e*mission of *r*adiation), a device for the production of coherent radiation in a very narrow frequency band. The maser was first developed in 1954 by Charles H. Townes (1915–) in association with Herbert J. Zeiger (1925–) and James P. Gordon (1928–). They succeeded in making microwave radiation induce transitions to upper energy levels of the ammonia molecule, with resulting emission of coherent radiation in a very narrow frequency range as the molecule fell back to the lower energy level. This has led to the production of many types of masers. The basic 1954 article by Gordon, Zeiger, and Townes, is reproduced here as Paper 44.

After the pioneer work of Townes and his collaborators, it became obvious that the radiation pumping technique they used in stimulating coherent more-or-less single-frequency radiation could be readily applied to the optical frequency domain. This led to the invention of the laser (acronym for *l*ight *a*mplification by *s*timulated *e*mission of *r*adiation). The first such laser was constructed in 1960 by Theodore H. Maiman (1927–), who succeeded in stimulating a piece of ruby to emit intense coherent light of wavelength 6943Å. Maiman's seminal article is reproduced here as Paper 45.

During the period covered by the present volume, much progress was made in the development of a satisfactory theory of superconductivity—that is, the practically entire loss of resistance to the flow of electric current in certain solids close to absolute zero in temperature. Early work on the theory dating back to the 1930s realized that a suitable basis would be found in some treatment of the interaction between the electrons and the vibrations of the lattice of the superconducting solid. But in the early attempts, the details failed to account for all the experimental facts. In 1957 John Bardeen, Leon N. Cooper (1930–), and John R. Schrieffer (1931–) finally developed a microscopic theory in which they were able to construct a wave function for the electrons for which only the important ones are paired. The results of this theory agree with experiment. This work was first set forth in a 1957 article, reproduced here as Paper 46. An illuminating, semipopular account of the essential features of the Bardeen, Cooper, and Schrieffer theory can be found in the McGraw-Hill Encyclopedia of Science and Technology.[2]

NOTES AND REFERENCES

1. L. de Forest, "The Audion—Its Action and Some Recent Applications," *Journal of the Franklin Institute* **190** (1) (1920):1-38. Reprinted as Paper 11 in R. B. Lindsay, ed., 1977, *The Control of Energy,* Benchmark Papers on Energy, vol. 6, Stroudsburg, Pa.: Dowden, Hutchinson & Ross pp. 81-118.
2. "Superconductivity," 1982, *Encyclopedia of Science and Technology,* 5th ed., vol. 13, New York: McGraw-Hill, pp. 355-356.

BIBLIOGRAPHY

Brotherton, M., 1964, *Masers and Lasers, How They Work, What They Do,* McGraw-Hill, New York.

Moll, J., 1964, *Physics of Semiconductors,* McGraw-Hill, New York.

Richardson, O. W., 1921, *The Emission of Electricity from Hot Bodies,* Longmans, Green, London.

Spitzer, L., Jr., 1956, *Physics of Fully Ionized Gases,* Interscience, New York.

36

Reprinted from pages 381-391 of *Rev. Mod. Phys.* **2**:381-394 (1930)

THERMIONIC EMISSION

By Saul Dushman
Research Laboratory of the General Electric Company

TABLE OF CONTENTS

Introductory Remarks

THERMIONIC devices have become such an essential factor in so many purely scientific and technical applications that it is difficult to realize at the present time that only sixteen years ago the very existence of a pure electron emission from incandescent solids was questioned by many physicists of good repute.[1] The apparently trivial observations made by Edison on the discharge of negative electricity from the carbon filament of an incandescent lamp to an auxiliary electrode in the bulb was the beginning of a series of scientific investigations carried out by O. W. Richardson, A. Wehnelt, I. Langmuir, W. Schottky and a large number of other investigators. Their observations have led to the development of a number of hot cathode devices, the application of which in the radio and electrical industry in general has been of far reaching importance.

The use of the hot cathode is becoming increasingly important not only in these high vacuum devices, but also in connection with various types of gas discharges and arcs. It has appeared, therefore, that it might be well to take stock, as it were, of the state of our knowledge of thermionic phenomena at the present time. There is a certain measure of reason for this as a considerable period has intervened since the publication of the classical work on the subject by O. W. Richardson,[2] while the most recent summaries are available only in German treatises, such as those of W. Schottky.[3] To these excellent discussions of the subject the writer wishes here to express his indebtedness in the preparation of the following paper.

I. Technique of Electron Emission

A. Equations for emission. On the basis of certain theoretical considerations, which will be discussed in a subsequent section, O. W. Richardson derived an equation for electron emission as a function of the temperature of the form

$$I = aT^{1/2}\epsilon^{-b/T} \tag{1}$$

where I = emission per unit area,

T = absolute temperature,

and a and b are constants characteristic of the emitting surface.

Although Richardson[4] and M. v. Laue[5] had also pointed out that an equation of the form

$$I = AT^2\epsilon^{-b_0/T} \tag{2}$$

would be just as valid theoretically, and would be in satisfactory agreement with observed data, it was shown by S. Dushman[6] that on the basis of the third law of thermodynamics, A in this equation should be a universal constant having the value 60.2 amps/cm² per deg² while b_0 should vary with the nature of the emitter.

As will be discussed more fully in subsequent sections, b or b_0 is a measure

[1] In this connection it is of historical interest to read the discussion on this point by I. Langmuir in Proc. Inst. Rad. Eng. **3**, 261 (1915), and G. E. Review **18**, 327 (1915).

[2] O. W. Richardson, The Emission of Electricity from Hot Bodies, Longmans, Green and Company (1921).

[3] W. Schottky, H. Rothe and H. Simon, Wien-Harms' Handbuch der Experimentalphysik **13**, Akadem. Verlagsgesellschaft, m.b.H. Leipzig (1928).

[4] O. W. Richardson, Phil. Mag. **28**, 633 (1914).

[5] M. v. Laue, Jahrb. d. Elektronik u Radioakt. **15**, 205, 257 (1918).

[6] S. Dushman, Phys. Rev. **21**, 623 (1923).

of the latent heat of evaporation of the electrons, i.e., the energy necessary to get the electrons through the surface. While these are expressed in degrees Kelvin in the above equations, it is also customary to express them in terms of volts by means of the relation

$$b_0 k = \Phi_0 e \tag{3a}$$

where k denotes Boltzmann's constant, e, the charge on the electron, and Φ_0 is known as the "work function." Substituting the well-known values for e and k, it is readily shown that

$$\Phi_0 = 8.62 \times 10^{-5} b_0 \text{ (volts)}. \tag{3b}$$

It has been the object of a large number of investigators to determine the values of these emission constants a and b(or A and b_0) as accurately as possible, and in the following sections the data thus obtained will be reviewed. Since it has been found actually that the observed value of A in equation (2) is not independent of the composition of the emitting surface, a great many theoretical investigations have been published with the object of explaining this apparent anomaly. While these views will be discussed in a subsequent section, it would seem at the present time that the reliable data available are insufficient to draw any definite conclusions on this point. Therefore more emphasis should be laid on experimental investigations to obtain accurate emission data on those substances for which data are at present unavailable, and furthermore it is necessary to obtain data, independent of emission measurements, on the values of the work functions themselves, since from such data combined with observations on emission more exact values of the constant A could be obtained.

B. General remarks on conditions requisite for accurate determinations of emission constants. From equation (2) it is seen that

$$\frac{dI/I}{dT/T} = \frac{b_0}{T} + 2.$$

For tungsten, $b_0 = 52{,}400$, and therefore for $T = 2400°$K, the temperature coefficient of emission is approximately 24. In general, the value of b_0 is lower the lower the range of temperatures at which emission data can be obtained under practical conditions, so that the ratio b_0/T does not vary within very wide limits.

Thus the observations on emission should be made under conditions which will make it possible to determine the temperature accurately. Furthermore, since the emitting source is either in the form of a filament, or an extended surface, care has to be taken to determine as accurately as possible the actual area at the maximum temperature, and to introduce proper corrections, where necessary, for temperature gradients along the emitting surface. The most important precautions, however, in connection with observations on emission are those regarding the removal of adsorbed and occluded gases from the emitter itself, and the necessity of maintaining as nearly perfect vacuum conditions as possible. Most of the observations on emission made up to 1914, and a considerable number of those made since then, are almost worthless because of the poor vacuum conditions under which they were made. As a result largely of the work of I. Langmuir and his associates it has become recognized that adsorbed gases may alter the emission profoundly, and, furthermore, that if there is sufficient residual gas in the hot cathode tube, positive ion bombardment tends to decrease the emission, especially from coated surfaces or those covered with monatomic films of "active" substances.

At the present time it is not necessary to describe in detail the methods available for obtaining ideal vacuum conditions and for removing adsorbed gases from the surface of the emitter. These are described in several treatises on this subject,[7] as well as in numerous papers. It is also hardly necessary to remark that all stop-cocks, greased joints, etc., should be avoided in connection with the exhaust and preparation of tubes containing cathodes for which it is desired to determine the electron emissivity.

While the evaporation of tungsten with the bulb immersed in liquid air was used by Langmuir in his first investigations in this field, other "getters" (clean-up agents for producing high vacuum) have come into use with the advent of vacuum devices for radio. For this purpose, magnesium, calcium, barium and alloys of rare-earth metals ("Mischmetal") have been utilized. According to investigations carried out in this laboratory by Mrs. M. R. Andrews, barium cleans up practically all residual gases at ordinary temperatures, while magnesium is ineffective in the case of hydrogen, and calcium does not take up nitrogen to any great extent.

Extremely low pressures may also be obtained by the use of a side tube containing charcoal (which has previously been well exhausted) immersed in liquid air. In this case, however, care has to be taken to see that the liquid air is maintained at constant level during the series of measurements.

C. Forms of hot cathode tubes for investigation of emission. One of the simplest forms of hot cathode devices is that used by I. Langmuir[8] in his first investigations on the emission from tungsten and thoriated tungsten. Two filaments, one of tungsten, and the other of the metal under investigation are sealed into a lamp bulb. Platinum flush seals are inserted in the walls of the bulb opposite to the filaments, and the tube is exhausted on a high vacuum system consisting of mercury condensation pump with liquid air trap between bulb and pump. During exhaust, the bulb is baked out

[7] *References on High Vacuum Technique*
(1) L. Dunoyer, Vacuum Practice, G. Bell and Sons, London, 1926.
(2) F. H. Newman, The Production and Measurement of Low Pressures, D. Van Nostrand Company, New York, 1925.
(3) G. W. C. Kaye, High Vacua, Longmans, Green and Company, 1927.
(4) S. Dushman, High Vacuum, G. E. Review, 1922, also translated by R. G. Berthold and E. Reimann, Julius Springer, Berlin (1926).
(5) A. Goetz, Physik u. Technik des Hochvakuumes, Friedr. Vieweg and Sohn, Akt. Ges., Braunschweig (1926).

[8] I. Langmuir, Phys. Rev. **2**, 450 (1913); also Phys. Zeits. **15**, 516 (1914).

for about one hour to the maximum temperature to which the particular glass may be subjected without collapsing. Then the filaments are flashed for a few minutes at very high temperatures (just below the melting point), and for a longer time at lower temperatures (2400°K approximately for tungsten), to eliminate occluded gases and to dissociate or evaporate any oxide films on the surface. This high temperature treatment should be continued until, as indicated by a proper type of low pressure gauge, the evolution of gas has practically ceased, and the pressure in the tube is as low as possible. The bulb is sealed off, immersed in liquid air, and the tungsten filament heated at the same time to a temperature of about 2900–3000°K. At these temperatures the tungsten evaporates fairly rapidly, "cleans up" residual gases, and forms a deposit on the walls which is subsequently used as anode in the emission observations. The evaporation should be continued until the resistance of the deposit as measured between the flush seals is less than 100 ohms.

The tungsten filament and deposit are used as anode in the subsequent emission measurements on the second filament, and it is advisable to carry out these measurements with the bulb immersed in liquid air in order to clean up any further traces of gases which may be evolved from the heated cathode.

The filaments used should preferably have comparatively low voltage drop, but should be long enough to minimize the effect of errors in correcting for lead losses (see below.) To determine the temperature of the filament, accurate observations should be made on the voltage drop and filament current.

A more convenient modification of Langmuir's arrangement has been used by S. Dushman[9] and his associates in precision measurements on electron emission from various metals. Instead of volatilizing tungsten on the walls, a piece of calcium wire inserted in a tungsten spiral is used and the metal evaporated while the tube is on the pump. Care should be taken in using this method to see that the calcium wire used is not badly oxidized and the tube should be exhausted as soon as possible after the wire has been sealed in.

K. H. Kingdon[10a] and K. H. Kingdon and I. Langmuir[10b] have used a form of tube which, while more complicated in construction, has the advantage that no corrections have to be made for lead losses. The filament under investigation is suspended along the axis of three co-axial cylinders of similar diameter and arranged with minimum distance of separation between them. The emission is measured to the central cylinder, while the two end cylinders are connected to the filament and act as "guard rings" to prevent any electrons emitted from the ends of the filament from reaching the central anode.

When the material for which it is desired to obtain emission data cannot be drawn in wire form and is available only in the form of sheets, both the

[9] S. Dushman *et al*, Phys. Rev. **25**, 338 (1925): **29**, 857 (1927).

[10a] K. H. Kingdon, Phys. Rev. **24**, 510 (1924).

[10b] K. H. Kingdon and I. Langmuir, Phys. Rev. **22**, 148 (1923).

problem of heating the surface to a uniform temperature and that of measuring the maximum temperature accurately becomes very difficult. In the case of molten metals, A. Goetz has overcome these difficulties by an ingenious method which will be described in connection with the discussion of his observations.

D. Temperature scale. For a number of substances data are now available by means of which very accurate determinations may be made of the temperature of the emitting surface. The results of the large number of investigations on this subject have been reviewed comprehensively by E. Lax and M. Pirani.[11] Data on the total radiation from a black body an various solids have also been summarized by W. W. Coblentz.[12] References to some of the more important materials are given in the following section.

Undoubtedly the most accurate method for determining the temperature consists in the measurement of the brilliancy (candles per cm^2). The temperature coefficient, that is, the ratio of dB/B to dT/T (where B denotes the brightness in international candles per cm^2) for tungsten varies from 22.75 at $T=1000°$K to 8.45 at $T=3000°$K.[13] Thus with the exception of the electron emission and the rate of evaporation, the candle power shows the greatest variation with temperature, and therefore any inaccuracy in the determination of this value involves less error in that of the temperature than is involved in the determination of any other property of the material, such as resistance or watts radiated.

While the values of B as a function of T have been determined for a number of substances, it is possible to obtain a fair approximation to the true temperature of a material for which the average luminous emissivity[14] is unknown by the following methods.

By means of a photometer it is possible to determine the temperature T_c, at which the material emits light of the *same color* as a black body. This gives a value for the temperature which is higher than the true temperature. For tungsten the color temperatures corresponding to different values of the true temperature as reported by Worthing and Forsythe are given in Table I.

TABLE I. *Relation between color temperature (T_c), brightness temperature (T_s) and true temperature (T) for tungsten.*

T_c for tungsten	T	$(T_c-T)/T$	T_s for tungsten
1006	1000	0.006	966
1517	1500	.011	1420
2033	2000	.0165	1857
2557	2500	.023	2274
3094	3000	.031	2673

[11] Handbuch der Physik **19,** 1–45, **21,** 190–272, Julius Springer, Berlin, (1929).
[12] International Critical Tables 5, 238–245 (1929).
[13] H. A. Jones and I. Langmuir, G. E. Review **30,** 310, 354, 408 (1927).
[14] By average luminous emissivity is meant the ratio of the total normal brightness to that of a black body at the same temperature. For tungsten this ratio varies from 0.464 at $T=1000°$K to 0.440 at $T=3000°$K (Forsythe and Worthing, Astrophys. J. **61,** 126 (1925)).

The *brightness* of the material may be compared with that of a standard lamp for a given wave-length. Usually the determination is made for red light ($\lambda = 0.665\mu$). The temperature thus determined is known as the brightness temperature, T_s, and it is *lower* than the true temperature by an amount which increases with decrease in e_λ, the spectral emissivity (for this wave-length) at any given temperature, and also increases with the temperature. Thus for tungsten, e_λ for $\lambda = 0.665\mu$ varies from 0.456 at $T = 1000°K$ to 0.415 at $T = 3000°K$ and the observed values of T_s are given in the last column of Table I.

Therefore from determinations of both T_c and T_s it is possible to deduce fairly approximate values to the true temperatures. On the other hand, it is of course possible to determine the actual value of e_λ by methods described by both Langmuir and Worthing and Forsythe, and thus to calculate T, the true temperature, from optical pyrometer measurements of T_s.

Further information on methods of determining temperatures by optical methods is given in these publications as well as in reviews by W. E. Forsythe.[15] Methods of measuring temperatures by optical methods have also been discussed by P. D. Foote and C. O. Fairchild.[16]

At temperatures below 1100–1200°K, optical methods become impracticable and it is then necessary to refer to some other property of the material such as the watts radiated per unit area or the resistance. The former method is more accurate whenever data are available on the total radiant emissivity. Naturally a correction has to be applied for the cooling effect of the leads (see below), which can be obtained by means of measurements on different lengths of the filaments. Data on the energy radiated as a function of the temperature have been summarized in the review by Lax and Pirani[17] which contains complete references to the literature. The metals mentioned in this summary are the following: W, Mo, Ta, Pt, Os, Au, Ni, Fe, C, Ag, Cu, Zr.

(1) *Tungsten*:—The temperature scale for tungsten has been determined accurately as a result of the work of I. Langmuir and H. A. Jones[13] in the Research Laboratory of the General Electric Company at Schenectady; W. E. Forsythe and A. G. Worthing[18] in the Research Laboratory of the National Lamp Works of the General Electric Company at Cleveland, and C. Zwikker[19] in the Research Laboratory of the Philips' Lamp Company in Eindhoven, Holland.

The best method for determining the temperature, as mentioned previously, is to make a vacuum lamp with a V-shaped filament from a sample of the same wire as used in the emission observations. This filament is treated during exhaust in the *same manner* as the filament in the emission tube. The

[15] W. E. Forsythe, J. Opt. Soc. Am. and Rev. Scientific Inst. **16**, 307 (1928); J. Am. Ceramic Soc. **12**, 780 (1929).

[16] P. D. Foote and C. O. Fairchild, Symposium on Pyrometry, Am. Inst. Mining and Metallurgical Engineers, **338** (1920), p. 324. See also in the same publication the papers by W. E. Forsythe (p. 291) A. G. Worthing (p. 367), and E. P. Hyde (p. 285).

[17] Lax and Pirani, Handbuch der Physik **21**, 236–240.

[18] W. E. Forsythe and A. G. Worthing, Astrophys. J. **61**, 126 (1925).

[19] C. Zwikker, Physica **5**, 249 (1925), Archives Neerlandaises des Sciences **9**, 207 (1925).

lamp is then either pyrometered against a standard lamp, or else the candle power emitted from a definite area is measured as a function of filament current and volts. Knowing the brightness (candles/cm²) it is then possible to determine from the tables published by the above investigators the temperature as a function of filament current.

However, a very convenient method, and one which yields results almost as accurate, is to determine the watts consumed by the actual cathode used, and after correcting this for lead losses to calculate the watts/cm². From the tables, the corresponding temperature may then be determined quite accurately. According to Worthing and Forsythe, the energy radiated from tungsten as a function of the absolute temperature is given in watts/cm² by the relation.

$$\log_{10} E = 3.680(\log_{10} T - 3.3) - 1040/T + 1.900.$$

For a filament of known length, but unknown diameter, the value of $VA^{1/3}/l$ is a function of the temperature (independent of diameter), where V = voltage drop corrected for lead loss, and A is the current. Values of this function have also been tabulated by the above mentioned investigators.

(2) *Molybdenum and tantalum*: The brightness (candles/cm²) and intensity of radiation (watts/cm²) for seasoned molybdenum and tantalum have been determined by A. G. Worthing.[20] Either function, therefore, may be used as a measure of the temperature.

(3) *Platinum*: The spectral emissivity of well aged platinum as a function of the temperature has been determined by Worthing[21] for the wave-length $\lambda = 0.665\mu$. The value varies linearly with temperature from 0.295 at T = 1200°K to 0.310 at T = 1850°K. The procedure used by L. A. DuBridge[22] in measuring electron emission from this metal consisted in determining the brightness temperature by comparison with a calibrated optical pyrometer in which a standard tungsten lamp was used and then correcting for the difference in emissivities of tungsten and platinum by means of Worthing's values given above.

W. Geiss[23] has determined the total radiation intensity for the metal and finds that this may be expressed as a function of the temperature by the relation

$$W = aT^n \cdot \sigma T^4 \text{ watts/cm}^2$$

where $a = 6.22 \times 10^{-4}$

$n = 0.767$

σ = Stefan-Boltzmann constant

$= 5.75 \times 10^{-12}$ watts·cm⁻² deg.⁻⁴

The total radiation emissivity from this metal as a function of temperature has also been investigated by C. Davisson and J. R. Weeks.[24]

[20] A. G. Worthing, Phys. Rev. **28**, 190 (1926).
[21] A. G. Worthing, Phys. Rev. **28**, 174 (1926).
[22] L. A. DuBridge, Phys. Rev. **31**, 236 (1928); **32**, 961 (1928).
[23] W. Geiss, Physica **5**, 203 (1925).
[24] C. Davisson and J. R. Weeks, J. Opt. Soc. and Rev. Scient. Inst. **8**, 581 (1924).

(4) *Carbon*: The spectral emissive power of graphite at $\lambda = 0.660\mu$ has been determined by C. H. Prescott and W. B. Hincke[25] for the temperature range 1250°K to 2700°K. From their data the true temperature may be calculated from values of the brightness temperature as determined by an optical pyrometer.

Assuming that carbon has an average total emissivity of 0.85, the energy radiated as a function of the temperature may also be calculated by means of the Stefan-Boltzmann law.

(5) *Nickel:* The spectral emissivity of this metal at three different wavelengths has been measured by A. G. Worthing.[26] The emissivity does not change with temperature, and for $\lambda = 0.665\mu$ the mean value 0.375 was obtained over the range 1200°K to 1650°K.

The total radiation from nickel and cobalt has been measured by C. L. Utterback.[27] The following equations were derived as expressions for the relation between energy and temperature. For nickel,

$$E = C_1 T^{5.29} \text{ for } 650° < T < 1400°\text{K}$$

$$E = C_2 T^{4.75} \text{ for } 1450° < T < 1600°\text{K}$$

For cobalt

$$E = C_3 T^{5.20} \text{ for } 672° < T < 1320°\text{K}$$

$$E = C_4 T^{4.62} \text{ for } 1380° < T < 1590°\text{K}$$

The exact values of the constants are not given.

(6) *Oxide coated filaments.* The temperature scale for these filaments differs both with the treatment and composition of the surface layer and with the nature of the coated material. Therefore this point is more conveniently treated in connection with the discussion of the emission data.

E. Lead loss correction. Owing to cooling effect by the leads, the length of filament from which the emission is obtained is less than the actually measured length. The problem of correcting for this effect of leads was first discussed by I. Langmuir,[28] and then by A. G. Worthing.[29] The paper by W. E. Forsythe and A. G. Worthing[18] gives tables from which the effect of temperature distribution along a filament may be calculated. A mechanical method for solving the differential equation involved in such problems has been developed by V-Bush and K. E. Gould.[30] However, a most comprehensive discussion of the whole problem has been published in a recent paper by I. Langmuir, S. MacLane and K. B. Blodgett,[31] and the following remarks represent a summary of the relations derived in this paper.

We consider a filament fastened to two leads which are presumably at a low temperature as compared with the maximum temperature T_m, at the

[25] C. H. Prescott and W. B. Hincke, Phys. Rev. **31**, 130 (1928).
[26] A. G. Worthing, Phys. Rev. **28**, 174 (1926).
[27] C. L. Utterback, Phys. Rev. **34**, 785 (1928).
[28] I. Langmuir, Trans. Faraday Soc. **17**, 21 (1921).
[29] A. G. Worthing, Jr. Frank. Inst. **194**, 597 (1922).
[30] V. Bush and K. E. Gould, Phys. Rev. **29**, 337 (1927).
[31] I. Langmuir, S. MacLane and K. B. Blodgett, Phys. Rev. **35**, 478 (1930).

center of the filaments. If the whole filament were at this temperature, the observed value of any property (e.g., watts radiated, electron emission), which we shall denote by H, would be H_m. The ratio $f = H_m/H$ therefore gives the correction factor by which the observed value has to be multiplied to correct for the effect of leads. Let ΔH denote the decrease in H due to one lead. Then

$$f = \frac{H_m \cdot}{H_m - 2\Delta H} \cdot \tag{4}$$

This decrease may be represented by a volt equivalent ΔV_H, and if ΔV denotes the actual decrease in voltage drop due to a single lead, it may readily be shown that

$$f = \frac{V + 2\Delta V}{V + 2\Delta V - 2\Delta V_H} \cdot \tag{5}$$

The magnitude of ΔV_H obviously depends upon the manner in which H varies with T, and also upon the ratio of the maximum temperatures T_m to that of the junction between filament and lead, T_0.

The accurate evaluation of these corrections is, therefore, a somewhat tedious matter. However, in many cases only approximate results are desired and for this purpose the following empirical type of equation may be used:

$$\Delta V_H = P(T_m/1000) - Q(T_0/1000) - R.$$

The values of the constants, P, Q and R for pure tungsten are given for different properties in Table II.

TABLE II. *Constants in formula for lead loss correction for tungsten filaments.*

H	P	Q	R	Range T_m	Range T_0
Voltage*	0.154*	0.081*	0.056*	1000–2500	any value
Candle Power	.338	.182	−.004	600–3500	300–1400
Electron Emission	.440	.158	.072	1000–3500	300– 900
Watts Radiated	.293	.160	.084	1100–3000	300– 900

* For voltage (watts input), a term, $-2.1 \times 10^{-8}\, T_0 \cdot T_m$, is to be added to the right-hand side of the above equation.

For electron emission from tungsten filaments covered with monatomic films of more electropositive elements, the values of ΔV_H will be different from those given in the table, but may be calculated by methods indicated in the original paper. For other materials than tungsten, which obey the Wiedemann-Franz law, the lead-loss correction may also be calculated approximately.

However, it would seem best in most cases either to use the guard ring principle and thus eliminate the necessity for lead-loss corrections completely, or else to measure emission on two different lengths of filaments over the same range of temperature values.[32]

[32] As, for instance, in the paper by S. Dushman and J. Ewald, Phys. Rev. 29, 857 (1927).

[*Editor's Note:* Material has been omitted at this point.]

37

Reprinted from pages 954–964 of *Phys. Rev.* **3**:954–968 (1929)

THE INTERACTION OF ELECTRON AND POSITIVE ION SPACE CHARGES IN CATHODE SHEATHS

BY IRVING LANGMUIR

ABSTRACT

Effect of positive ions generated at a plane anode upon the space charge limitation of electron currents from a parallel cathode.—Mathematical analysis shows that single ions emitted with negligible velocity permit 0.378 $(m_p/m_e)^{1/2}$ additional electrons to pass; but with an unlimited supply of ions the electron current approaches a limiting value 1.860 times that which flows when no ions are present, and the electron current is then $(m_p/m_e)^{1/2}$ times the ion current, both currents thus being limited by space charge and the electric field being symmetrically distributed between the electrodes. Single ions introduced into a pure electron discharge at a point 4/9ths of the distance from cathode to anode produce a maximum effect, 0.582 $(m_p/m_e)^{1/2}$, in increasing the electron current. These conditions apply to a cathode emitting a surplus of electrons surrounded by ionized gas. The cathode sheath is then a double layer with an inner negative space charge and an equal outer positive charge, the field being zero at the cathode and at the sheath edge. The electron current is thus limited to $(m_p/m_e)^{1/2}$ times the rate at which ions reach the sheath edge. If ions are generated without initial velocities uniformly throughout the space between two plane electrodes, a parabolic potential distribution results. If the total ion generation exceeds 2.86 times the ion current that could flow from the more positive to the more negative electrode, a potential maximum develops in the space. Electrons produced by ionization are trapped within this region and their accumulation modifies the potential distribution yielding a region (named plasma) in which only weak fields exist and where the space charge is nearly zero. The *potential distribution in the plasma*, given by the Boltzmann equation from the electron temperature and the electron concentrations, *determines the motions of the ions* and thus fixes the rate at which the ions arrive at the cathode sheath. The *anode sheath* is usually also a positive ion sheath, but with anodes of small size a detached double-sheath may exist at the boundary of the anode glow. In discharges from *hot cathodes in gases* where the current is limited by resistance in series with the anode, the electron current is space-charge-limited, being fixed by the rate of arrival of ions at the cathode sheath. Thus the cathode drop is fixed by the necessity of supplying the requisite number of ions to the cathode. The *effect of the initial velocities of the ions and electrons* that enter a double-sheath from the gas is to decrease the electron current by an amount that varies with the voltage drop in the sheath. A nearly complete theory of this effect is worked out for plane electrodes. A detailed study is made of the *potential distribution in the plasma and near the sheath edge* for a particular case and the conclusion is drawn that the velocities of the ions that enter the sheath can be calculated from the electron temperature if the geometry of the source of ionization is given.

Experiments with double sheaths.—With large cathodes coated with barium oxide in low pressure mercury vapor, simultaneous measurements showed that the electron current density was independent of the cathode temperature and was from 140 to 200 times the ion current density, this ratio being independent of the intensity of ionization and of the gas pressure but varying slowly with the voltage drop in the cathode sheath, in good accord with the theory. The observed ratio, however, was about 40 percent of that calculated, this discrepancy being probably due to non-uniformity in the cathode coating. Similar results were obtained with double sheaths on wire type cathodes, the ratio of the electron current to the ion current through the sheath ranging from 450:1 at high current densities to 2000:1 and more at very low currents, this variation being in agreement with the approximate theory developed for cylindrical sheaths. In these experiments *two cathodes* were used; one at rather large negative voltage to produce any desired intensity of ionization, while from the volt-ampere characteristics of the other cathode the space-charge-limited electron currents were measured. The ion currents were measured either by cooling the test cathode so that it emitted no electrons, or by the use of an auxiliary ion collector.

THE maximum electron current that can flow from a given hot cathode to an anode in high vacuum is limited by the space charge of the electrons.[1] If even a small amount of gas is present and the applied anode voltage is appreciably higher than the ionizing potential, the positive ions formed tend to neutralize the electron space charge and thus allow the current to increase until, with sufficient gas, the current becomes limited only by the electron emission from the cathode as determined by its temperature.

Since in a given electric field the ions move hundreds of times slower than the electrons, the rate at which the ions need to be produced in order to neutralize the space charge is usually less than one percent of the rate at which the electrons flow from the cathode. In formulating a quantitative theory for calculating the increase in electron current produced by a given amount of ionization we meet the difficulty that the ions are produced at different points within the gas and therefore are not all moving with the same velocity. Then, too, the probability of ionization as a function of the electron velocity must be known. The problem thus becomes so complicated that it seems hardly worth while to attempt a general solution.

About 13 years ago the writer derived the equations, given in the present paper, for the space charge problem between parallel planes where the cathode emits a surplus of electrons and the anode emits positive ions without initial velocities. The results, although interesting, did not seem to be applicable directly to experimental conditions and therefore the results were not published. Some years later,[2] however, by the discovery that all caesium atoms which strike a tungsten surface at 1300°K are converted into ions, it became practicable to generate positive ions at the anode in any desired number, and thus obtain the conditions which were assumed in the theory.

Still more recently in a study of gaseous discharges at low pressures[3] it was found that the space charge equations could be applied to the positive ion currents flowing to negatively charged collectors.

If we consider a negatively charged hot collector or in fact any hot cathode in a gas we see that there are present in the positive ion sheath electrons as well as ions. In the present paper it will be shown that the theory which was developed 13 years ago is now applicable to the calculation of the properties of these double sheaths.

[1] Langmuir, Phys. Rev. **2**, 450 (1913); Phys. Zeits. **15**, 348, 516 (1914).
[2] Langmuir and Kingdon, Proc. Royal Soc. **A107**, 61 (1925).
[3] Langmuir and Mott-Smith, Gen. Elec. Rev. **27**, 449, 538, 616, 762, 810 (1924), and Phys. Rev. **28**, 727 (1926).

Jaffe[4] has attempted to develop a theory of the effect of small amounts of gas ionization on currents limited by space charge in gases at low pressures. However, he based his entire treatment upon the inadmissible assumption that as many ions recombine in each element of volume as are produced by ionization within that volume. We now know that even with the high current densities in a mercury vapor arc carrying amperes, recombination of ions in the gas is negligible, compared to the removal of the ions by diffusion to the walls and to the electrodes. Thus the equations which Jaffe derived are not even approximately correct.

Theory of the Effect of Ions on Space Charge Currents Between Parallel Planes

Consider an infinite plane cathode C at zero potential, and a similar parallel plane anode A at the potential V_A and at a distance a from C. Let the cathode emit a surplus of electrons without appreciable initial velocities. There will thus be an infinite concentration of electrons at the cathode surface and the potential gradient will be zero, but a finite electron current, I_0 per unit area, will flow to the anode, this current limited by space charge being given by the equation[1]

$$I_0=\frac{(2)^{1/2}}{9\pi}\left(\frac{e}{m_e}\right)^{1/2}\frac{V_A{}^{3/2}}{a^2} \tag{1}$$

where e is the charge and m_e the mass of the electrons.

Let us now consider the effect of introducing positive ions without initial velocities, uniformly distributed over a plane B which is at a distance b from C. Between B and C there is thus an ion current I_p per unit area. Because of the partial neutralization of the electron space charge, the electron current from the cathode will increase to a new value, say I_e per unit area. We assume that the ions and electrons do not collide with gas molecules nor with each other and that no appreciable number of ions or electrons is lost by recombination during the passage between the electrodes. Our problem is to determine how the electron current I_e depends on I_p the positive ion emission and on b the location of the source of ions.

Let v_e be the velocity of the electrons at any point P which is at a distance x from the cathode, ρ_e the electron space charge density at P. The corresponding quantities for the positive ions are denoted by the subscript p. The signs of all these quantities will be taken as positive.

Then

$$\rho_e v_e=I_e \qquad \text{and} \qquad \rho_p v_p=I_p \tag{2}$$

and assuming the ions have unit charge

$$\tfrac{1}{2}m_e v_e{}^2=V_e \qquad \text{and} \qquad \tfrac{1}{2}m_p v_p{}^2=(V_B-V)e \tag{3}$$

where V is the potential at the point P and V_B the potential at B.

[4] George Jaffe, Ann. d. Physik 63, 145–174 (1920).

Poisson's equation gives

$$d^2V/dx^2 = 4\pi(\rho_e - \rho_p) \tag{4}$$

We may eliminate ρ and v by Eqs. (2) and (3) and then after substituting

$$\alpha = (I_p/I_e)(m_p/m_e)^{1/2} \quad \text{and} \quad \phi = V/V_A \tag{5}$$

we obtain

$$\frac{d^2\phi}{dx^2} = 2(2)^{1/2}\pi\left(\frac{m_e}{e}\right)^{1/2}\frac{I_e}{V_A{}^{3/2}}\left[\phi^{-1/2} - \alpha(\phi_B - \phi)^{-1/2}\right] \tag{6}$$

Combining Eqs. (1) and (6) and substituting

$$\lambda = x/a \tag{7}$$

we have

$$\frac{d^2\phi}{d\lambda^2} = \frac{4}{9}\frac{I_e}{I_0}\left[\phi^{-1/2} - \alpha(\phi_B - \phi)^{-1/2}\right] \tag{8}$$

Since the electron current is limited by space charge, we impose the condition $dV/dx = 0$ when $x = 0$, or in other words

$$d\phi/d\lambda = 0 \quad \text{when} \quad \lambda = 0 \tag{9}$$

Integration of Eq. (8) then gives

$$d\phi/d\lambda = (4/3)(I_e/I_0)^{1/2}\left[\phi^{1/2} + \alpha\left\{(\phi_B - \phi)^{1/2} - \phi_B{}^{1/2}\right\}\right]^{1/2} \tag{10}$$

Ions Emitted From Anode, $\phi_B = 1$

According to Eq. (10) the potential gradient at the surface of the anode, $(\phi = \phi_B = 1)$, is proportional to $(1-\alpha)^{1/2}$ and so becomes imaginary if $\alpha > 1$. When $\alpha = 1$ the potential gradient at the anode is zero and the positive ion current as well as the electron current is thus limited by space charge.

It appears, therefore, that even an unlimited supply of positive ions available at the anode is not capable of neutralizing the electron space charge, for the positive ion current cannot become more than a definite fraction of the electron current, this fraction (according to Eq. 5, when $\alpha = 1$) being equal to the square root of the ratio of the mass of the electron to that of the ion.

Examination of Eqs. (8) and (10) shows that when $\phi_B = 1$ and $\alpha = 1$ the equations remain unchanged in form if we substitute $1 - \phi$ in place of ϕ. Thus the curve representing the potential distribution between the cathode and anode is symmetrical about its central point ($\lambda = \frac{1}{2}$, $\phi = \frac{1}{2}$). Between the cathode and this central point there is an excess of negative space charge, while from the central point to the anode there is an excess of positive charge.

To calculate the potential distribution we integrate Eq. (10) after placing $\phi_B = 1$

$$\lambda=(3/4)(I_0/I_e)^{1/2}\int_0^{\phi}[\phi^{1/2}+\alpha\{(1-\phi)^{1/2}-1\}]^{-1/2}d\phi. \tag{11}$$

The ratio I_e/I_0 is found by observing that $\phi=1$ when $\lambda=1$, thus

$$(I_e/I_0)^{1/2}=(3/4)\int_0^{1}[\phi^{1/2}+\alpha\{(1-\phi)^{1/2}-1\}]^{-1/2}d\phi \tag{12}$$

Table I gives values of λ obtained from Eq. (11) by numerical integration. The values of I_e/I_0 which were used in these calculations as found by Eq. (12) are given for various values of α in the next to last horizontal line of the table.[5]

From the values of I_e/I_0 we see that the electron current increases as more positive ions are emitted from the anode until the positive ion current also becomes limited by space charge. When this occurs the electron current and the positive ion current are each 1.860 times as great as the currents of electrons or ions that could flow (with the same applied potentials) if carriers of the opposite sign were absent.

It is interesting to inquire how large is the effect of single positive ions emitted from the anode, in causing an increased electron flow from the cathode. By differentiating Eq. (12) with respect to α and then placing $\alpha=0$, and $I_e=I_0$ we find in terms of Gamma functions

$$dI_e/d\alpha=[3-3\Gamma(1.25)\Gamma(1.5)/\Gamma(1.75)]I_0=0.378I_0 \tag{13}$$

or by Eq. (5)

$$dI_e/dI_p=0.378(m_p/m_e)^{1/2} \qquad \text{for} \qquad \alpha=0 \tag{14}$$

A similar calculation for the case $\alpha=1$ involves a numerical evaluation of the resulting integral giving

$$dI_e/dI_p=3.455(m_p/m_e)^{1/2} \qquad \text{for} \qquad \alpha=1$$

A plot of I_e/I_0 as function of α from the data of Table I shows that the slope of the curve increases gradually from 0.378 at $\alpha=0$ up to 3.455 at $\alpha=1$. Thus the effectiveness of the ions in raising the electron current increases as the field strength decreases in the region where they originate, but only up to a certain limiting value. Of course when $\alpha=1$ the further increase in the electron current is stopped by the space charge limitation of the ion current.

The square root of the ratio of the masses of the ions and the electrons is 607 for mercury vapor, 271 for argon, and 60.8 for hydrogen, and there-

[5] In carrying out these calculations it was found convenient to replace ϕ by a new variable σ such that $\phi=\sigma^{4/3}$. By so doing the infinite value of the integrand that occurs when $\phi=0$ is avoided. When $\alpha=1$ and $\phi=1$ a similar difficulty still occurs but the value of λ in the range $\phi=\frac{1}{2}$ to 1 can be obtained from those calculated in the range $\phi=0$ to $\frac{1}{2}$ by making use of the fact already noted that in this case λ is symmetrical about the point $\phi=\frac{1}{2}$.

fore each positive ion of these gases liberated at the anode will increase the number of electrons that cross the space by 229, 102 or 23 respectively in the case of a pure electron discharge ($\alpha=0$).

TABLE I. *Potential distribution between plane cathode emitting surplus of electrons and parallel plane anode which emits given numbers of ions. Table of values of λ, the fraction of the distance to the anode, at which the potential is a given fraction φ of the anode potential (zero potential at cathode).*

	$\alpha=(I_p/I_e)\ (m_p/m_e)^{1/2};\ \lambda=x/a$						
ϕ	$\alpha=0$	$\alpha=0.2$	$\alpha=0.4$	$\alpha=0.6$	$\alpha=0.8$	$\alpha=0.9$	$\alpha=1.0$
0	0	0	0	0	0	0	0
0.02	0.0532	0.0513	0.0491	0.0467	0.0438	0.0419	0.0396
0.05	.1057	.1022	.0981	.0934	.0879	.0842	.0798
0.1	.1778	.1723	.1661	.1588	.1498	.1437	.1367
0.2	.2991	.2911	.2823	.2714	.2573	.2477	.2363
0.3	.4054	.3962	.3855	.3721	.3546	.3423	.3274
0.4	.5030	.4932	.4815	.4667	.4467	.4324	.4146
0.5	.5946	.5847	.5731	.5580	.5371	.5218	.5000
0.6	.6817	.6723	.6612	.6461	.6245	.6080	.5854
0.7	.7653	.7570	.7471	.7332	.7123	.6958	.6726
0.8	.8459	.8395	.8314	.8198	.8016	.7861	.7637
0.9	.9240	.9201	.9149	.9074	.8940	.8813	.8633
1.0	1.0000	1.0000	1.0000	1.0000	1.0000	1.0000	1.0000
I_e/I_0	1.0000	1.0839	1.1872	1.3237	1.5186	1.6644	1.8605
a/a_0	1.0000	1.0411	1.0896	1.1505	1.2323	1.2901	1.3640

THE SOURCE OF IONS IS AT A PLANE BETWEEN CATHODE AND ANODE $\phi_B<1$.

Let us consider the effect produced by ions that start from a plane B which lies between C and A. In the region between C and B, Eqs. (6) and (10) are applicable and thus the value of $d\phi/d\lambda$ at B is found by putting $\phi=\phi_B$ in Eq. (10). Equation (6) is applicable also in the region between B and A but here $\alpha=0$ for there is no ion current. Thus by integration we obtain the two equations

$$\left.\begin{aligned} &4(I_e/I_0)^{1/2}\lambda_B=3\phi_B^{3/4}\int_0^1[\mu^{1/2}+\alpha\{(1-\mu)^{1/2}-1\}]^{-1/2}d\mu \\ \text{and}\quad &4(I_e/I_0)^{1/2}(1-\lambda_B)=3\int_{\phi_B}^1[\phi^{1/2}-\alpha\phi_B^{1/2}]^{-1/2}d\phi \end{aligned}\right\} \tag{15}$$

The constant of the first integration for the second equation was chosen to make $d\phi/d\lambda$ at $\phi=\phi_B$ the same as for the first equation.

Differentiating Eqs. (15) with respect to α, putting $\alpha=0$, adding the resulting equations and combining with Eq. (5) gives rigorously[6]

$$\frac{dI_e}{dI_p}\left(\frac{m_e}{m_p}\right)^{1/2}=3\phi_B^{1/2}-2.622\phi_B^{3/4}. \tag{16}$$

[6] The coefficient 2.622 is equal to 3 minus the coefficient 0.378 as given in Eq. (14).

This equation, which reduces to Eq. (14) if $\phi_B=1$, allows us to calculate the number of additional electrons that can flow in a pure electron discharge ($\alpha=0$) if single positive ions are introduced *at any point* in the space between the electrodes. The values given in Column 3 of Table II were calculated by Eq. (16); the second column represents λ_B the fraction of the distance from cathode to anode at which the ions originate. From Eq. (1) we see that $\lambda_B=\phi_B^{3/4}$.

TABLE II. *The increase in electron current caused by ions originating at various positions between cathode and anode, $\alpha=0$. Initial velocities neglected.*

ϕ_B	λ_B	$\frac{dI_e}{dI_p}\cdot\left(\frac{m_e}{m_p}\right)^{1/2}$
0.00	0.00	0.0
0.001	0.0056	0.080
0.01	0.0316	0.217
0.1	0.177	0.483
0.2	0.300	0.554
0.338	0.444	0.582 max.
0.5	0.595	0.561
0.6	0.683	0.534
0.7	0.765	0.502
0.8	0.846	0.464
0.9	0.925	0.419
1.0	1.000	0.378

The ions have a maximum effect in increasing the electron current when they are introduced at a point which is 4/9 of the distance from cathode to anode. If a trace of gas is present and the voltage is so high that we may assume the probability of ionization per cm of electron path is uniform, we find readily from Eq. (16) (integrating with respect to λ) that the average value of dI_e/dI_p is 0.489 $(m_p/m_e)^{1/2}$.

If the probability of ionization is greater near the end of the path, as in the case of low anode voltages, the coefficient will lie between 0.489 and 0.378.

Low Pressure Discharges With Hot Cathode

Let us consider for a moment the phenomena that are observed as we pass from a pure electron discharge to a discharge at low gas pressure in which there is abundant gas ionization. When the ions originate at the anode or at a definite plane between the electrodes we have been able to follow through the effects produced by even an unlimited supply of ions. But when the ions are produced throughout all or a large part of the space between the electrodes, we have been able to analyze only the effects produced by a very small total number of ions ($\alpha=0$). To understand the typical characteristics of gas discharges, however, we must devise methods of treating the problem involving a large intensity of ionization throughout a volume. The nature of the problem will best be realized by considering briefly some experimental observations.

Suppose for example, we have a hot tungsten cathode (at zero volts) capable of emitting 50 ma in a bulb containing mercury vapor saturated at room temperature. With an anode at 10 volts the current is limited by electron space charge and is practically the same as in the absence of mercury vapor. Beginning at the ionization voltage (10.4 volts) the current increases with the anode voltage more rapidly than in good vacuum until at a voltage of 15 to 25 volts depending on the vapor pressure, and the geometry of the tube, the current rises abruptly to the saturation value corresponding to the cathode temperature.

In this second state of the discharge practically the whole of the voltage difference from cathode to anode is concentrated in a cathode sheath in which there is a positive ion space charge; the rest of the volume is nearly field-free, the space charge of the positive ions being neutralized by low-velocity or "ultimate electrons" which accumulate in this space until their concentration is hundreds or thousands of times greater than that of the primary electrons from the cathode.

By means of a sufficiently large resistance in series with the anode it is possible to observe points on the negative resistance part of the current voltage curve that lies between the parts corresponding to the two regions we have just considered. In this transition region the current is limited by the electron and ion space charges in a double layer or double sheath on the cathode.

In order to understand the formation of this double sheath, the final neutralization of space charge and the accumulation of the ultimate electrons, we will first consider the following problem.

Potential Distribution and Current Flow Resulting from the Production of Ions Uniformly throughout the Volume Between Two Planes

Consider two parallel plane electrodes A and C separated by the distance a and let V_c be the potential of C taking that of A to be zero. We assume that S ions of charge e are generated per unit time in each unit volume, and for the present will assume that no electrons are generated by this ionization.

We will first consider the case in which there is no maximum potential in the space between A and C. We will take V_c negative so that the ions move towards C. Then the ion current density I at any plane at a distance x from A is

$$I = Sex \qquad (17)$$

but this current is composed of ions having widely different velocities depending on the potentials of their points of origin. The space charge ρ at any point at a distance x from A where the potential is V_1 is thus given by the integral $\int(1/v)dI$ where v, the velocity of an ion which originated at a point of potential V, is found from

$$\tfrac{1}{2}m_p v^2 = (V - V_1)e \tag{18}$$

Thus

$$\rho = S(em_p/2)^{1/2} \int_0^{x_1} (V - V_1)^{-1/2} dx. \tag{19}$$

The problem is now to find V as a function of x which satisfies this integral equation simultaneously with Poisson's equation. By trial the following is found to be a particular solution

$$V = V_c x^2/a^2 \tag{20}$$

which we may now prove as follows.
Substituting this value of V and the corresponding value of V_1 in Eq. (19) We obtain

$$\rho = \tfrac{1}{2}\pi I_a [m_p/(-2eV_c)]^{1/2} \tag{21}$$

where

$$I_a = S_1 ea \tag{22}$$

is the ion current which reaches C, and S_1 is the particular value of S which corresponds to the potential distribution assumed in Eq. (20).
Differentiation of Eq. (20) gives

$$dV/dx = 2V_c x/a^2 \tag{23}$$

and a second differentiation and combination with Poisson's equation gives

$$d^2V/dx^2 = 2V_c/a^2 = -4\pi\rho \tag{24}$$

whence

$$\rho = -V_c/2\pi a^2 \tag{25}$$

We see that both Eqs. (21) and (25) give ρ independent of x or a uniform space charge between the planes. Equating the two expressions for ρ and solving the resulting equation for I_a gives

$$I_a = (2e/m_p)^{1/2}(-V)^{3/2}/\pi^2 a^2. \tag{26}$$

If, as before, I_0 is the current calculated by the ordinary space charge equation, Eq. (1), we see that

$$I_a = (9/\pi)I_0 = 2.865 I_0 \tag{27}$$

We now recognize that the parabolic potential distribution assumed in Eq. (20) is a solution of our problem only when the rate of ionization S has the particular value S_1 given by Eq. (22) and where I_a is given by Eq. (26). Curve 1 in Fig. 1 illustrates this parabolic distribution. It will be noted by Eq. (23) that the potential gradient is zero at A, i.e. at $x = 0$.

If $S<S_1$ the potential distribution curve will lie between the straight line 0 and Curve 1 in Fig. 1, but will not be parabolic. On the other hand, if $S>S_1$ it is evident that there will be a potential maximum between A and C and that Eq. (26) can then be applied separately to the two branches of

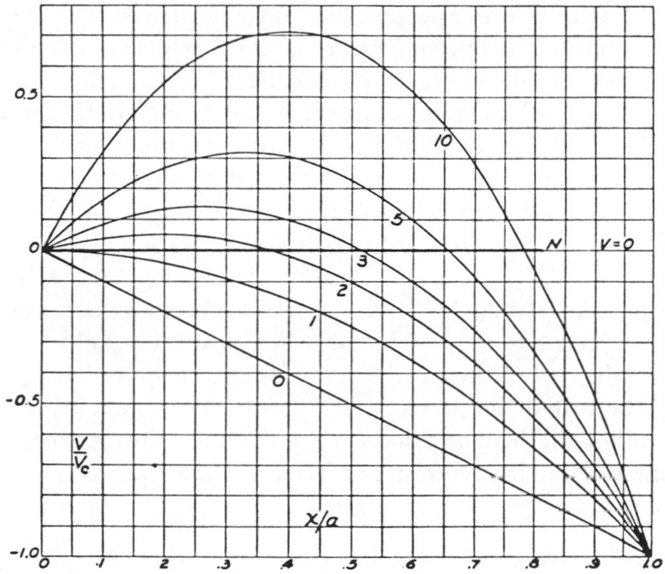

Fig. 1. Potential distributions between plane electrodes when ions are generated uniformly between them.

the curve on the opposite sides of the maximum. The potential distribution curve is thus still a parabola but the origin is no longer at A. The curves marked, 2, 3, 5 and 10 in Fig. 1 have been calculated for values of S equal respectively to 2, 3, 5 and 10 times S_1. The equation of these parabolas is

$$\frac{V}{V_c}=\frac{x}{a}\left[1-\left(\frac{S}{S_1}\right)^{2/3}\left(1-\frac{x}{a}\right)\right]. \tag{28}$$

Effect of the Electrons Generated by Ionization

If the positive ions in the foregoing problem are produced by the ionization of a gas an equal number of electrons will be generated simultaneously. If $S<S_1$ there will be no potential maximum between the electrodes so that these electrons will flow to the electrode A without having appreciable effect on the space charge, for it would take an electron current hundreds of times greater than I_a to neutralize the positive ion space charge due to I_a.

The situation is very different, however, if $S>S_1$ for there is then a tendency to develop a potential maximum as illustrated in Fig. 1. In any region at a potential higher than that of both electrodes the low velocity electrons produced by ionization will accumulate until they nearly neutralize the positive space charge. The potential of the region which would otherwise be above anode potential (above line ON in Fig. 1) is thus lowered to a value at which the electrons, in virtue of their initial velocities, can just escape to the anode as fast as they are produced. The accumulation of the ultimate or low velocity electrons is greatly favored by the smallness of the field available for drawing away the positive ions. In general there will still be a maximum potential in the space but this usually exceeds the anode potential by not more than a volt or so, and thus the ions flow in nearly equal numbers to anode and cathode, while the electrons go to the anode only. Any calculation of the exact potential distribution must involve some knowledge of the velocity distribution of the electrons and ions.

We see from Fig. 1 that when S/S_1 is as great as 10, the fields at the cathode and anode which are necessary to draw these ion currents, are very large.

These regions of strong field due to space charge which cover the electrodes will be referred to as the *sheaths*. The relatively field-free regions between the sheaths where the positive and negative space charges are nearly balanced will be called the *plasma*. We shall find in general that these two regions are rather distinct and have very different properties. Let us first consider some of the characteristics of the plasma.

[*Editor's Note:* Material has been omitted at this point.]

38

Reprinted from *Ark. Mat. Astron. Fys.* **29B**:1-7 (1942)

On the existence of electromagnetic-hydrodynamic waves.

By

HANNES ALFVÉN.

With 2 figures in the text.

Communicated September 9th 1942 by C. W. OSEEN.

§ 1. If a conducting liquid is placed in a constant magnetic field, a mechanical motion in the liquid will in general give rise to an e. m. f., which produces electric currents. The interaction between the magnetic field and these currents causes mechanical forces which change the state of motion of the liquid.

Thus the application of a magnetic field to a conducting liquid causes a mutual interaction between hydrodynamic motion and electric current. Thus kinetic energy can be converted into electromagnetic energy and *vice versa.* This mechanism makes possible the existence of a kind of combined *electromagnetic-hydrodynamic wave*, which — as far as I know — has as yet attracted no attention.

For the electromagnetic vectors we have

$$\text{rot } H = \frac{4\pi}{c} i \tag{1}$$

$$\text{rot } E = -\frac{1}{c}\frac{\partial B}{\partial t} \tag{2}$$

$$B = \mu H \tag{3}$$

$$i = \sigma\left(E + \frac{v}{c} \times B\right) \tag{4}$$

where E is the electric and H the magnetic field, i the electric current density, v the velocity of the liquid, σ the electric conductivity, μ the permeability, and c the velocity of light.

These equations must be combined with the hydrodynamic equation

$$\vartheta \frac{dv}{dt} = \frac{1}{c}(i \times B) - \text{grad } p \tag{5}$$

where ϑ denotes the mass density and p the hydrostatic pressure. If we suppose that the liquid is uncompressible, we have

$$\text{div } v = 0. \tag{6}$$

§ 2. In order to study the phenomenon under as simple conditions as possible, let us suppose that the primary magnetic field H_0 is homogeneous and parallel to the z-axis of an orthogonal coordinate system, the conductivity σ is infinite, and ϑ is constant.

The magnetic field consists of the primary field H_0 and the field H' which is caused by the current i. In order to study a *plane wave* in the direction of H_0, we assume that all vectors are independent of x and y (but depend upon z and the time t).

This implies that according to (1) and (2) we have $i_z = 0$ and $H_z = \text{const} = H_0$. Further, according to (6) we may put $v_z = 0$.

If we turn the coordinate system in such a way that $i_y = 0$, we obtain from (1)

$$\begin{aligned} i_x &= -\frac{c}{4\pi}\frac{\partial H_y}{\partial z} \\ i_y &= i_z = 0 \end{aligned} \tag{7}$$

$$\begin{aligned} H_x &= \text{const} = 0 \\ H_z &= H_0 . \end{aligned} \tag{8}$$

We introduce these values into (5). As according to our assumptions grad p can have no components perpendicular to the z-axis, we obtain

$$\frac{\partial v_x}{\partial t} = 0; \quad v_x = \text{const} = 0$$

$$\frac{\partial v_y}{\partial t} = \frac{\mu H_0}{4\pi\vartheta} \frac{d H_y}{d z} \tag{9}$$

$$v_z = 0$$

and further

$$\frac{d p}{d z} = -\frac{\mu}{8\pi} \frac{d(H_y^2)}{d z}. \tag{10}$$

Because i is finite, equation (4) gives

$$E = -\mu \frac{v}{c} \times H$$

or with (8) and (9)

$$E_x = -\mu \frac{v_y}{c} H_0$$

$$E_y = E_z = 0. \tag{11}$$

Equation (2) gives

$$\mu \frac{d H_y}{d t} = -c \frac{d E_x}{d z}. \tag{12}$$

Combining (12), (11) and (9) we obtain

$$\frac{d^2 H_y}{d t^2} = \frac{\mu H_0^2}{4\pi\vartheta} \frac{d^2 H_y}{d z^2} \tag{13}$$

which means a wave in the direction of the z-axis with the velocity

$$V = \frac{H_0 \sqrt{\mu}}{\sqrt{4\pi\vartheta}}. \tag{14}$$

The velocity of the electromagnetic-hydrodynamic wave is independent of the frequency as well as of the amplitude.

§ 3. If we put

$$H_y = A \sin \omega \left(t - \frac{z}{V}\right) \tag{15}$$

we find

$$v_y = -\frac{A \sqrt{\mu}}{\sqrt{4 \pi \vartheta}} \sin \omega \left(t - \frac{z}{V}\right) \tag{16}$$

$$i_x = A \frac{c\,\omega}{H_0} \sqrt{\frac{\vartheta}{4 \pi \mu}} \cos \omega \left(t - \frac{z}{V}\right) \tag{17}$$

$$E_x = A \frac{H_0 \sqrt{\mu^3}}{c \sqrt{4 \pi \vartheta}} \sin \omega \left(t - \frac{z}{V}\right) \tag{18}$$

$$p = p_0 - \frac{\mu A^2}{8 \pi} \sin^2 \omega \left(t - \frac{z}{V}\right). \tag{19}$$

The magnetic lines of force, which with no waves were straight lines

$$x = x_0, \quad y = y_0 \tag{20}$$

change their shape into sine curves:

$$x = x_0$$

$$y = y_0 + \frac{A \sqrt{\mu}}{\omega \sqrt{4 \pi \vartheta}} \cos \omega \left(t - \frac{z}{V}\right). \tag{21}$$

Differentiating (21) we find that the magnetic lines of force oscillate with the same velocity (given by (16)) as the liquid.

§ 4. The phenomenon can also be treated along quite different lines. Suppose that we have a homogeneous magnetic field in a perfectly conducting liquid. The magnetic lines of force can be considered as elastic strings according to the usual mechanical picture of electrodynamical phenomena. In view of the infinite conductivity, every motion (perpendicular to the field) of the liquid in relation to the lines of force is forbidden because it would give infinite eddy currents. Thus the matter of the liquid is »fastened» to the lines of force, constituting a series of strings.

A swinging string obeys the equation

$$m \frac{d^2 y}{d t^2} = S \frac{d^2 y}{d z^2} \tag{22}$$

if the string has the direction of the z-axis and swings parallel to the y-xis. S means the tension and m the mass per unit length.

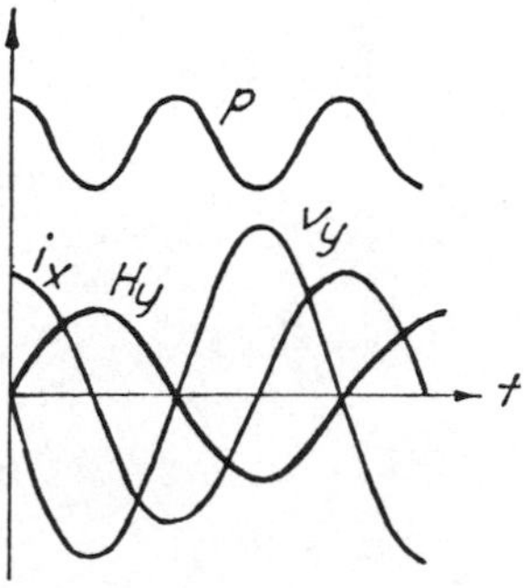

Fig. 1. Current i_x, velocity v_y, variable component of magnetic field H_y, and pressure p, as functions of the time t.

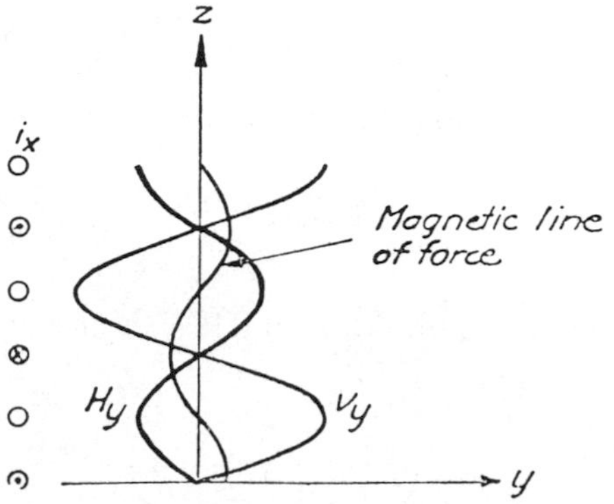

Fig. 2. Velocity v_y and variable component of magnetic field H_y as functions of z. Geometrical form of a magnetic line of force. To the left is shown where i_x is zero (○), directed upwards (⊙) or downwards (⊕) through the paper.

In such a string the velocity of transverse waves is

$$V = \sqrt{\frac{S}{m}}. \tag{23}$$

In our case we substitute for the string a filament of the liquid with unit cross-section. Thus m corresponds to ϑ. If the string in a certain moment has the shape

$$y = f(z) \tag{24}$$

its length has increased by

$$\int \left[\sqrt{1 + \left(\frac{d y}{d z} \right)^2} - 1 \right] d z,$$

which means that its potential energy is

$$W = S\int\left[\sqrt{1+\left(\frac{dy}{dz}\right)^2}-1\right]dz \approx \frac{S}{2}\int\left(\frac{dy}{dz}\right)^2 dz \qquad (25)$$

The magnetic lines of force have the same shape as the string if, to the original field H_0 in the z-direction, we add

$$H_y = H_0\frac{dy}{dz}. \qquad (26)$$

When doing so we increase the energy by the amount

$$W = \frac{\mu}{8\pi}\int H_y^2 dz = \frac{\mu H_0^2}{8\pi}\int\left(\frac{dy}{dz}\right)^2 dz. \qquad (27)$$

The energies (25) and (27) become equal if we put

$$S = \frac{\mu H_0^2}{4\pi}. \qquad (28)$$

Introducing this expression in (23) we obtain

$$V = \frac{H_0\sqrt{\mu}}{\sqrt{4\pi\vartheta}}.$$

which is in accordance with (14).

§ 5. Electromagnetic-hydrodynamic waves are probably very important in solar physics. The sun's general magnetic field constitutes the primary field H_0 in which the waves move. Owing to its ionization, solar matter is a good electrical conductor. The fact that mechanical motions as well as strong magnetic fields are observed in sunspots indicate that they may be associated with waves of this type, although more complicated than the plane waves we have studied.

During the 11-year period the sunspot zone moves from a latitude of about 30° towards the equator with a velocity of the order of 100 cm sec^{-1}. As the general magnetic field is of the order of $H_0 = 15$ gauss, the velocity of an electromagnetic-hydrodynamic wave amounts to 100 cm sec^{-1} if $\mu = 1$ and

$$\vartheta = \frac{H_0^2}{4\pi V^2} = 0.002 \text{ g cm}^{-3}.$$

The solar density has this value at about one tenth of the solar radius below the surface. The original cause of the sunspots may very well be situated at that depth. Thus the cause of sunspots may be electromagnetic-hydrodynamic waves probably originating in the sun's interior and reaching the surface in the sunspot zones.

The problem of electromagnetic-hydrodynamic waves in the sun will be treated in a later publication.

As the term »electromagnetic-hydrodynamic waves» is somewhat complicated, it may be convenient to call the phenomenon »*magneto-hydrodynamic*» waves. (The term »hydromagnetic» is still shorter but not quite adequate.)

Stockholm Aug. 1942.

K. Tekniska Högskolan.

39

Reprinted from *Roy. Soc. (London) Proc.*, ser. A, **130:**499–513 (1931).

Quantum Mechanics of Electrons in Crystal Lattices.

By R. de L. Kronig and W. G. Penney, University of Groningen.

(Communicated by R. H. Fowler, F.R.S.—Received November 13, 1930.)

Introduction.—Through the work of Bloch our understanding of the behaviour of electrons in crystal lattices has been much advanced. The principal idea of Bloch's theory is the assumption that the interaction of a given electron with the other particles of the lattice may be replaced in first approximation by a periodic field of potential. With this model an interpretation of the specific heat,* the electrical and thermal conductivity,† the magnetic susceptibility,‡ the Hall effect,§ and the optical properties|| of metals could be obtained. The advantages and limitations inherent in the assumption of Bloch will be much the same as those encountered when replacing the interaction of the electrons in an atom by a suitable central shielding of the nuclear field, as in the work of Thomas and Hartree.

* Bloch, 'Z. Physik,' vol. 52, p. 555 (1928).

† Bloch, 'Z. Physik,' vol. 52, p. 555 (1928), vol. 53, p. 216 (1929), and vol. 59, p. 208 (1930); Peierls, 'Ann. Physik,' vol. 4, p. 121 (1930), and vol. 5, p. 244 (1930).

‡ Bloch, 'Z. Physik,' vol. 53, p. 216 (1929).

§ Peierls, 'Z. Physik,' vol. 53, p. 255 (1929).

|| Kronig, 'Proc. Roy. Soc.,' A, vol. 124, p. 409 (1929).

In the papers quoted a number of general results were given regarding the behaviour of electrons in any periodic field of potential. To obtain a clearer idea of the details of this behaviour with a view to the application in special problems, however, it appeared worth while to investigate the mechanics of electrons in periodic fields of potential somewhat similar to those met with in practice and of such nature that the energy values W and eigenfunctions ψ of the wave-equation can actually be computed. It is the purpose of this article to discuss a case where the integration is possible. In Section 1 the energy values and in Section 2 the wave-functions in their dependence on the binding introduced by the potential field are discussed for the one dimensional problem. In Section 3 the matrix elements of the linear momentum, which furnish the electric current associated with the various stationary states as well as the probability of radiative transitions between these states, are evaluated. In Section 4 the results are extended to the three dimensional case and those features considered which one may expect to find in the case of more general periodic fields of potential. Section 5 deals with some applications to physical problems.

1. *The Energy Values.*—The potential field which we shall consider is essentially that shown in fig. 1. Later on we shall pass to the limit $b = 0$

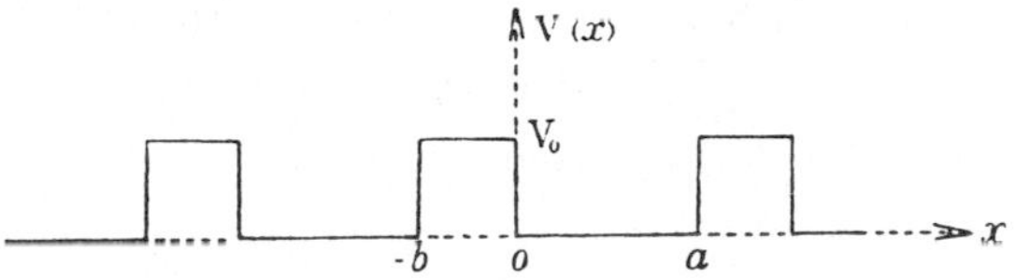

Fig. 1.

and $V_0 = \infty$ as this makes the results mathematically simpler without essentially altering their character, but to begin with we leave both b and V_0 finite. The wave-equation of the problem is given by

$$\frac{d^2\psi}{dx^2} + \kappa^2 [W - V(x)]\, \psi = 0, \qquad \kappa^2 = \frac{8\pi^2 m}{h^2}, \tag{1}$$

and following Bloch* we enquire after the solutions ψ which are periodic over a distance $L = G(a + b)$, where G is a large integer. As he has shown, these solutions must be of the form

$$\psi(x) = u(x)\, e^{i\alpha x}, \qquad \alpha = 2\pi k/L, \tag{2}$$

where k is any integer and $u(x)$ a function periodic in x with the period $(a + b)$. To find $u(x)$ we can hence confine ourselves to one period of the lattice, say

* Bloch, 'Z. Physik,' vol. 52, p. 555 (1928).

from $x = -b$ to $x = a$, and moreover, since we later shall make $V_0 = \infty$, we may assume $0 < W < V_0$. Substituting the expression (2) in equation (1) imposes on u in the range $-b \leqq x \leqq 0$ the condition

$$\frac{d^2u}{dx^2} + 2i\alpha\frac{du}{dx} - (\alpha^2 + \gamma^2)\,u = 0$$

with the solution

$$u = Ae^{(-i\alpha+\gamma)x} + Be^{(-i\alpha-\gamma)x}, \tag{3}$$

while in the region $0 \leqq x \leqq a$ we have

$$\frac{d^2u}{dx^2} + 2i\alpha\frac{du}{dx} - (\alpha^2 - \beta^2)\,u = 0$$

with the solution

$$u = Ce^{i(-\alpha+\beta)x} + De^{i(-\alpha-\beta)x}. \tag{4}$$

Here β and γ are given by

$$\beta = \kappa\sqrt{W}, \qquad \gamma = \kappa\sqrt{V_0 - W}. \tag{5}$$

In particular β and γ are real quantities and may be assumed positive without loss of generality.

The constants A, B, C, D are to be so chosen that the solutions in the two regions have the same value and the same first derivative for $x = 0$, while from the periodic nature of u it follows that the solution (3) and its first derivative for $x = -b$ shall be equal respectively to the solution (4) and its first derivative for $x = a$. The four linear homogeneous equations resulting for the constants from these requirements are

$$\begin{aligned}
A+B &= C+D,\\
(-i\alpha+\gamma)A+(-i\alpha-\gamma)B &= i(-\alpha+\beta)C+i(-\alpha-\beta)D,\\
Ae^{(i\alpha-\gamma)b}+Be^{(i\alpha+\gamma)b} &= Ce^{i(-\alpha+\beta)a}+De^{i(-\alpha-\beta)a}\\
(-i\alpha+\gamma)Ae^{(i\alpha-\gamma)b}+(-i\alpha-\gamma)Be^{(i\alpha+\gamma)b} &= i(-\alpha+\beta)Ce^{i(-\alpha+\beta)a}\\
&\quad +i(-\alpha-\beta)De^{i(-\alpha-\beta)a},
\end{aligned}$$

and can be satisfied only if the quantities β and γ, which according to equation (5) are directly related to the energy value W, satisfy the relation

$$\frac{\gamma^2-\beta^2}{2\beta\gamma}\sinh\gamma b\sin\beta a + \cosh\gamma b\cos\beta a = \cos\alpha\,(a+b).$$

Passing now to the limit where $b = 0$ and $V_0 = \infty$ in such a way that $\gamma^2 b$ stays finite and calling

$$\lim_{\substack{b\to 0\\ \gamma\to\infty}} \frac{\gamma^2 ab}{2} = P,$$

this relation becomes

$$P \sin \beta a / \beta a + \cos \beta a = \cos \alpha a, \tag{6}$$

a transcendental equation for βa.

To discuss the roots of this equation we have plotted its left-hand side as a function of βa in fig. 2, assuming for P the value $3\pi/2$. The values of βa

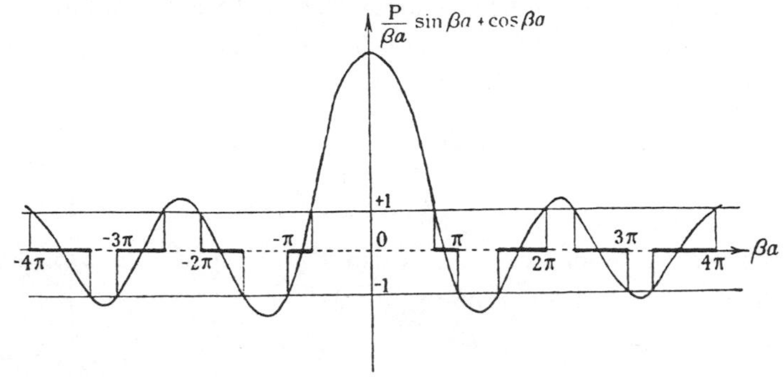

Fig. 2.

satisfying equation (6) are obtained as the projections on the βa-axis of the intersections of the curve with a straight line drawn at the distance $\cos \alpha a$ parallel to this axis. Since $\cos \alpha a$ lies between -1 and $+1$, and since the ordinates of the maxima of the curve have an absolute magnitude greater than 1, then, upon varying α by giving k different values, the curve is found to consist of the portions between the parallels ± 1 on which there lie intersections and the remaining portions outside these parallels on which there lie no intersections. By projection the βa-axis too is divided into portions containing permissible values of βa (drawn heavily in fig. 2) and portions not containing such values. Letting L approach infinity allows us according to equation (2) to vary α continuously, and the permissible values of βa will then fill the heavily drawn portions of the βa-axis continuously. *The energy values which an electron moving through the lattice may have, hence form a spectrum consisting of continuous pieces separated by finite intervals.*

It is of interest to study the influence of P upon this spectrum. If P vanishes, the curve of fig. 2 goes over into $\cos \beta a$, the forbidden intervals of the βa-axis disappear, and we have one continuous spectrum of all energy values from 0 to ∞. This is the limiting case of *free electrons.* Increasing P we obtain the forbidden intervals, the ratio of the length of these intervals to that of the adjoining allowed portions decreasing as we pass to large values of βa. We may say that fast electrons have less trouble to pass the potential barriers than slow electrons and can be considered more nearly as free. Letting P approach infinity reduces the allowed portions of the βa-axis to the points $n\pi$ $(n = \pm 1, \pm 2, \ldots)$. The energy spectrum now becomes discrete, the energy values

$$W = n^2h^2/8ma^2, \qquad (n = 1, 2, \ldots)$$

being those of an electron confined to move between two impenetrable potential barriers at a distance a apart. *The electrons* are caught between the potential walls, they *have become bound.*

Fig. 3 shows the change in the energy spectrum during the transition from the case of free to that of bound electrons. In the region $0 \leqq P \leqq 4\pi$, $P/4\pi$ has been chosen as abscissa, while for $4\pi \leqq P \leqq \infty$ $4\pi/P$ is used. The shaded area represents the allowed values of $(\beta a/\pi)^2$.

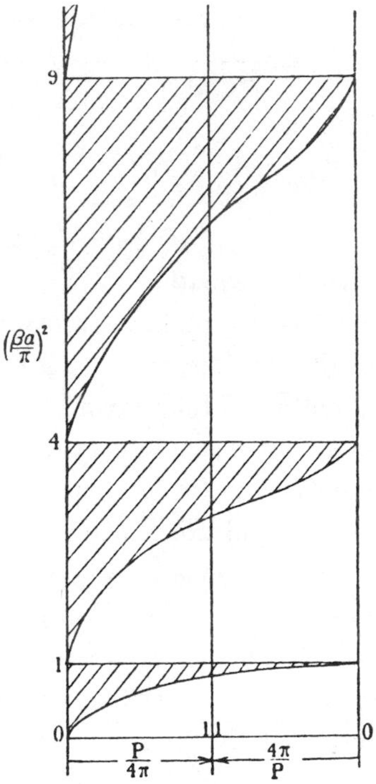

FIG. 3.

2. *The Wave-functions.*—We come now to consider the *wave-functions* of our problem. Since b in the limit is reduced to zero, we are only concerned with the solution (4). Solving the linear equations for the constants A, B, C, D in this limiting case gives us

$$D = -C\frac{1 - e^{-i(\alpha-\beta)a}}{1 - e^{-i(\alpha+\beta)a}}. \qquad (7)$$

u is given by equation (4) in the cell extending from 0 to a, while in the cell extending from ra to $(r + 1)\,a$ it will be given by

$$u = Ce^{i(-\alpha+\beta)(x-ra)} + De^{i(-\alpha-\beta)(x-ra)},$$

being periodic with the period a. According to equation (2) ψ in this cell will then be given by

$$\psi = Ce^{i\beta x + ir(\alpha-\beta)a} + De^{-i\beta x + ir(\alpha+\beta)a}, \qquad (8)$$

with C and D related by equation (7). If we require that ψ be normalised over a distance equal to its period L, we must have

$$CC^* = \frac{1}{2L}\frac{1 - \cos(\alpha+\beta)a}{1 - \cos\alpha a\cos\beta a + \frac{1}{P}(\cos\alpha a - \cos\beta a)^2}, \qquad (9)$$

the asterisk denoting the conjugate.

From equation (6) it appears that if with a given permissible value of βa a certain value of αa satisfies this equation, then also $\alpha a + 2n\pi$ and $-\alpha a + 2n\pi$ (n an integer) will satisfy it. We see, however, from equations (7) and (8) that by substituting $\alpha a + 2n\pi$ for αa we do not obtain any new wave-functions, while substituting $-\alpha a + 2n\pi$ for αa we get the same wave-function as that belonging to $-\beta a$, $-\alpha a$. We may hence, without loss of generality, accept the convention that if $n\pi \leqq \beta a \leqq (n+1)\pi$, we shall associate with it that value of αa for which $n\pi \leqq \alpha a \leqq (n+1)\pi$. In this way there is associated with every permissible value of βa one and only one value of αa, and we may use α to distinguish the stationary states.

We shall next investigate the influence of the value of P on the wave-functions. If we make P vanishingly small, then according to the convention just introduced βa will equal αa, and according to equation (7) D will vanish excepting for $\alpha a = n\pi$, in which case $D = \pm C$. The wave-functions according to equation (8) will go over into $e^{\pm i\alpha x}$ excepting for the special case just mentioned, in which they become equal to $\sin \alpha x$ and $\cos \alpha x$ so that we obtain the *wave-functions of free electrons*. At the same time the significance of α becomes apparent. For P approaching infinity, β takes the values $n\pi$, $D = -C$ according to equation (7), and we get from equation (8) the *wave-functions of electrons confined in their motion between impenetrable potential walls* at a distance a apart.

If we let L increase, the energy values, as mentioned before, come closer together, and for very large values of L we may enquire after the *density distribution of the energy values* in the allowed regions of βa. According to equation (2) this is evidently given by

$$\rho(\beta a) = \frac{L}{2\pi}\frac{d\alpha}{d\beta} = \frac{L}{2\pi \sin \alpha a}\left(\frac{P}{(\beta a)^2}\sin \beta a - \frac{P}{\beta a}\cos \beta a + \sin \beta a\right)$$

with $\rho(\beta a)$ so normalised that

$$\int_{(a\beta)\,\text{min.}}^{(\beta a)\,\text{max.}} \rho(\beta a)\, d\beta a = \frac{L}{2},$$

the integration extending over any one of the allowed regions of βa. Fig. 4 shows $\rho(\beta a)/L$ in the first o fthese for a value $P = 3\pi/2$, while for free electrons $(P = 0)$ one obtains instead a horizontal line with an ordinate $1/2\pi$ for all values of βa. The binding has thus the effect of concentrating the stationary states at the limits of an allowed region.

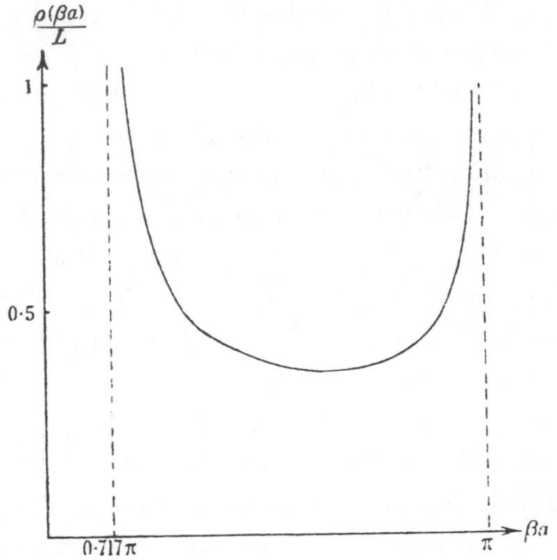

FIG. 4.

3. *The Linear Momentum.*—According to quantum mechanics the matrix elements of the linear momentum of an electron belonging to two stationary states α and α' are given by

$$p\,(\alpha,\,\alpha') = \frac{h}{2\pi i}\int \psi_\alpha{}^* \frac{\partial \psi_{\alpha'}}{\partial x}\,dx,$$

the integration extending over one period L of the functions ψ_α and $\psi_{\alpha'}$. Introducing for ψ_α and $\psi_{\alpha'}$ their expressions given by equations (7), (8), (9) and performing the integration it is found that $p\,(\alpha,\,\alpha')$ vanishes unless αa and $\alpha' a$ differ by an integral multiple of 2π. Thus if the state α lies in the first allowed region of the positive βa-axis, then $p\,(\alpha, \alpha')$ is different from zero for only one state α' in the first, one in the third, one in the fifth positive allowed region, etc., and one in the second, one in the fourth, one in the sixth negative allowed region, etc., viz., those states for which $\alpha' a = \alpha a + 2n\pi$. A simple calculation gives us for the square of the absolute value of these non-vanishing matrix elements :

$$p\,(\alpha,\,\alpha')\,p\,(\alpha',\,\alpha) = \left(\frac{h}{\pi a}\,\frac{\beta\beta'}{\beta'^2-\beta^2}\right)^2$$

$$\times \frac{\sin^2 \alpha a\,(\cos \beta a - \cos \beta' a)^2}{\left[1-\cos \alpha a \cos \beta a + \frac{1}{P}(\cos \alpha a - \cos \beta a)^2\right]\left[1-\cos \alpha a \cos \beta' a + \frac{1}{P}(\cos \alpha a - \cos \beta' a)^2\right]}. \quad (10)$$

To make the significance of equation (10) clearer we have computed from it in the first place $\pi a p(\alpha, \alpha)/h$ and plotted it in fig. 5 as a function of βa for the

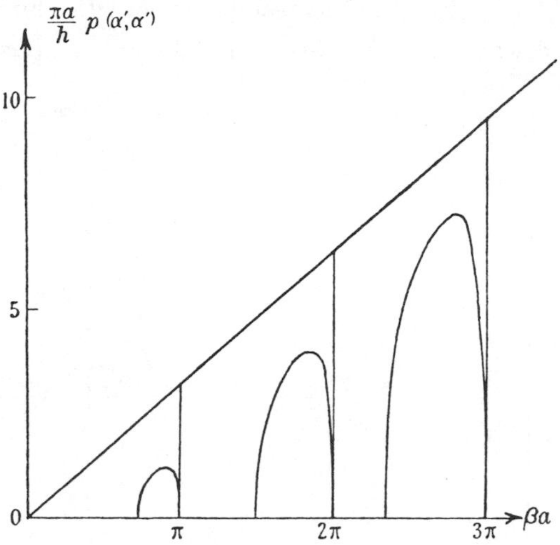

FIG. 5.

values of βa lying in the first, second and third allowed regions of the positive βa-axis, taking again P equal to $3\pi/2$. $p(\alpha, \alpha)$ represents the *time average of the linear momentum* of an electron in the stationary state α. As one easily sees

$$p(-\alpha, -\alpha) = -p(\alpha, \alpha).$$

Since the energy of the stationary states α and $-\alpha$ is the same, one may enquire if to a linear combination of the wave-functions ψ_α and $\psi_{-\alpha}$ there might correspond a linear momentum of greater absolute value, but from a simple calculation one finds that p lies then between $p(-\alpha, -\alpha)$ and $p(\alpha, \alpha)$. The straight line going through the origin in fig. 5 gives us $\pi a p(\alpha, \alpha)/h$ for the case of free electrons, showing that the potential barriers have the effect of reducing the linear momentum for a given value of the energy. Indeed they make it vanish when P approaches infinity. $p(\alpha, \alpha)$ also furnishes us the *electric current* associated with the stationary state α, whose time average is obtained by multiplying $p(\alpha, \alpha)$ by e/m, the ratio of the charge of the electron to its mass.

Furthermore we have plotted in fig. 6 the values of $\pi^2 a^2 p(\alpha, \alpha')\, p(\alpha', \alpha)/h^2$ as a function of βa when α is a state in the first allowed region while α' is that state in the second negative, third positive and fourth negative allowed region (curves 1, 2 and 3 respectively) for which the above quantity is different from zero. Again P has been taken equal to $3\pi/2$. For vanishing P all the curves are reduced to zero. For infinite P the results agree with those for an electron confined to move between impenetrable walls a distance a apart. The quantities

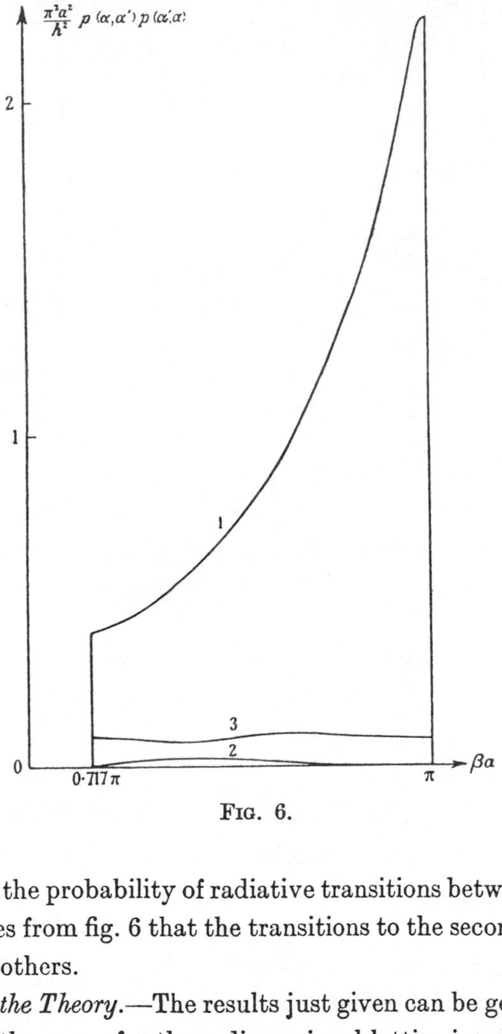

FIG. 6.

$p(\alpha,\alpha')$ determine the probability of radiative transitions between the stationary states, and one sees from fig. 6 that the transitions to the second negative region far outweigh all others.

4. *Extension of the Theory.*—The results just given can be generalised directly so as to apply to the case of a three-dimensional lattice in which the potential is the sum of three terms V (x), V (y), V (z), each of which depends upon its co-ordinate in the same way as V (x) on x in the one dimensional case. This means that the space is divided by infinitely thin potential barriers into cubical cells with an edge a. The wave-equation can then be separated

into three equations of the form (1), each involving only a single co-ordinate, viz.,

$$\left.\begin{aligned}\frac{d^2\psi_1}{dx^2}+\kappa^2[W_1-V(x)]\psi_1=0\\ \frac{d^2\psi_2}{dy^2}+\kappa^2[W_2-V(y)]\psi_2=0\\ \frac{d^2\psi_3}{dz^2}+\kappa^2[W_3-V(z)]\psi_3=0\end{aligned}\right\}, \qquad (11)$$

the total energy of the system and its wave-function being given by

$$W = W_1 + W_2 + W_3, \qquad (12)$$

$$\psi = \psi_1\,\psi_2\,\psi_3 \qquad (13)$$

respectively. Each of these equations can be treated in exactly the same way as equation (1). The stationary states will now be characterised by three quantities β_1, β_2, β_3, related to W_1, W_2, W_3 through equations analogous to equation (5), or by the corresponding quantities α_1, α_2, α_3.

In the one dimensional case we have seen that for $P = 0$ the energy W is a continuous function of αa, while for $P \neq 0$ W has discontinuities for $\alpha a = n\pi$ ($n = \pm 1, \pm 2, \ldots$). Similarly we shall have here that for $P = 0$, W is a continuous function of $\alpha_1 a$, $\alpha_2 a$, $\alpha_3 a$, while for $P \neq 0$ the function becomes discontinuous on the surfaces $\alpha_1 a = n_1\pi$, $\alpha_2 a = n_2\pi$, $\alpha_3 a = n_3\pi$ ($n_1, n_2, n_3 = \pm 1, \pm 2, \ldots$). The detailed behaviour of the energy values as well as of the wave-functions when P is varied follows directly from equations (12) and (13) and the results of the previous sections. The same remark applies to the matrix elements of the linear momentum. It is interesting to note in this connection that $p_x(\alpha_1, \alpha_2, \alpha_3\,;\,\alpha_1', \alpha_2', \alpha_3')$ is different from zero only provided $\alpha_1' a = \alpha_1 a + 2n\pi$, $\alpha_2' a = \alpha_2 a$, $\alpha_3' a = \alpha_3 a$ (n integer) with similar results holding for p_y and p_z.

A number of general qualitative features encountered in the special problem discussed may be expected to occur also for other periodic fields of potential. The falling apart of the energy spectrum into continuous regions separated by finite intervals has been met with previously in the case of a potential given by $V(x) = A\cos 2ax$, discussed by Strutt,* who also mentions that a similar

* M. J. O. Strutt, 'Ann. Physik,' vol. 86, p. 319 (1928). Attention may be called here to the fact that the energy regions considered by him as allowed are just the ones excluded for translatory motion through the lattice, and *vice versa*. The energy spectra of the equations corresponding to the potential A cos $2ax$ and to a potential as shown in fig. 1 for the particular case $a = b$ have also been discussed in connection with problems

effect arises for any one dimensional periodic field of potential according to theorems proved by Haupt.† The stationary states, just as in our problem, can be characterised by a value α, the coefficient of the exponent in the function $e^{i\alpha x}$ which results from the wave-function when the potential field is reduced to zero. The energy W of the stationary states again has discontinuities at the points $\alpha a = n\pi$ ($n = \pm 1, \pm 2, \ldots$). In an analogous manner the surfaces $\alpha_1 a = n_1\pi$, $\alpha_2 a = n_2\pi$, $\alpha_3 a = n_3\pi$ ($n_1, n_2, n_3 = \pm 1, \pm 2, \ldots$) will be surfaces of discontinuity for the function $W(\alpha_1 a, \alpha_2 a, \alpha_3 a)$ in the general three-dimensional case.

Another result, valid for any periodic field of potential, is concerned with the matrix elements of the linear momentum $p(\alpha_1, \alpha_2, \alpha_3; \alpha_1', \alpha_2', \alpha_3')$‡. It may be shown that these can be different from zero only if $\alpha_1' a = \alpha_1 a + 2n_1\pi$, $\alpha_2' a = \alpha_2 a + 2n_2\pi$, $\alpha_3' a = \alpha_3 a + 2n_3\pi$ (n_1, n_2, n_3 integers). For the wave-functions in the two states $\alpha_1, \alpha_2, \alpha_3$ and $\alpha_1', \alpha_2', \alpha_3'$ according to Bloch (*loc. cit.*) are given by

$$\psi_{\alpha_1\alpha_2\alpha_3} = u_{\alpha_1\alpha_2\alpha_3} e^{i(\alpha_1 x + \alpha_2 y + \alpha_3 z)},$$
$$\psi_{\alpha_1'\alpha_2'\alpha_3'} = u_{\alpha_1'\alpha_2'\alpha_3'} e^{i(\alpha_1' x + \alpha_2' y + \alpha_3' z)},$$

where the u's are periodic in x, y, z with the period a. Introducing these functions in the expression for the matrix element of p_x, say, gives us

$$p_x(\alpha_1, \alpha_2, \alpha_3; \alpha_1', \alpha_2', \alpha_3') = \frac{h}{2\pi i}\int \psi^*_{\alpha_1\alpha_2\alpha_3} \frac{\partial \psi_{\alpha_1'\alpha_2'\alpha_3'}}{\partial x}\, dx\, dy\, dz$$
$$= \frac{h}{2\pi i}\int u^*_{\alpha_1\alpha_2\alpha_3}\left[\frac{\partial u_{\alpha_1'\alpha_2'\alpha_3'}}{\partial x} + i u_{\alpha_1'\alpha_2'\alpha_3'}\right] e^{i[(\alpha_1'-\alpha_1)x + (\alpha_2'-\alpha_2)y + (\alpha_3'-\alpha_3)z]}\, dx\, dy\, dz$$
$$= \frac{h}{2\pi i}\int u^*_{\alpha_1\alpha_2\alpha_3} \frac{\partial u_{\alpha_1'\alpha_2'\alpha_3'}}{\partial x} e^{i[(\alpha_1'-\alpha_1)x + (\alpha_2'-\alpha_2)y + (\alpha_3'-\alpha_3)z]}\, dx\, dy\, dz,$$

the second term in the bracket not contributing anything on account of the orthogonality of the two wave-functions. From the periodicity of the u's it follows that we may write

$$u^*_{\alpha_1\alpha_2\alpha_3} \frac{\partial u_{\alpha_1'\alpha_2'\alpha_3'}}{\partial x} = \sum_{n_1 n_2 n_3 = -\infty}^{\infty} A_{n_1 n_2 n_3} e^{\frac{2\pi i}{a}(n_1 x + n_2 y + n_3 z)},$$

of classical physics by van der Pol and Strutt, "Phil. Mag.," vol. 5, p. 18 (1928). The advantage of the potential field considered in our article as compared with the field A cos $2dx$ lies in the fact that only elementary functions occur, making the evaluation of the various matrix elements very easy.

† O. Haupt, 'Math. Ann.,' vol. 79, p. 281 (1919).

‡ See also Kronig, *loc. cit.*

and introducing this in p_x leads immediately to the conclusion stated above. For two stationary states not separated by a surface of discontinuity of W the matrix element p hence always vanishes.

5. *Applications.*—A problem which we shall investigate here with the help of the one-dimensional model is the reflection of electrons of a given velocity falling from vacuum on to the lattice. The potential will then be as shown in fig. 7, for if our model is to represent an actual lattice, the potential of the

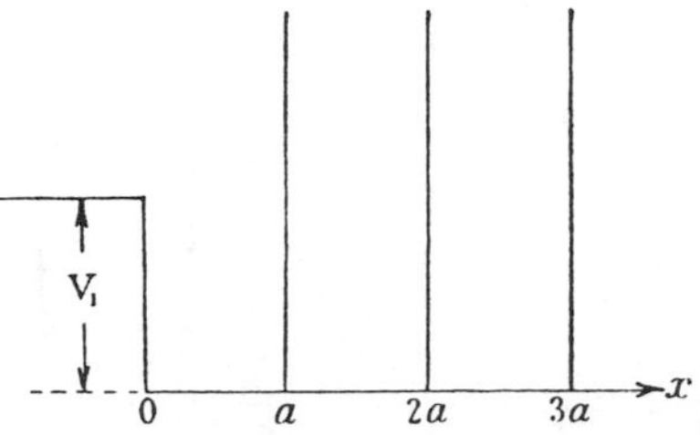

FIG. 7.

bottom of the lattice must be assumed to lie below that outside by a certain amount V_1. We enquire after a solution representing an incident and a reflected beam of electrons for negative values of x and a transmitted beam for positive values of x. For $x < 0$ we shall have thus:

$$\psi = e^{i\alpha_0 x} + Ae^{-i\alpha_0 x}, \tag{14}$$

while for $x > 0$ the solution is given by equation (8). α_0 and β are related by

$$\alpha_0^2 + \kappa^2 V_1 = \beta^2,$$

since the right and left-hand sides of this equation represent respectively the energies of the electron on the left and right of the point $x = 0$, which must be equal. The requirement that the function and its first derivative shall be continuous for $x = 0$ gives us the values of A in equation (14) and of C in equation (8).

AA* measures the intensity of the reflected beam, that of the incident beam being equal to 1, and is hence equal to the coefficient of reflection R. We find that

$$R = \frac{(\cos \alpha a - \cos \beta a)^2 + (\sin \alpha a - \frac{\alpha_0}{\beta} \sin \beta a)^2}{(\cos \alpha a - \cos \beta a)^2 + (\sin \alpha a + \frac{\alpha_0}{\beta} \sin \beta a)^2} \tag{15}$$

in the allowed regions of βa, while $R = 1$ for the forbidden regions of βa. The quantity α in equation (15) is the one associated with β in the manner described in the earlier sections. In fig. 8 R is represented as a function of the velocity

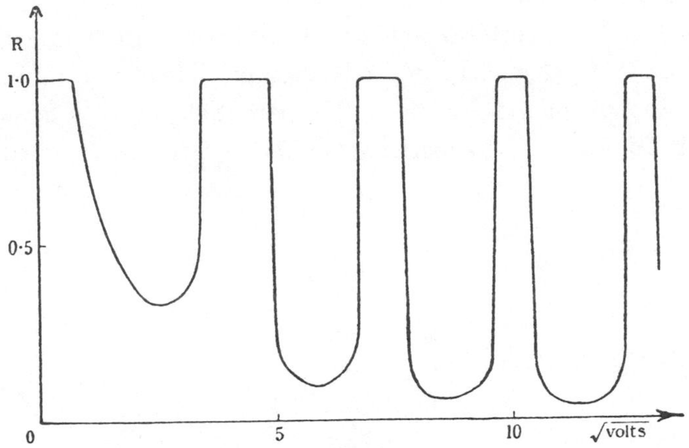

FIG. 8.

of the incident beam measured in $\sqrt{\text{volts}}$, which is proportional to α_0. For the binding constant P we have again taken $3\pi/2$, while for V_1 and a we have assumed the values $V_1 = 15$ volts, $a = 2{\cdot}2 \,.\, 10^{-8}$ cm. (suggested by the work of Rupp quoted below). The most essential feature of the diagram is, that regions of partial reflection alternate with regions of total reflection, and that these latter have a finite breadth, decreasing as the velocity of the particles gets greater. For very large values of the velocity total reflection takes place only when the lattice constant a is very nearly an integral multiple of the de Broglie wave-length of the incident particles, the regions of total reflection then becoming quite narrow.*

Measurements of the reflection and transmission of electrons in crystal lattices have been performed by various observers.† A quantitative comparison of their results with those obtained here is made difficult by the following circumstances: (1) If we had made our calculation with the three-dimensional

* Morse, 'Phys. Rev.,' vol. 35, p. 1310 (1930) has investigated the reflection and scattering of electrons by a crystal in which the potential depends upon the three co-ordinates through terms of the type A cos $2ax$. He obtains for the reflection coefficient at perpendicular incidence a curve essentially of the same type as that shown in fig. 8.

† Davisson and Germer, 'Phys. Rev.,' vol. 30, p. 705 (1927); 'Proc. Nat. Acad.,' vol. 14, pp. 317, 619 (1928); Rupp, 'Ann. Physik,' vol. 85, p. 981 (1928), vol. 1, p. 801 (1929), vol. 3, p. 497 (1929), and vol. 5, p. 453 (1930); 'Z. Physik,' vol. 61, p. 587 (1930); Thomson, 'Proc. Roy. Soc.,' A, vol. 117, p. 600 (1928), vol. 119, p. 651 (1928), and vol. 125, p. 352 (1929).

lattice, we would have obtained, besides the reflected and transmitted beams, diffracted beams which reduce the intensities of the former; (2) the periodic field of potential considered here differs from that actually present in a metal; (3) in the experiments it is found that some of the incident electrons suffer energy losses due to inelastic collisions with the electrons in the lattice, a phenomenon not provided for in our model. This effect also reduces the intensities of the reflected and transmitted beams. Nevertheless the principal features which appeared in our investigation, viz., the finite breadth of the reflection maxima, the decrease of this breadth with increasing velocity of the incident electrons and the decrease in the ordinates of the reflection minima, can be recognised in the experimental data (see in particular the work of Rupp).

An explanation can be given here also of a phenomenon recently observed by Rupp.* He finds that when the velocity of the incident electrons is gradually increased, new radiations appear in the soft X-ray spectrum of the substance bombarded at about the same velocities at which the reflection coefficient has a maximum. This can be understood if it be remembered that when the velocity begins to exceed that corresponding to the upper limit of a forbidden interval in the energy spectrum of the crystal, the impinging electron can enter into a new region of allowed energy values, giving it new possibilities for radiative transitions. According to this view the excitation threshold should differ from the reflection maximum by an amount equal to half the top-breadth of the maximum (lying toward higher velocities), but since in Rupp's measurements the maxima are already rather sharp the difference is probably obliterated by effects such as the inhomogeneity in the velocities, the imperfections of the crystal, etc. Perhaps, also, the constant energy losses of electrons impinging upon incandescent metals as observed by Rudberg† may be interpreted as corresponding to the transfer of the conduction electrons to the higher allowed regions of energy.

It is hoped to investigate later other physical properties of the model discussed as they appear in the phenomena mentioned at the beginning of this article.

Summary.

1. It is shown that the wave-equation representing the motion of an electron in a periodic field of potential can be integrated in terms of elementary functions when the potential takes the form of a series of equidistant rectangular barriers.

* 'Naturwiss.,' vol. 18, p. 880 (1930).

† Rudberg, 'Proc. Roy. Soc.,' A, vol. 127, p. 111 (1930).

When the breadth b of these barriers is made infinitely small and their height V_0 infinitely large, the results become particularly simple, the influence of the barriers depending then only on the product bV_0.

2. In the one dimensional problem for this limiting case the spectrum of permissible energy values is found to consist of continuous regions separated by finite intervals. By varying the quantity bV_0 from zero to infinity we pass from the case of free to that of bound electrons, and can thus study the changes in the allowed and forbidden ranges of the energy and in the wave-functions during this transition.

3. An investigation of the matrix elements of the linear momentum shows that the electrons can pass through the lattice and that there exists for them the possibility of transition to other stationary states under emission or absorption of radiation, the electrons thus having at the same time the characteristic properties of free and of bound electrons.

4. An investigation of the reflection of electrons by a crystal represented by the field of potential considered leads to results in qualitative agreement with the experimental facts. An explanation of a phenomenon recently observed by Rupp is given, and the possible connection of the theory with measurements of Rudberg is pointed out.

40

Reprinted from pages 553–559 of *J. Appl. Phys.* **16:**553–562 (1945)

The Basic Principles of Semi-Conductors[1]

BY FREDERICK SEITZ

Carnegie Institute of Technology, Pittsburgh, Pennsylvania

A review is given of typical deviations from ordinary valence rules by alloys and inorganic compounds. The effect on the electrical properties of a compound of a deviation from stoichiometric proportions is discussed. The value of the information obtained from electrical measurements is illustrated by a discussion of the properties of several semi-conducting compounds.

1. INTRODUCTION[2]

IT is a commonly known fact that the best insulating materials, such as organic resins and hydrocarbons and the high melting inorganic materials like the silicates and light-metal oxides, obey the rules of combining proportions very accurately. On the other hand, there is a large number of non-metallic solids which do not obey the rules of combining proportions very accurately and which, interestingly enough, are usually not very good insulators. In this category, for example, fall materials such as copper oxide, lead sulfide, iron sulfide, and other heavy-metal oxides, sulfides, and selenides. Careful examination shows that these materials are almost inevitably electronic conductors in the vicinity of room temperature, so that they represent a class of substance intermediate between the ideal metals and ideal insulators. We shall follow convention by calling this class *semi-conductors*. It should be added that several monatomic materials, such as selenium and tellurium, fall in the same category in the sense that they are electronic conductors that have conductivities intermediate between the values usually associated with the pure metals and the ideal insulators. These monatomic materials evidently cannot possess compositions which deviate from the rules of combining proportions; however, their electrical properties usually are greatly affected by impurities, so that we may conclude that their electronic conductivity is intimately connected with their chemical behavior.

In the present paper the relationship between the chemical and electrical properties of semi-conductors will be reviewed with the purpose of showing that the two are inextricably interwoven, and actually are manifestations of one set of principles. In addition, it will be shown that, as in many other fields of chemistry, purely physical measurements make it possible to explore with comparative ease a domain that would otherwise have been opened only with the greatest difficutly.

2. THE LAWS OF COMBINING PROPORTIONS

It is an interesting fact that in a large part of the conventional field of inorganic chemistry the

[1] Based on a paper presented at a symposium on semi-conductors at the Detroit meeting of the American Chemical Society, April, 1943.

[2] More detailed surveys of this topic may be found in the following books: F. Seitz, *The Modern Theory of Solids* (McGraw-Hill Book Company, Inc., New York, 1940); N. F. Mott and R. W. Gurney, *Electronic Processes in Ionic Crystals* (Oxford University Press, New York, 1940); A. H. Wilson, *Semi-Conductors and Metals* (Cambridge University Press, New York, 1939).

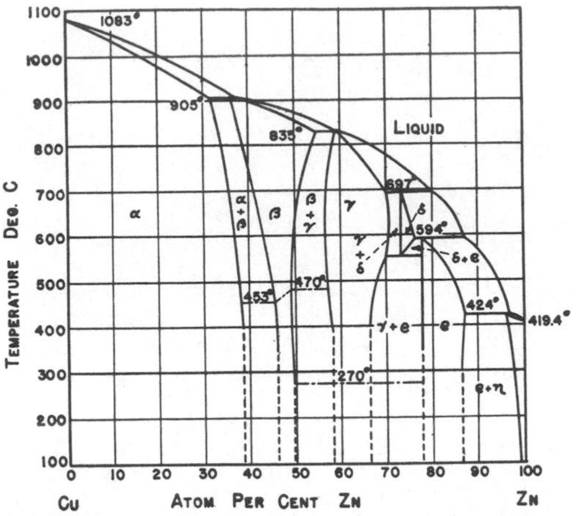

FIG. 1. The phase diagram of the brass system (Cu-Zn). The phases of this typical alloy system form over wide ranges of composition and for combining ratios that have no meaning from the standpoint of conventional valence theory.

amount of each element present in a given compound may be uniquely specified by giving the valence states of each of the elements in the compound. The valences which a given element may exhibit are in turn uniquely limited by its electronic structure or, more practically, by its position in the periodic chart. In this category fall, for example, such simple materials as the common halides and oxides of the lighter metals. The field of organic chemistry, on the other hand, is set apart not only by the fact that the compounds with which it deals are those found in living matter, but also by the fact that the combining ratios are not uniquely determined by stating the valence of the elements contained in these compounds. We know, as a result of the work of Kekule and his followers, that the principles of valency actually do apply in this field; however, it is necessary to augment the principles of valency with a theory of molecular structure before the combining proportions of a compound can be specified. Once the structure pattern of a compound has been settled upon, the laws of valency operate as effectively as in the part of the field of inorganic chemistry referred to above. In fact, once the importance of structure has been realized, we are able to correlate a much larger part of the field of inorganic chemistry with the use of simple principles of valency. For example, a clear understanding of the nature and composition of the many complex crystalline silicates and similar "chain" compounds can be then obtained.

It is a notable fact that the rules of combination obeyed by ionic and homopolar compounds are identical insofar as obedience to the laws of valency are concerned. The principal difference between the two types of compound shows up in the nature of the structure found in each. The great similarity goes back, of course, to the fact that the ability of an atom to become charged and its ability to form homopolar bonds are intimately related to the structure of its outer shell, which in turn determines its valence.

In spite of the degree to which the simple theory of valency is successful in determining combining proportions in large parts of the field of chemistry, when coupled with the rules of structure, we need not look very far to find an important class of simple crystalline compound in which the theory fails completely. It is evident (Fig. 1) that such alloys as the brasses (Cu-Zn alloys) and the bronzes (Cu-Sn alloys) obey an entirely different set of laws of combining proportions. For, in the first place, many of the metallic elements combine over a wide range of composition; and, in the second, even if we overlook the range of solubility, we must introduce formulae of the type Cu_5Zn_8, which are meaningless from ordinary valence theory, in attempting to describe the composition of the alloys. It is true that liquids and glasses usually show a similar range of solubility; however, these materials are in a somewhat different category since they are not crystalline and since they are formed by dissolving a compound which obeys the ordinary rules of combining proportions in another which does the same.

A significant clue to the irregular position occupied by the alloys is provided by the fact that the metals which alloy well, in the sense that their phases have well-separated solubility limits, are those which lie close to one another in the electromotive series. This means that they have very nearly the same chemical affinity, and as a result do not become strongly charged when in combination with one another. Thus, the ion-ion forces which play a very important role in most inorganic compounds are weak in the alloys. A careful analysis shows that in ionic substances which are composed of elements lying far apart in the electromotive series, these forces actually are responsible for the ordinary rules of combining proportions. For they require that the system be electrostatically neutral and, as a result, require that there be fixed ratios of the electropositive and electronegative elements. The same is undoubtedly true to a large extent in many of the homopolar compounds in which the homopolar bonds occur between different elements; that is, the displacement of charge from one atom to another that takes place in such compounds is probably sufficient to compel them to obey the rules followed by ionic substances. In addition, of course, the properties of the homopolar bond are such that the rules of valence are closely satisfied. We may conclude, then, that the constituents of ideal alloys disobey the ordinary rules of valence chemistry because (*a*) they lie so close together in the electromotive series that they do not obtain sufficient ionic charge to behave like ionic crystals and (*b*) the bond between them is sufficiently unlike the homopolar bond that they need not follow the same rules as homopolar substances.

Further insight into the rules obeyed by ionic and homopolar compounds on the one hand and by the alloys on the other may be obtained by consideration of the principles that determine the stability of any chemical system. According to thermodynamics the stable state of any system is determined by the condition that its free energy is a minimum. Since we are primarily interested in solids, we may take the free energy in the form

$$A = E - TS. \qquad (1)$$

Here E is the mechanical energy of the system, which is equal to the sum of the kinetic and potential energy of its constituent particles; S is its entropy; and T is the temperature, which for convenience we may take to be the absolute temperature. It is well known that the entropy of any given energy state of a system may be related to the number of possible ways N in which it may have that energy by means of the equation:

$$S = kT \log N, \qquad (2)$$

in which k is Boltzmann's constant. Thus S increases as the number of ways in which a given energy state may be achieved increases.

We may now note two extreme cases in which the free energy is used to determine the equilibrium state of a system. In the first place, let us suppose we are at the absolute zero of temperature; then T is zero and A reduces to E. Thus, in this circumstance, the condition for chemical equilibrium is identical with the condition for mechanical equilibrium. On the other hand, let us suppose that the temperature is so high that E in Eq. (1) is negligible in comparison with TS. Then A is a minimum when S is a maximum; that is, the system tends to that energy state which has the greatest value of N associated with it or which has the greatest degree of randomness. At any intermediate temperature in which E and TS are comparable, the equilibrium state will be determined by a compromise between the tendency of E to become as small as possible and S to become as large as possible. This competition is usually very genuine, for the states of low energy usually have small values of N or S associated with them, whereas, the states of large S and N usually have large values of E.

We may now apply these basic principles to the case of solid compounds, restricting our attention at first to the absolute zero of temperature where E alone determines the state of equilibrium. If we form an ionic compound in a case in which we have an excess of one of the constituent elements, experience tells us that the system is most stable, that is, has the lowest energy when "ideal" combining proportions of the two unite to form as much of the salt as possible, and the excess constituent remains uncombined. We know from the theory of ionic crystals that this occurs because the salt has its lowest energy when the ions are very precisely arranged in a perfect lattice. If the excess constituent was forced into the lattice as a solution of neutral atoms or molecules, so much work would have to be done that there would be a resultant rise in energy. Exactly the same situation occurs when we combine the constituents which make up an ideal homopolar compound.

On the other hand, when we form an ideal alloy, the atoms are relatively unconcerned about the precise nature of their neighbors since all of the constituent atoms are close together in the electromotive series and since none exhibit homopolar tendencies. Thus, in this case there is a net decrease in energy when all of the constituents are in the solid, regardless of the composition.

Now we may note that in general as we leave the absolute zero of temperature, the change that takes place in ionic and homopolar solids should push the state of these materials in the direction of that found in the ideal alloys at the absolute zero of temperature. The reason for this is that the TS term in (1), which requires randomness, will begin to exert its effect as soon as T is no longer zero. It is evident that the state of order in the ideal alloy is less than that found

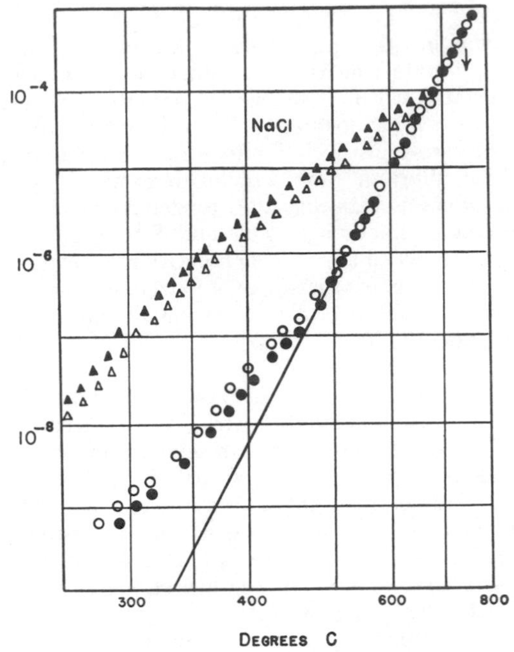

FIG. 2. The electrical conductivity of several specimens of sodium chloride (after Lehfeldt). The vertical scale is in units of $ohm^{-1}\ cm^{-1}$. It may be noted that the conductivities of all specimens are the same just below the melting point, but deviate at lower temperatures.

in the well-disciplined arrangements of ionic and homopolar substances. Thus we may begin to expect deviations from ideal stoichiometric relations to take place in these materials at sufficiently high temperatures under proper circumstances, such as if they are prepared with an excess of one of the constituents.

It is interesting to note in passing that under proper conditions even ideal alloys may show an opposite tendency at absolute zero; that is, show a tendency to obey the rules of valency. For example, there is a phase in the brass system (Fig. 1) known as β-brass, which has a body-centered cubic lattice and possesses the approximate composition CuZn. At elevated temperatures this phase forms over a fairly wide range of composition. Moreover, the arrangement of atoms among the various lattice sites of the lattice is random. However, as the temperature is lowered, two changes occur. In the first place, the solubility limits become very small; in the second, the arrangement of atoms becomes *ordered* in the sense that each copper atom prefers to have its eight neighbors be zinc atoms and *vice versa*. Thus, at low temperatures the lattice resembles very closely that of cesium chloride. It has been shown by Mott that this change is a result of the charges which the atoms possess as a result of their relatively small difference in position in the electromotive series; that is, the zinc atom behaves like a positive ion and the copper like a negative ion. The ionic charge is of the order of 0.1 electronic unit.

A more extreme example occurs in the magnesium-antimony system. Both these metals form ideal alloys when combined with metals which are close to them in the electromotive series. However, they do not lie close to one another in this series and as a result form only a single compound having the composition Mg_3Sb_2, which has very narrow solubility limits. It is evident that this compound, though an electronic conductor, has much in common with the ionic substances. In it magnesium behaves like a bivalent metal and antimony like a trivalent electronegative atom.

We shall now examine the types of deviation from stoichiometric proportions that occur in "ideal" ionic and homopolar substances and shall illustrate the way in which these deviations affect the electronic conductivity of these materials.

3. THE ALKALI HALIDES[3]

The alkali halides are the prototypes of the large and important class of ionic crystals and as such merit particular attention. In addition, their stoichiometric properties have been investigated extensively by Pohl and his associates under just the conditions in which we are interested in this article.

Ordinary specimens of the alkali halides which have been prepared under conditions of high chemical purity show no significant deviations from ideal stoichiometric combination. For this reason it might seem at first sight that their lattices are perfect representatives of the ideal ionic structures. If, however, we examine their electrical conductivity, we find good evidence to show that at any finite temperature these salts actually are the seat of an important effect which has bearing on their composition, as was first pointed out by Frenkel. Figure 2 shows the logarithm of the ohmic electrical conductivity of sodium chloride as a function of the reciprocal

[3] The experimental research on the properties of the alkali-halide crystals is described in the books of reference 2 and in the paper by R. W. Pohl, Physik. Zeits. **39**, 36 (1938).

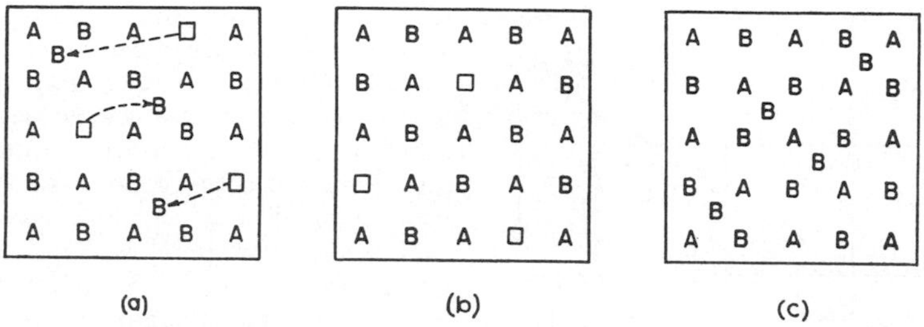

FIG. 3. Three types of lattice defects. (a) represents interstitial atoms and vacancies that are produced by atoms leaving normal positions in the lattice. (b) represents vacancies that are produced as a result of migration of one of the constituents out of the lattice. (c) represents interstitial atoms that are produced as a result of vaporization of one of the constituents from the surface of the lattice and the subsequent diffusion of the extra constituent into the interior of the crystal. It is believed that the normal alkali halides possess equal numbers of vacancies of the type shown in (b). In this case, the atoms that diffuse out of the lattice unite on the surface to form additional layers of the crystal.

of the absolute temperature. Transport experiments show that this conductivity is entirely ionic and is due to the migration of both positive and negative ions in variable ratio. Frenkel emphasized that the occurrence of this ionic conductivity in the solid can only mean that there must be present a degree of disorder in the lattice. For if the lattice were perfect and all ions were rigidly attached to fixed positions, except for thermal oscillations, we would expect no current to flow until the applied electric field reaches a value sufficient to tear the ions from the lattice and drive them to the electrodes. This type of conductivity evidently would not be ohmic, as the observed conductivity is, but would be analogous to dielectric breakdown, in which a large current is generated suddenly when the field reaches a definite value. Frenkel concluded that the ions must be migrating at any temperature at which the ionic conductivity is observed and that the field must play the role of directing this migration. It may readily be shown that such a mechanism will lead to Ohm's law.

It was originally suggested that the ions in the alkali halides occasionally leave their normal lattice positions as the result of thermal fluctuations and force their way into interstitial positions where they are more or less free to wander. (Fig. 3a). It is now known, however, as a result of extensive research, that the mechanism actually is somewhat different. Under the influence of thermal fluctuations, ions in the surface layers jump to the outer surface and leave *vacancies* behind (Figs. 3b, 5a). These vacancies wander through the lattice and permit a degree of mobility to the normal ions in the salt. The higher the temperature, the larger is the number of vacancies and the greater is the conductivity. This situation leads to the type of conductivity shown in Fig. 2 which increases with increasing temperature. It may be seen that the conductivity curve illustrated in the figure is perfectly linear near the melting point. This means that in this range of temperature the conductivity C obeys the equation:

$$C = Ae^{-Q/RT}, \tag{3}$$

in which A and Q are constants that are dependent upon the material, R is the gas constant, and T is the absolute temperature. This equation resembles closely that obeyed by the vapor pressure of a liquid or solid in which case Q is the heat of sublimation. The analogy actually is very close, for in the electrical case C is proportional to the density of vacancies in the lattice; whereas, in the case of vaporization the vapor pressure is proportional to the density of vapor molecules in the gas. It follows that Q in Eq. (3) is related to the energy required to form the vacancies in the crystal, and that these vacancies may be regarded as a type of "vapor" within the solid.

It may be observed that as the temperature is lowered the curve shown in Fig. 2 shows deviations from the linear relation that is valid just below the melting point. The temperature at which this deviation occurs depends upon the previous history of the specimen, although the conductivity at high temperature is not dependent upon this history. It is now known that this deviation is related to the "freezing-in" of vacancies as the temperature is lowered. Impurities may play an important role in aiding

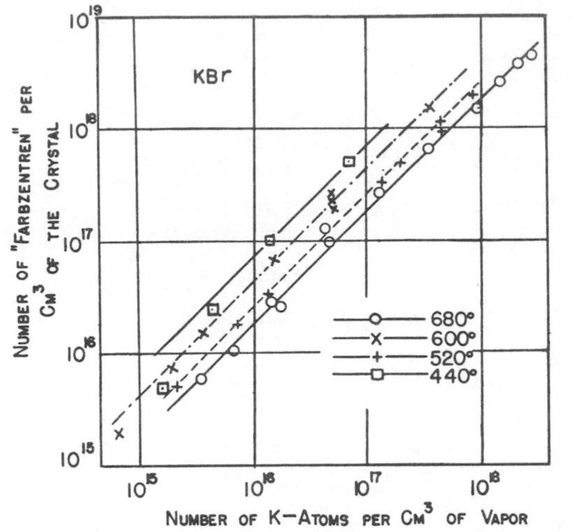

FIG. 4. Relationship between the number of excess potassium atoms dissolved in KBr and the density of atoms in the vapor surrounding the crystal. (After Pohl.)

this freezing, as is shown by the fact that the temperature at which the deviation occurs is lower the higher the purity. In any case, there is good reason to believe that if sufficient time were allowed for true equilibrium to occur at any given temperature, the conductivity of a pure specimen would follow accurately the linear relation observed at high temperatures.

Under normal conditions the numbers of positive-ion vacancies and negative-ion vacancies are equal. Thus, even though the ratio of vacancies to normal sites becomes as large as 0.1 percent near the melting point, no deviation from ideal stoichiometric relations is observed, as would be the case if there were more vacancies of one type than of the other. The reason for this equality in the number of vacancies is clear. If more positive ions than negative ones diffused to the surface, the surface would obtain a resultant positive charge and the interior would obtain a negative charge. This development of electrostatic potential would halt the diffusion process before it had gone far enough to lead to measurable differences in the ratio. It is conceivable that each of the excess halogen ions that would be left in the lattice under these conditions could give up an electron to the alkali metal ions at the surface and thereby prevent the development of potential. In this case the crystals would contain a layer of alkali metal on the surface and an excess of neutral halogen in the interior. Since the layer of alkali metal would be volatile at the temperatures of interest, it would disappear, and the crystals would be left with a stoichiometric excess of halogen. Simple consideration of energy, however, shows that this process could not occur below the melting point in the alkali halides, although, as we shall see later, a process very much like it occurs in other materials.

Suppose now that we heat an alkali-halide crystal in the presence of either the alkali metal vapor or the halogen gas. Then we might expect the gas atoms to take advantage of the porosity of the crystal that results from the presence of the vacancies and diffuse into it. This actually occurs. In other words, it is possible to produce alkali-halide crystals containing a stoichiometric excess of one of the constituents by heating in the presence of that constituent. Crystals containing an excess of the metal have been studied very extensively by Pohl and his co-workers, and a great deal is known about them at present. Figure 4 shows the relationship between the number of potassium atoms in the vapor and the excess in the crystal when KCl is heated in the presence of potassium at several temperatures. It is possible to determine the equilibrium constant for the reaction:

$$K(\text{gas}) \rightleftarrows K(\text{crystal}) \qquad (4)$$

from curves of this type. It turns out that the alkali atom becomes ionized after striking the surface of the crystal and that its valence electron enters the crystal and occupies a halogen-ion vacancy (Fig. 5). Other halogen ions diffuse to the surface to maintain a balance of electrical charge and form layers of the salt. If the temperature is sufficiently high, these electrons may migrate about in the crystal and transform it into an electronic conductor; that is, to a semi-conductor. Thus, we see that in this case there is a

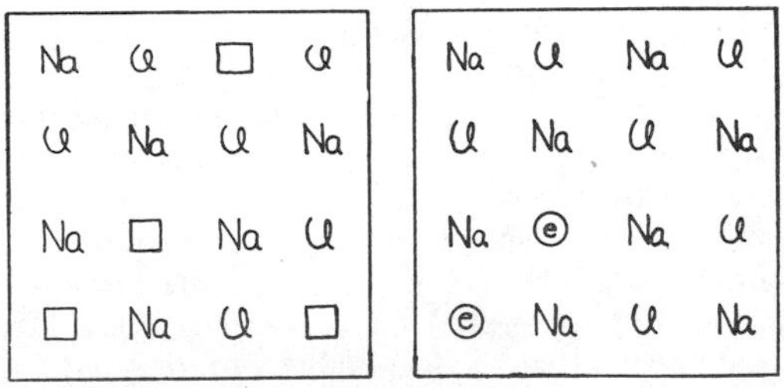

FIG. 5. The figure on the left represents an uncolored sodium chloride crystal in which there are equal numbers of vacancies of each sign. The right-hand figure represents the change produced when excess sodium is added. The positive ion vacancies become filled with sodium ions and the halogen vacancies become filled with electrons.

very intimate relationship between the electronic conductivity and the deviations from ideal proportions that occur when the crystal is heated in the presence of the metal.

The electrons bound at halogen-ion vacancies may absorb optical radiation that lies in the visible part of the spectrum. As a result, they discolor the crystal. When a colored crystal is placed between electrodes at a temperature sufficiently high to promote diffusion, the discoloration diffuses to the anode, and the crystal reverts to its original condition. During this process the electrons diffuse out of the crystal via the anode, and the excess alkali ions are converted to neutral atoms at the cathode. The removal of the stoichiometric excess of metal is accompanied by the decrease in electrical conductivity shown in Fig. 6. The shaded portion corresponds to the electronic part of the current.

It is evident from the observations on the alkali halides that if we were constructing a temperature *versus* composition-phase diagram for this system, it would be improper to represent the salt by an infinitely narrow line. Instead, the true relation should be that shown in Fig. 7 in an exaggerated degree. The boundary limits of the NaCl phase are shown to be fairly symmetrical since it is known that excess halogen can be added to the crystal as well as excess alkali metal.

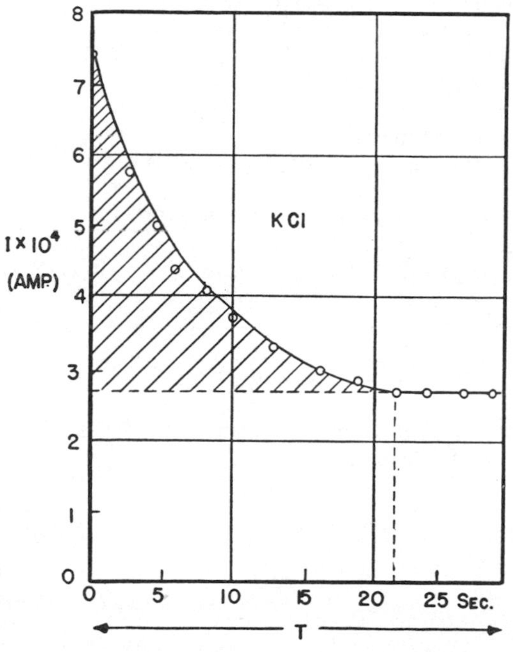

FIG. 6. Decrease in conductivity of colored potassium chloride as the color centers are removed from the lattice by electrolysis. (After Pohl.)

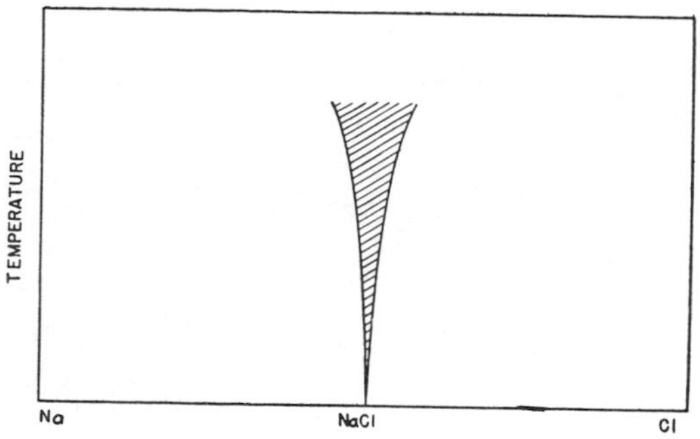

FIG. 7. Schematic phase diagram of the Na-Cl system. The cross-hatched region indicates the range of composition over which the crystal may be formed. The width of this region actually is about 0.1 percent near the melting point instead of the large value shown.

4. THE SILVER HALIDES

Silver chloride and silver bromide resemble the alkali halides closely in that they possess the sodium-chloride structure and are composed of singly charged ions. Moreover, at room temperature they possess a relatively large ionic conductivity which is due almost entirely to the mobility of silver ions. Careful analysis shows, however, that the lattice defects which occur in these salts probably are of a different nature from those found in the alkali halides. It seems most likely that in these cases a fraction of the silver ions force their way into interstitial positions, so that the ionic conductivity is due to the migration, both of these ions and of the vacancies left behind. As a result of this difference, it does not appear to be possible to convert the silver halides into electronic conductors by the methods used for the alkali halides.

It should be added that the high ionic mobility of the silver halides observed at room temperature plays an important role in the processes that take place in the darkening of the photographic emulsion.

[*Editor's Note:* Material has been omitted at this point.]

41

Reprinted from *J. Appl. Phys.* **16:**563–570 (1945)

The Electrical Properties of Semi-Conductors*

BY ROBERT J. MAURER

Carnegie Institute of Technology, Pittsburgh, Pennsylvania

The electrical properties of semi-conductors are briefly reviewed. The information concerning the structure of these materials which is obtained by electrical measurements is discussed. The most important of the physical-chemical methods available for the investigation of deviations from stoichiometric proportions in semi-conducting compounds are illustrated by a discussion of several typical examples.

INTRODUCTION

SEMI-CONDUCTORS are a class of solids whose electronic conductivity falls in the range 10^{-5} to $10^{3}\Omega^{-1}-\text{cm}^{-1}$. This is to be compared with the typical conductivities of metals which are of the order of $10^{5}\Omega^{-1}-\text{cm}^{-1}$. The significant difference between semi-conductors and metals is not, however, the difference in the order of magnitude of their conductivities but the tremendous variation of the conductivity of semi-conductors with the chemical and thermal treatment given them.

Our knowledge of the fundamental electrical properties of semi-conductors is small by comparison with that of metals. The quantum mechanical theory of solids has progressed to a point where it is possible to make a quantitative comparison of the calculated and measured conductivities of a specific metal. In discussing semi-conductors we must still be content with general models. It is the purpose of this paper to review some of the important experiments which have yielded information regarding the behavior of electrons in semi-conductors. Because of the peculiar nature of the origin of the conductivity of these materials, it is impossible to do this and neglect their chemistry.

ELECTRICAL CONDUCTIVITY

Our introductory discussion has tacitly assumed that the conductivity of a semi-conductor is its most important electrical property. Figure 1 shows a typical experimental arrangement for the measurement of conductivity and its temperature variation. The potential drop across a portion of the specimen is measured with a known current passing through it. The purpose of this method is to avoid errors due to a large contact resistance which is frequently present at the junction of a metal and a semi-conductor. It must be recognized that the conductivity may not be due entirely to electrons. The ionic conductivity may be very small, as in cuprous oxide at 1000°C, where it amounts to only 4×10^{-4} of the total conductivity.[1] On the other

* Based on a paper presented at a symposium on semi-conductors at the Detroit meeting of the American Chemical Society, April, 1943.

[1] J. Gunderman and C. Wagner, Zeits. f. physik. Chemie **B37**, 155 (1937).

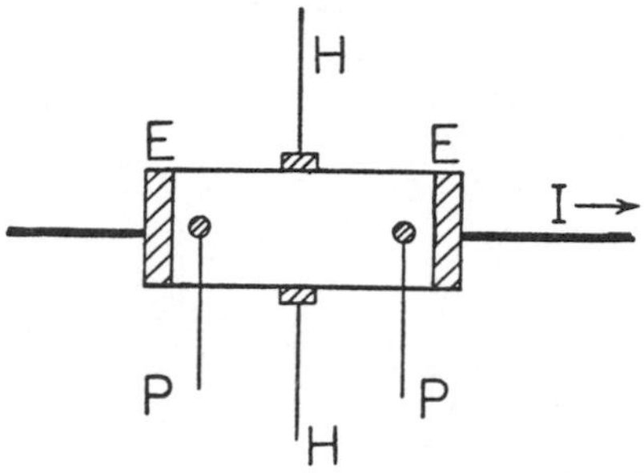

FIG. 1. Measurement of conductivity and Hall effect; E,E are electrodes for passing the constant current, I, through sample, P,P and H,H are electrodes for potentiometric measurement of voltage drop and Hall e.m.f., respectively. Magnetic field is assumed perpendicular to plane of paper.

hand, the electrolytic conductivity of β-Ag_2S at 170°C is 81 percent of the total conductivity.[2]

The electronic conductivity of a semi-conductor can be expressed in terms of an equation which was originally proposed by Lorentz[3] to describe the conductivity of metals.

$$\sigma = \frac{4}{3} \frac{e^2 nl}{(2\pi mkT)^{\frac{1}{2}}}, \tag{1}$$

where e is the electronic charge, k is Boltzmann's constant, T is the absolute temperature, n is the density of conducting electrons, and l is their mean free path. The density of the free electron gas in a metal is too large to permit the use of the Maxwell-Boltzmann statistics as was done by Lorentz in the derivation of this equation. In semi-conductors, the density of conducting electrons is seldom greater than $10^{18}/cm^3$ and Lorentz' equation is valid. The mass, m, which appears in the equation is an effective mass which may be greater or less than the electronic mass. It will be noted that the product of n and l appears in (1). The determination of the conductivity fixes neither of these important quantities uniquely.

Over rather wide ranges of temperature the conductivity of a semi-conductor can usually be expressed by an empirical equation of the form

$$\sigma = Ae^{-E/kT}, \tag{2}$$

where A and E are constants. This means that either the density of free electrons or the mean free path, or both, are strongly temperature dependent. It has been frequently assumed in the past that the mean free path of an electron in a semi-conductor is determined by the same mechanisms as in a metal. In a metal the most important factor is the interaction of the electron with the ions of the lattice as a result of the thermal motion of the latter. This results in a transfer of the energy, which the electron has gained from the applied electric field, to the lattice with the stimulation of lattice vibrations. At ordinary and high temperatures it is well known that this process leads to the approximately linear increase of resistance of metals with temperature. The exponential increase of the conductivity of semi-conductors with temperature has been ascribed to an exponential increase in the number of free electrons. Furthermore, since the presence of traces of impurities in metals leads to an increase in resistance which is small and independent of the temperature, the great dependence of the resistance of semi-conductors upon their impurity content has been ascribed to the effect of these impurities upon the number of free electrons. In fact, it is believed that in the pure state at ordinary temperatures practically all semi-conductors are good insulators. The impurities are a source of potentially free electrons. To free an electron from an impurity center requires an activation energy, E, and the density of free electrons follows from Boltzmann's principle

$$n = n_0 e^{-E/kT}. \tag{3}$$

Wilson[4] has been able to fit these ideas into the quantum-mechanical theory of solids. We shall not pursue this phase of the topic but will point out one important result. An impurity center may possess no bound electrons which can be thermally freed but may be able to trap one of the "normal" electrons of the semi-conductor. This leaves a "hole" in the electron distribution of the solid. Such "holes" will give rise to a conductivity described by Eqs. (1)–(3) but the sign of the charge carriers will appear to be positive. This peculiar "hole" conduction, which occurs in many metals and is predicted by the theoretical treatment of the motion of an electron in the periodic potential field of a lattice, can be detected by measurement of the Hall effect.

[2] C. Tubandt and H. Reinhold, Zeits. f. anorg. allgem. Chemie **160**, 222 (1927).

[3] F. Seitz, *Modern Theory of Solids* (McGraw-Hill Book Company, Inc., New York, 1940), p. 190.

[4] A. H. Wilson, *Semi-conductors and Metals* (Cambridge University Press, New York, 1939).

THE HALL EFFECT

When an electronic current is present in a material and there is applied a uniform magnetic field in a direction perpendicular to that of the current, an e.m.f. appears across the sample in a direction perpendicular to the plane defined by the current and magnetic field vectors. This e.m.f. is proportional to the product of the current and the magnetic field strength and inversely proportional to the thickness t of the specimen.

$$\epsilon = R(Hi/t). \quad (4)$$

An extension[5] of Lorentz' treatment of an electron gas to the case where electric and magnetic fields are present gives the following equation for the constant R:

$$R = (-3\pi/8)(1/nec). \quad (5)$$

The measurement of the Hall constant, R, yields information of two kinds. The sign of R tells us whether the conduction is due to electrons or to "holes." The numerical magnitude of R allows the calculation of the density of conducting electrons or "holes." If the Hall constant and the conductivity are measured the mean free path can be obtained from Eqs. (1) and (5)

$$l = \left(\frac{8c^2mkT}{\pi e^2}\right)^{\frac{1}{2}} \sigma R. \quad (6)$$

The experimental arrangement for the measurement of the Hall e.m.f. is shown in Fig. 1. The Hall e.m.f. is not easy to determine because it is usually of the order of a fraction of a millivolt and at high temperatures is masked by thermoelectric e.m.f.'s which occur in the circuit as a result of unavoidable thermal gradients across the sample.

Figure 2 shows the mean free path in cuprous oxide at low temperatures. These results are calculated by Engelhard[6] from Hall effect and conductivity data taken by him. The experimental temperature dependence agrees well with the theory of Fröhlich and Mott[7] which predicts the rapid increase of the mean free path at low temperatures.

Miller's data upon the conductivity of ZnO have been described by Seitz[8] in the preceding paper. The conduction electrons result from the thermal excitation of electrons from interstitial zinc atoms. In Miller's work the experiments were conducted in a time short compared with that necessary for equilibrium to be established so that the number of interstitial zinc atoms was kept constant. Miller used the slope of the high temperature portion of the curve to determine the activation energy for the release of an electron from an interstitial zinc atom with the assumption of a temperature independent mean free path. The change in slope of the low temperature part of the curve he ascribed to the pairing of zinc atoms with the production of interstitial zinc molecules. On purely statistical grounds there will be a small fraction of these pairs present. Since the activation energy for the pairs will be lower than that of the atoms, the behavior of the pairs will dominate the low temperature part of the curve where the number of electrons released from atoms will be small as compared with the number released from the pairs. Figure 3 gives the data of Miller[9] on the

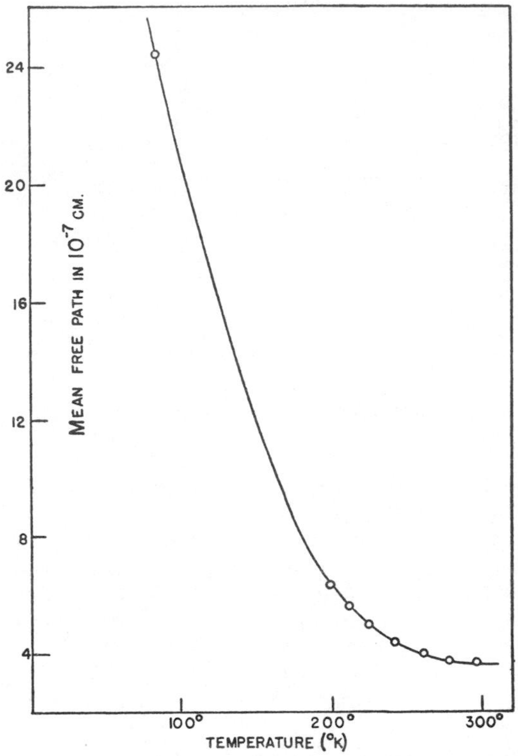

FIG. 2. The mean free path in cuprous oxide as a function of temperature (after Engelhard).

[5] F. Seitz, see reference 3.
[6] E. Engelhard, Ann. d. Physik **17**, 501 (1933).
[7] H. Fröhlich and N. Mott, Proc. Roy. Soc. **A171**, 496 (1939).
[8] F. Seitz, J. App. Phys. **16**, 553 (1945).
[9] P. Miller, Jr., Phys. Rev. **60**, 890 (1941).

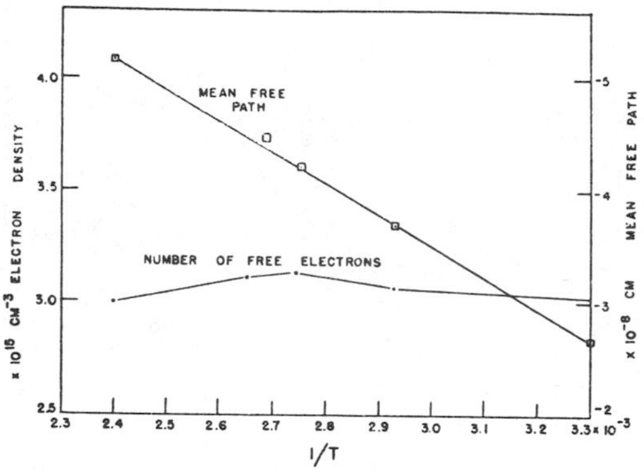

FIG. 3. The mean free path and density of conduction electrons in zinc oxide (after Miller).

number of electrons and mean free path in the low temperature region. It was impossible to measure the Hall e.m.f. at high temperatures. It is to be noted that at low temperatures the number of free electrons is constant while the mean free path increases by a factor of two over a temperature range of 100°C. The constant number of free electrons can be explained by the assumption that at these temperatures each interstitial zinc molecule has lost an electron, and the contribution of the interstitial zinc atoms is still negligible. The number of molecules, determined from the electrical measurements in this manner, is what one would expect from the density of interstitial zinc atoms and the statistical probability of a certain fraction being nearest neighbors in the lattice. The rapid change in mean free path is anomalous. This behavior is not unique, however. Feldman[10] found that the mean free path in cuprous oxide was a strong function of the temperature in the neighborhood of 300°C. It must be realized that in the interpretation of conductivity data the assumption of a mean free path which is independent of temperature and concentration of impurity is extremely dangerous.

THE THERMOELECTRIC EFFECTS

The thermoelectric effects in solids are rather complicated second-order effects so that little precise information can be obtained from them. The sign of the Seebach e.m.f. gives, however, the sign of the charges whose motion constitutes the current and its determination is a relatively simple matter. It is known that the reason for the fact that the Seebach e.m.f. at a semi-conductor-metal junction is a thousand times greater than at metal-metal junctions is the result of the great difference in the density of conduction electrons in semi-conductor and metal but differences cannot be determined precisely in this manner.

DEVIATIONS FROM STOICHIOMETRIC PROPORTIONS

The experiments which have been described have all been performed upon semi-conductors which possessed an unknown constant deviation from stoichiometric proportions. Obviously, one of the most important problems is the correlation of the purely electrical properties with the excess or deficiency of a given component in the compound. In addition there is the purely chemical problem of determining how the extent of the deviation from stoichiometric proportions depends upon the equilibrium conditions.

The types of deviations from stoichiometric proportions which are to be expected in semi-conducting compounds have been reviewed by Seitz.[8] First, there is the wandering of one type of ion into an interstitial position while the other type of ion leaves the crystal for the gas phase. This is usually referred to as Frenkel disorder. Another type of disorder is the Schottky type where the ions wander to the surface and build an extension of the lattice with the simultaneous formation of vacancies. In principle, the type of disorder can be established by purely experimental methods. The combination of density measurement and x-ray determination of the lattice constant gives the number of molecules per unit cell,

$$\nu = Na^3d/M,$$

where ν is the number of molecules per unit cell, N is Avogadro's number, a is the lattice constant, d is the density of the sample, and M is the molecular weight. In the case of Frenkel disorder, the average number of molecules per unit cell becomes larger as the degree of disorder increases. With Schottky disorder the density of molecules decreases as the degree of disorder increases. The number of vacancies per unit volume is usually in the range from 10^{-7} to 10^{-4} so that the precision of the measurements is ordinarily insufficient to make the method of practical value. The Wüstite phase of FeO is an exception. Wüstite has the sodium chloride type of lattice, and the iron content can be

[10] W. Feldman, Thesis, University of Pennsylvania (1942).

varied from 76.08 to 76.72 wt. percent of iron. Jette and Foote[11] prepared a large number of samples with different iron contents and determined the iron content by chemical analysis. Density and x-ray measurements then enabled them to show that the disorder was of the Schottky type. One iron ion diffuses to the surface for each excess oxygen atom present in the solid phase. The oxygen atom becomes an ion by obtaining two electrons from the solid. The result is the creation of iron ion vacancies and electron "holes." In this case the degree of disorder is so great that it is measurable by chemical analysis.

Juse and Kurtschatow[12] were able to demonstrate the presence of excess oxygen in cuprous oxide by chemical analysis. The cuprous oxide was dissolved in acid potassium iodide solution and the oxidized iodine titrated with hyposulfite solution. Samples of cuprous oxide which had been heated in a vacuum showed no detectable excess oxygen. Samples heated in oxygen possessed excess oxygen which increased with the time of heating and reduction of the annealing temperature. Their results are tabulated in Table I. These authors determined also the conductivity *versus* temperature curves for their samples. The activation energies were the same for all of the samples possessing excess oxygen, but the magnitude of the constant A in Eq. (2) increased with the amount of excess oxygen. Dünwald and Wagner[13] also attempted to determine the amount of excess oxygen in cuprous oxide by determining the change in oxygen pressure in a sealed flask when the oxygen was exposed to the solid. They established that the amount of oxygen absorbed varied approximately as the seventh root of the oxygen pressure, which is the same pressure dependence as the conductivity. Baumbach and Wagner[14] attempted to detect a change in weight of nickelous oxide with change of oxygen pressure at 1000°C. Any change which occurred was less than that corresponding to an absorption of 2×10^{-3} g-atom of O per mol of NiO, which was the limit of sensitivity of their apparatus.

TABLE I. Excess oxygen in cuprous oxide as determined by chemical analysis. (Juse and Kurtschatow.)

Remarks	Excess oxygen in wt. percent
Original material	0.053
Vacuum heated	not detectable
1 hour in air at 1000°C	0.060
6 hours in air at 700°C	0.084
6 hours in air at 600°C	0.090
6 hours in air at 500°C	0.101

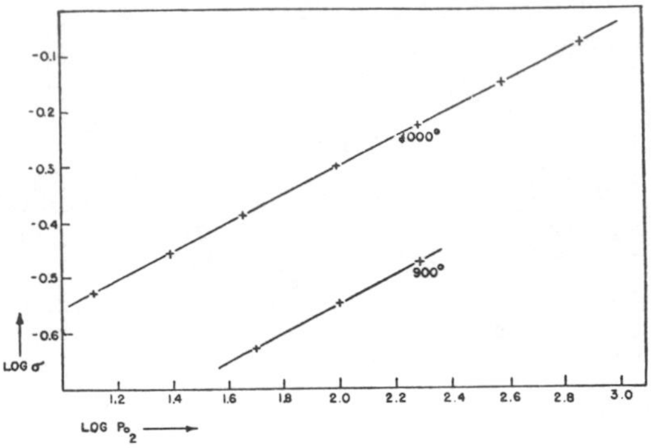

FIG. 4. The logarithm of the conductivity of nickelous oxide *versus* the logarithm of the oxygen pressure (after Baumbach and Wagner).

The dependence of the conductivity upon the vapor pressure of the gas phase of one component of a semi-conducting compound has frequently been used to give information regarding the course of the reaction which results in the production of disorder. Several examples have been given by Seitz in the preceding paper.

Another example is given in Fig. 4 which shows the dependence of the conductivity of nickelous oxide upon the pressure of oxygen in equilibrium with the solid.[14] The conductivity varies as the fourth root of the oxygen pressure. A mechanism which will yield this behavior is expressed in the following equation:

$$Ni^{++}+O^{=}+\tfrac{1}{2}O_2(g)\rightleftharpoons Ni^{++}+2O^{=}+[Ni^{++}]^{+}+h.$$

The reaction can be described as the diffusion of a nickelous ion to the surface where it builds an extension of the crystal lattice with the atom of oxygen from the gas phase. Two electrons diffuse to the surface to convert the oxygen atom into a double charged ion. One of the two "holes" which result is trapped at the vacancy left by the nickelous ion in the lattice and the symbol $[Ni^{++}]^{+}$ indicates this vacancy plus trapped "hole." The other "hole" is free and is

[11] E. Jette and F. Foote, J. Chem. Phys. 1, 29 (1933).
[12] W. Juse and B. Kurtschatow, Physik. Zeits. Sowjetunion 2, 453 (1932).
[13] H. Dünwald and C. Wagner, Zeits. f. physik. Chemie B22, 212 (1933).
[14] H. Baumbach and C. Wagner, Zeits. f. physik. Chemie B22, 59 (1934).

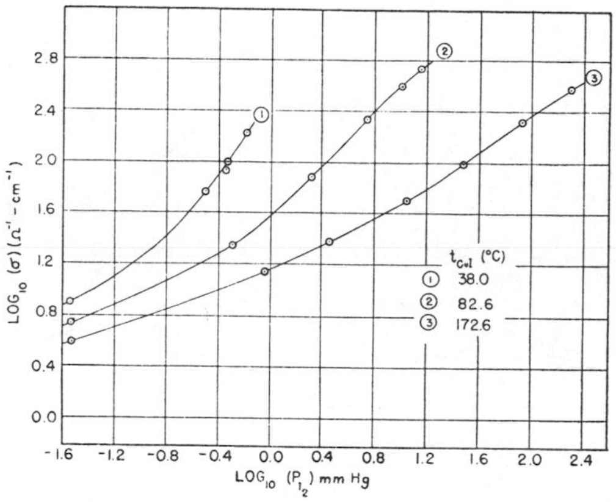

FIG. 5. The logarithm of the conductivity of cuprous iodide *versus* the logarithm of the iodine pressure.

the source of the electrical conductivity. The application of the mass action law to the above equation and the assumption that the conductivity is proportional to the concentration of holes leads to the observed dependence of the conductivity on the pressure.

When the conductivity does not vary as some simple power of the pressure of the gas in equilibrium with the solid, it is usually impossible to deduce the mechanism of the reaction from the pressure-conductivity curves. As an example of this type of behavior, we may refer to cuprous oxide below 800°C; and to cuprous iodide.

The conductivity of cuprous oxide varies according to the following equation:

$$\sigma = \sigma_0 p_{O_2}{}^{1/7}$$

for temperatures between 1000°C and 800°C, while the theory predicts an eighth root dependence upon the pressure of oxygen. Below 800°C there is no simple relationship between pressure and conductivity.

Cuprous iodide is a semi-conductor whose electrical properties depend upon the amount of excess iodine in a very complicated fashion. Nagel and Wagner[15] have proposed the following reaction:

$$Cu^+ + I^- + \tfrac{1}{2}I_2(g) \rightleftharpoons Cu^+ + 2I^- + [Cu^+] + h.$$

$[Cu^+]$ denotes a cuprous ion vacancy and h an electron hole. A cuprous ion and an electron are supposed to wander to the surface and combine with an iodine atom from the gas phase. The result is the formation of cuprous ion vacancies and holes. At a fixed temperature, the fraction of the holes which are free is assumed constant. An application of the mass action law to the above reaction, in which the vacancies and holes are treated as chemical entities, then leads to a fourth root dependence of the conductivity upon the pressure of the iodine vapor in equilibrium with cuprous iodide.

$$\sigma = \sigma_0 p_{I_2}{}^{1/4}.$$

It is assumed that the mean free path is independent of and the conductivity proportional to the density of holes. Figure 5 shows the experimental dependence of the conductivity upon the pressure of iodine vapor.[16] There is no simple power law relation between conductivity and pressure.

The weight of excess iodine absorbed by cuprous iodide varies, however, as the square root of the iodine pressure as shown in Fig. 6. This data was obtained with the quartz microbalance illustrated in Fig. 7. Magnetic control of the balance in the manner described by Blewett[17] allowed the balance to be operated in an atmosphere of iodine. The increase in weight of the sample due to absorption of iodine was balanced by the torque upon the iron cylinder caused by the magnetic field of a pair of Helmholtz coils. The iron cylinder and platinum counterweight

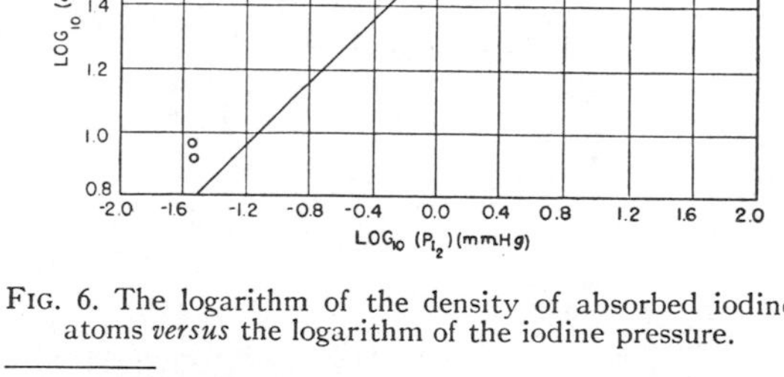

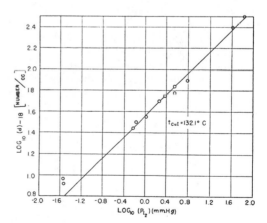

FIG. 6. The logarithm of the density of absorbed iodine atoms *versus* the logarithm of the iodine pressure.

[15] K. Nagel and C. Wagner, Zeits. f. physik. Chemie **B25**, 71 (1934).

[16] This work was supported by the Solar Energy Research Fund of Massachusetts Institute of Technology and carried out in the laboratory of Professor A. von Hippel. A detailed account will be published soon.

[17] J. P. Blewett, Rev. Sci. Inst. **10**, 231 (1939).

were sealed in quartz to protect them from the iodine vapor.

That the dependence of the conductivity of cuprous iodide upon the amount of excess iodine is very complicated is further illustrated by Fig. 8. Here we have the mean free path as a function of the density of holes as calculated from conductivity and Hall effect data of Steinberg.[18] The mean free path drops rapidly as the density of holes becomes greater than 10^{18} cc^{-1}.

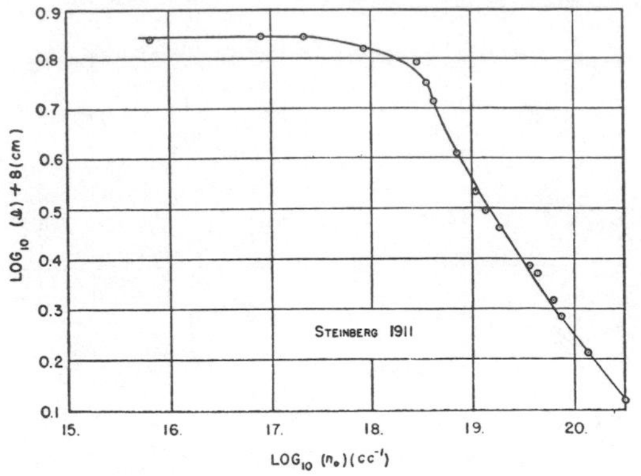

FIG. 8. The logarithm of the mean free path in cuprous iodide *versus* the logarithm of the density of conduction "holes" (after Steinberg).

The alkali halide semi-conductors have been extensively investigated by Pohl and his co-workers who have developed a very interesting optical method for the determination of the amount of excess alkali metal present. The reaction proposed for the absorption of excess potassium in potassium chloride is

$$K^+ + Cl^- + K(g) \rightleftharpoons 2K^+ + Cl^- + [Cl^-]^-.$$

When the crystal is heated at a high temperature in the presence of potassium vapor chlorine ions wander to the surface of the crystal and build on to the lattice with potassium atoms from the vapor. The potassium atoms lose electrons which diffuse into the crystal and are trapped at the vacant lattice sites where chlorine ions are missing. These sites will be the most stable positions in the crystal for the electrons. An electron trapped at a chlorine vacancy can absorb light and be excited to a higher energy level. The absorption band corresponding to this process is shown in Fig. 9.

The reason for the presence of an absorption band rather than an absorption line is the interaction of the electron with the lattice. If the half-width and height of the absorption band are measured, the number of absorbing centers can be calculated from dispersion theory.[19] There is one unknown factor, the oscillator strength which is proportional to the probability that a trapped electron will absorb a photon, but there are very good theoretical grounds for believing that this strength is close to unity. Kleinschrod[20] compared the amount of excess potassium in potassium chloride crystals as determined by the optical method and chemical analysis. He dissolved his crystals in water and determined the resultant *p*H of the solution using methyl red as an indicator. The absorption curves of the methyl red were determined before and after solution of the potassium chloride and the excess potassium obtained from the change in *p*H. For concentrations of excess potassium atoms between $1.0 \times 10^{+17}$ and $2.0 \times 10^{+18}$ cc^{-1}, the chemical procedure gave an increase of 19 percent over the concentrations determined optically. The interpretation is that the oscillator strength of an electron trapped at a halogen ion vacancy is not 1.0 but 0.8. The usefulness of the optical determination depends upon the occurrence of the absorption band due to the vacancies in an accessible portion of the spectrum where it does not overlap the characteristic absorption of the pure crystal. In addition, it is necessary that the absorption band have the simple bell shape predicted by optical dispersion theory.

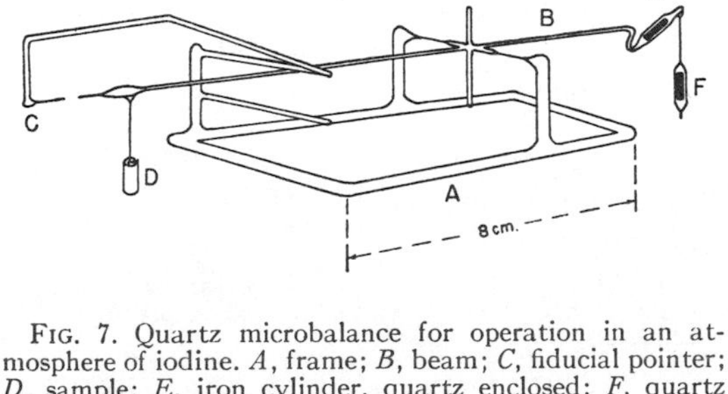

FIG. 7. Quartz microbalance for operation in an atmosphere of iodine. *A*, frame; *B*, beam; *C*, fiducial pointer; *D*, sample; *E*, iron cylinder, quartz enclosed; *F*, quartz enclosed platinum counterweight.

CONCLUSION

Deviations from stoichiometric proportions are the most important type of "impurity" which

[18] K. Steinberg, Ann. d. Physik **35**, 1009 (1911).

[19] A. Smakula, Zeits. f. Physik **59**, 603 (1930).
[20] F. Kleinschrod, Ann. d. Physik **27**, 97 (1936).

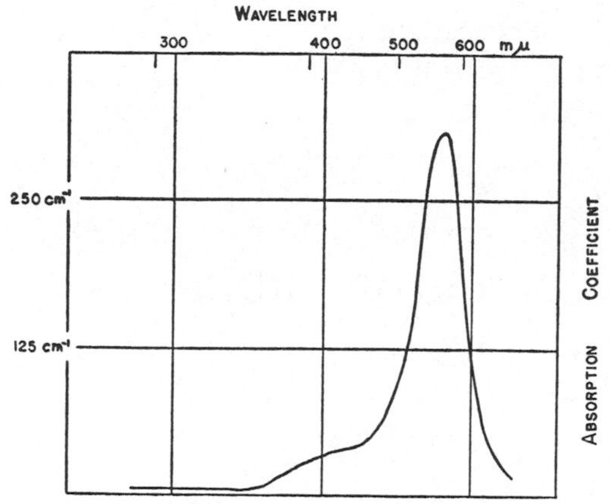

FIG. 9. Absorption coefficient *versus* wave-length for trapped electrons in KCl (after Pohl).

one has to consider when dealing with the electrical properties of semi-conducting compounds There are available physical-chemical methods by which, in favorable cases, the degree of disorder may be determined as a function of the equilibrium conditions. The combination of these measurements with the determination of the electrical conductivity, σ, and the Hall constant, R, enables one to deduce the mean free path and the density of conducting particles as a function of the degree of disorder and of the temperature. The absolute value of the mean free path is not uniquely fixed by these measurements because there is no reliable method of determining the effective mass of the particles, which occurs in Eq. 4 for the mean free path. It is assumed, ordinarily, that the effective mass is the electronic mass.

42

Reprinted from *Phys. Rev.* **74**:230–231 (1948)

THE TRANSISTOR, A SEMI-CONDUCTOR TRIODE

J. Bardeen and W. H. Brattain
Bell Telephone Laboratories, Murray Hill, New Jersey

A THREE–ELEMENT electronic device which utilizes a newly discovered principle involving a semi-conductor as the basic element is described. It may be employed as an amplifier, oscillator, and for other purposes for which vacuum tubes are ordinarily used. The device consists of three electrodes placed on a block of germanium[1] as shown schematically in Fig. 1. Two, called the emitter and collector, are of the point-contact rectifier type and are placed in close proximity (separation ∼.005 to .025 cm) on the upper surface. The third is a large area low resistance contact on the base.

The germanium is prepared in the same way as that used for high back-voltage rectifiers.[2] In this form it is an *N*-type or excess semi-conductor with a resistivity of the order of 10 ohm cm. In the original studies, the upper surface was subjected to an additional anodic oxidation in a glycol borate solution[3] after it had been ground and etched in the usual way. The oxide is washed off and plays no direct role. It has since been found that other surface treatments are equally effective. Both tungsten and phosphor bronze points have been used. The collector point may be electrically formed by passing large currents in the reverse direction.

Each point, when connected separately with the base electrode, has characteristics similar to those of the high back-voltage rectifier. Of critical importance for the operation of the device is the nature of the current in the forward direction. We believe, for reasons discussed in detail in the accompanying letter,[4] that there is a thin layer next

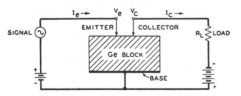

FIG. 1. Schematic of semi-conductor triode.

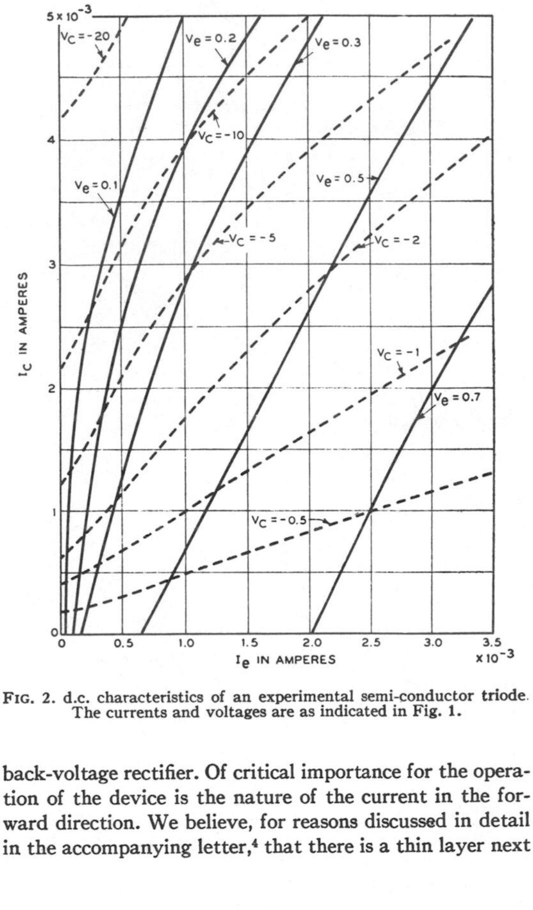

FIG. 2. d.c. characteristics of an experimental semi-conductor triode. The currents and voltages are as indicated in Fig. 1.

to the surface of P-type (defect) conductivity. As a result, the current in the forward direction with respect to the block is composed in large part of holes, i.e., of carriers of sign opposite to those normally in excess in the body of the block.

When the two point contacts are placed close together on the surface and d.c. bias potentials are applied, there is a mutual influence which makes it possible to use the device to amplify a.c. signals. A circuit by which this may be accomplished in shown in Fig. 1. There is a small forward (positive) bias on the emitter, which causes a current of a few milliamperes to flow into the surface. A reverse (negative) bias is applied to the collector, large enough to make the collector current of the same order or greater than the emitter current. The sign of the collector bias is such as to attract the holes which flow from the emitter so that a large part of the emitter current flows to and enters the collector. While the collector has a high impedance for flow of electrons into the semi-conductor, there is little impediment to the flow of holes into the point. If now the emitter current is varied by a signal voltage, there will be a corresponding variation in collector current. It has been found that the flow of holes from the emitter into the collector may alter the normal current flow from the base to the collector in such a way that the change in collector current is larger than the change in emitter current. Furthermore, the collector, being operated in the reverse direction as a rectifier, has a high impedance (10^4 to 10^5 ohms) and may be matched to a high impedance load. A large ratio of output to input voltage, of the same order as the ratio of the reverse to the forward impedance of the point, is obtained. There is a corresponding power amplification of the input signal.

The d.c. characteristics of a typical experimental unit are shown in Fig. 2. There are four variables, two currents and two voltages, with a functional relation between them. If two are specified the other two are determined. In the plot of Fig. 2 the emitter and collector currents I_e and I_c are taken as the independent variables and the corresponding voltages, V_e and V_c, measured relative to the base electrode, as the dependent variables. The conventional directions for the currents are as shown in Fig. 1. In normal operation, I_e, I_c, and V_e are positive, and V_c is negative.

The emitter current, I_e, is simply related to V_e and I_c. To a close approximation:

$$I_e = f(V_e + R_F I_c), \tag{1}$$

where R_F is a constant independent of bias. The interpretation is that the collector current lowers the potential of the surface in the vicinity of the emitter by $R_F I_c$, and thus increases the effective bias voltage on the emitter by an equivalent amount. The term $R_F I_c$ represents a positive feedback, which under some operating conditions is sufficient to cause instability.

The current amplification factor α is defined as

$$\alpha = (\partial I_c / \partial I_e)_{V_c = \mathrm{const.}}$$

This factor depends on the operating biases. For the unit shown in Fig. 2, α lies between one and two if $V_c < -2$.

Using the circuit of Fig. 1, power gains of over 20 db have been obtained. Units have been operated as amplifiers at frequencies up to 10 megacycles.

We wish to acknowledge our debt to W. Shockley for initiating and directing the research program that led to the discovery on which this development is based. We are also indebted to many other of our colleagues at these Laboratories for material assistance and valuable suggestions.

[1] While the effect has been found with both silicon and germanium, we describe only the use of the latter.

[2] The germanium was furnished by J. H. Scaff and H. C. Theuerer. For methods of preparation and information on the rectifier, see H. C. Torrey and C. A. Whitmer, *Crystal Rectifiers* (McGraw-Hill Book Company, Inc., New York, New York, 1948), Chap. 12.

[3] This surface treatment is due to R. B. Gibney, formerly of Bell Telephone Laboratories, now at Los Alamos Scientific Laboratory.

[4] W. H. Brattain and J. Bardeen, Phys. Rev., this issue.

Reprinted from *Phys. Rev.* **69:**37–38 (1946)

Resonance Absorption by Nuclear Magnetic Moments in a Solid

E. M. PURCELL, H. C. TORREY, AND R. V. POUND*
Radiation Laboratory, Massachusetts Institute of Technology, Cambridge, Massachusetts
December 24, 1945

In the well-known magnetic resonance method for the determination of nuclear magnetic moments by molecular beams,[1] transitions are induced between energy levels which correspond to different orientations of the nuclear spin in a strong, constant, applied magnetic field. We have observed the absorption of radiofrequency energy, due to such transitions, in a *solid* material (paraffin) containing protons. In this case there are two levels, the separation of which corresponds to a frequency, ν, near 30 megacycles/sec., at the magnetic field strength, H, used in our experiment, according to the relation $h\nu = 2\mu H$. Although the difference in population of the two levels is very slight at room temperature ($h\nu/kT \sim 10^{-5}$), the number of nuclei taking part is so large that a measurable effect is to be expected providing thermal equilibrium can be established. If one assumes that the only local fields of importance are caused by the moments of neighboring nuclei, one can show that the imaginary part of the magnetic permeability, at resonance, should be of the order $h\nu/kT$. The absence from this expression of the nuclear moment and the internuclear distance is explained by the fact that the influence of these factors upon absorption cross section per nucleus and density of nuclei is just cancelled by their influence on the width of the observed resonance.

A crucial question concerns the time required for the establishment of thermal equilibrium between spins and lattice. A difference in the populations of the two levels is a prerequisite for the observed absorption, because of the relation between absorption and stimulated emission. Moreover, unless the relaxation time is very short the absorption of energy from the radiofrequency field will equalize the population of the levels, more or less rapidly, depending on the strength of this r-f field. In the expectation of a long relaxation time (several hours), we chose to use so weak an oscillating field that the absorption would persist for hours regardless of the relaxation time, once thermal equilibrium had been established.

A resonant cavity was made in the form of a short section of coaxial line loaded heavily by the capacity of an end plate. It was adjusted to resonate at about 30 mc/sec. Input and output coupling loops were provided. The inductive part of the cavity was filled with 850 cm^3 of paraffin, which remained at room temperature throughout the experiment. The resonator was placed in the gap of the large cosmic-ray magnet in the Research Laboratory of Physics, at Harvard. Radiofrequency power was introduced into the cavity at a level of about 10^{-11} watts. The radiofrequency magnetic field in the cavity was everywhere perpendicular to the steady field. The cavity output was balanced in phase and amplitude against another portion of the signal generator output. Any residual signal, after amplification and detection, was indicated by a microammeter.

With the r-f circuit balanced the strong magnetic field was slowly varied. An extremely sharp resonance absorption was observed. At the peak of the absorption the deflection of the output meter was roughly 20 times the magnitude of fluctuations due to noise, frequency, instability, etc. The absorption reduced the cavity output by 0.4 percent, and as the loaded Q of the cavity was 670, the imaginary part of the permeability of paraffin, at resonance, was about $3\cdot 10^{-6}$, as predicted.

Resonance occurred at a field of 7100 oersteds, and a frequency of 29.8 mc/sec., according to our rather rough calibration. We did not attempt a precise calibration of the field and frequency, and the value of the proton magnetic moment inferred from the above numbers, 2.75 nuclear magnetons, agrees satisfactorily with the accepted value, 2.7896, established by the molecular beam method.

The full width of the resonance, at half value, is about 10 oersteds, which may be caused in part by inhomogeneities in the magnetic field which were known to be of this order. The width due to local fields from neighboring nuclei had been estimated at about 4 oersteds.

The relaxation time was apparently shorter than the time ($\sim$ one minute) required to bring the field up to the resonance value. The types of spin-lattice coupling suggested by I. Waller[2] fail by a factor of several hundred to account for a time so short.

The method can be refined in both sensitivity and precision. In particular, it appears feasible to increase the sensitivity by a factor of several hundred through a change in detection technique. The method seems applicable to the precise measurement of magnetic moments (strictly, gyromagnetic ratios) of most moderately abundant nuclei. It provides a way to investigate the interesting question of spin-lattice coupling. Incidentally, as the apparatus required is rather simple, the method should be useful for standardization of magnetic fields. An extension of the method in which the r-f field has a rotating component should make possible the determination of the sign of the moment.

The effect here described was sought previously by Gorter and Broer, whose experiments are described in a paper[3] which came to our attention during the course of this work. Actually, they looked for dispersion, rather than absorption, in LiCl and KF. Their negative result is perhaps to be attributed to one of the following circumstances: (a) the applied oscillating field may have been so strong, and the relaxation time so long, that thermal equilibrium was destroyed before the effect could be observed—(b) at the low temperatures required to make the change in permeability easily detectable by their procedure, the relaxation time may have been so long that thermal equilibrium was never established.

* Harvard University, Society of Fellows (on leave).
[1] Rabi, Zacharias, Millmann, and Kusch, Phys. Rev. **53**, 318 (1938).
[2] I. Waller, Zeits. f. Physik **79**, 370 (1932).
[3] Gorter and Broer, *Physica* **9**, 591 (1942).

44

Reprinted from *Phys. Rev.* **95:**282–284 (1954)

Molecular Microwave Oscillator and New Hyperfine Structure in the Microwave Spectrum of NH_3†

J. P. Gordon, H. J. Zeiger,* and C. H. Townes
Department of Physics, Columbia University, New York, New York
(Received May 5, 1954)

An experimental device, which can be used as a very high resolution microwave spectrometer, a microwave amplifier, or a very stable oscillator, has been built and operated. The device, as used on the ammonia inversion spectrum, depends on the emission of energy inside a high-Q cavity by a beam of ammonia molecules. Lines whose total width at half-maximum is six to eight kilocycles have been observed with the device operated as a spectrometer. As an oscillator, the apparatus promises to be a rather simple source of a very stable frequency.

A block diagram of the apparatus is shown in Fig. 1. A beam of ammonia molecules emerges from the source and enters a system of focusing electrodes. These electrodes establish a quadrupolar cylindrical electrostatic field whose axis is in the direction of the beam. Of the inversion levels, the upper states experience a radial inward (focusing) force, while the lower states see a radial outward force. The molecules arriving at the cavity are then virtually all in the upper states. Transitions are induced in the cavity, resulting in a change in the cavity power level when the beam of molecules is present. Power of varying frequency is transmitted through the cavity, and an emission line is seen when the klystron frequency goes through the molecular transition frequency.

If the power emitted from the beam is enough to maintain the field strength in the cavity at a sufficiently high level to induce transitions in the following beam, then self-sustained oscillations will result. Such oscillations have been produced. Although the power level has not yet been directly measured, it is estimated at about 10^{-8} watt. The frequency stability of the oscillation promises to compare favorably with that of other possible varieties of "atomic clocks."

Under conditions such that oscillations are not maintained, the device acts like an amplifier of micro-

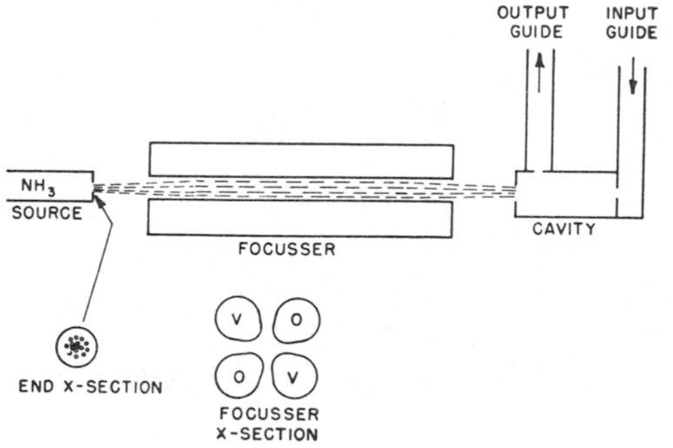

Fig. 1. Block diagram of the molecular beam spectrometer and oscillator.

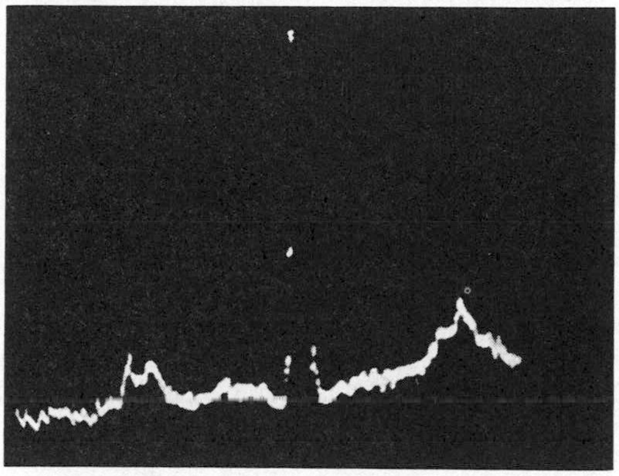

Fig. 2. A typical oscilloscope photograph of the NH_3, $J=K=3$ inversion line at 23 870 Mc/sec, showing the resolved magnetic satellites. Frequency increases to the left.

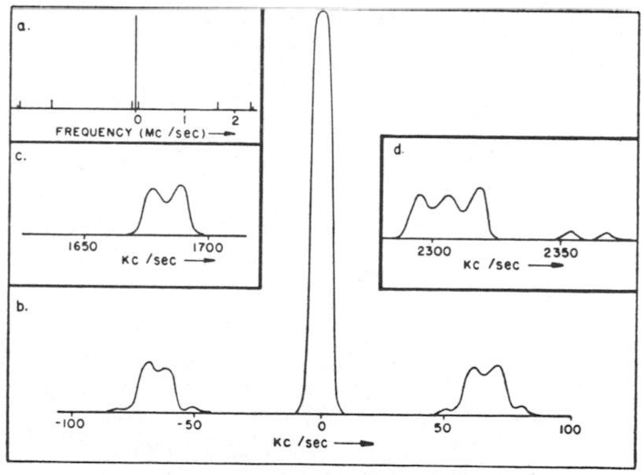

Fig. 3. The observed hyperfine spectrum of the 3,3 inversion line. (a) Complete spectrum, showing the spacings of the quadrupole satellites. (b) Main line with magnetic satellites. (c) Structure of the inner quadrupole satellites. (d) Structure of the outer quadrupole satellites. The quadrupole satellites on the low-frequency side of the main line are the mirror images of those shown, which are the ones on the high-frequency side.

wave power near a molecular resonance. Such an amplifier may have a noise figure very near unity.

High resolution is obtained with the apparatus by utilizing the directivity of the molecules in the beam. A cylindrical copper cavity was used, operating in the $TE011$ mode. The molecules, which travel parallel to the axis of the cylinder, then see a field which varies in amplitude as $\sin(\pi x/L)$, where x varies from 0 to L. In particular, a molecule traveling with a velocity v sees a field varying with time as $\sin(\pi vt/L)\sin(\Omega t)$, where Ω is the frequency of the rf field in the cavity. A Fourier analysis of this field, which the molecule sees from $t=0$ to $t=L/v$, gives a frequency distribution whose amplitude drops to 0.707 of its maximum at points separated by a $\Delta\nu$ of $1.2v/L$. The cavity used was twelve centimeters long, and the most probable velocity of ammonia molecules in a beam at room temperature is 4×10^4 cm/sec. Since the transition probability is proportional to the square of the field amplitude, the resulting line should have a total width at half-maximum given by the above expression, which in the present case is 4 kc/sec. The observed line width of 6–8 kc/sec is close to this value.

The hyperfine structure of the ammonia inversion transitions for $J=K=2$ and $J=K=3$ has been examined, and previously unresolved structure due to the reorientation of the hydrogen spins has been observed. Figure 2 is a typical scope photograph of these new magnetic satellites on the 3,3 line. The observed spectra for the 3,3 line is shown in Fig. 3, which contains all the observed hyperfine structure components, including the quadrupole reorientation transitions of the nitrogen nucleus, which have been previously observed as single lines.

Within the resolution of the apparatus, the hyperfine structures of the upper and lower inversion levels are identical, as evidenced by the fact that the main line is not split. Symmetry considerations require that the hydrogen spins be in a symmetric state under 120-degree rotations about the molecular axis. Thus for the 3,3 state, $I_H=3/2$, and one expects each of the quadrupole levels to be further split into four components by the interaction of the hydrogen magnetic moments with the various magnetic fields of the molecule. At the present writing, the finer details of the expected magnetic splittings have not been worked out.

This type of apparatus has considerable potentialities as a more general spectrometer. Since the effective dipole moments of molecules depend on their rotational state, some selection of rotational states could be effected by such a focuser. Similarly, a focuser using magnetic fields would allow spectroscopy of atoms. Sizable dipole moments are required for a strong focusing action, but within this limitation, the device may prove to have a fairly general applicability for the detection of transitions in the microwave region.

The authors would like to acknowledge the expert help of Mr. T. C. Wang during the latter stages of this experiment.

† Work supported jointly by the Signal Corps, the U. S. Office of Naval Research, and the Air Force.

* Carbide and Carbon post-doctoral Fellow in Physics, now at Project Lincoln, Massachusetts Institute of Technology, Cambridge, Massachusetts.

45

Reprinted from *Nature* **187:**493–494 (1960)

STIMULATED OPTICAL RADIATION IN RUBY

T. H. Maiman
Hughes Research Laboratories, A Division of Hughes Aircraft Co., Malibu, California

Schawlow and Townes[1] have proposed a technique for the generation of very monochromatic radiation in the infra-red optical region of the spectrum using an alkali vapour as the active medium. Javan[2] and Sanders[3] have discussed proposals involving electron-excited gaseous systems. In this laboratory an optical pumping technique has been successfully applied to a fluorescent solid resulting in the attainment of negative temperatures and stimulated optical emission at a wave-length of 6943 Å.; the active material used was ruby (chromium in corundum).

A simplified energy-level diagram for triply ionized chromium in this crystal is shown in Fig. 1. When this material is irradiated with energy at a wave-length of about 5500 Å., chromium ions are excited to the $^4F_2$ state and then quickly lose some of their excitation energy through non-radiative transitions to the ^2E state[4]. This state then slowly decays by spontaneously emitting a sharp doublet the components of which at 300° K. are at 6943 Å. and 6929 Å. (Fig. 2*a*). Under very intense excitation the population of this metastable state (^2E) can become greater than that of the ground-state; this is the condition for negative temperatures and consequently amplification via stimulated emission.

To demonstrate the above effect a ruby crystal of 1-cm. dimensions coated on two parallel faces with silver was irradiated by a high-power flash lamp;

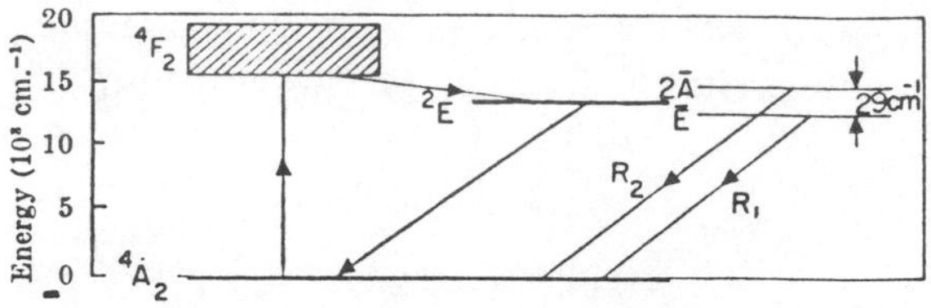

Fig. 1. Energy-level diagram of Cr^{3+} in corundum, showing pertinent processes

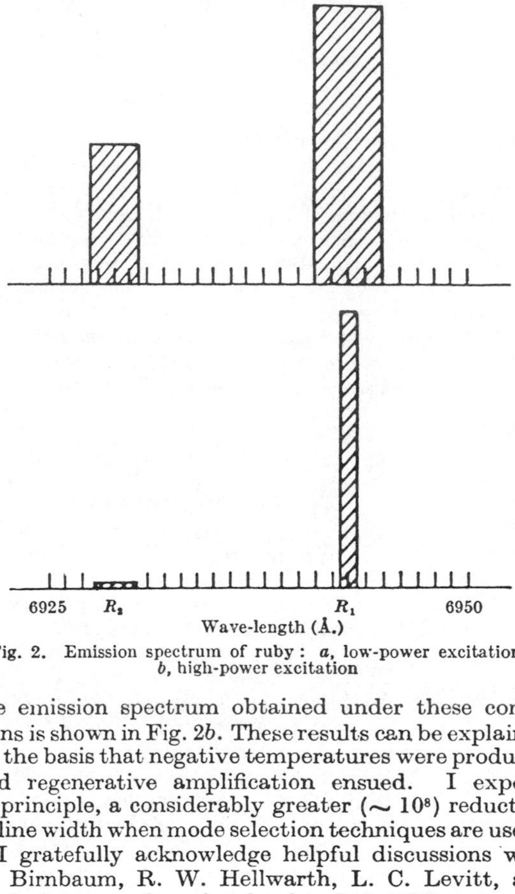

Fig. 2. Emission spectrum of ruby: *a*, low-power excitation; *b*, high-power excitation

the emission spectrum obtained under these conditions is shown in Fig. 2*b*. These results can be explained on the basis that negative temperatures were produced and regenerative amplification ensued. I expect, in principle, a considerably greater ($\sim 10^8$) reduction in line width when mode selection techniques are used[1].

I gratefully acknowledge helpful discussions with G. Birnbaum, R. W. Hellwarth, L. C. Levitt, and R. A. Satten and am indebted to I. J. D'Haenens and C. K. Asawa for technical assistance in obtaining the measurements.

[1] Schawlow, A. L., and Townes, C. H., *Phys. Rev.*, **112**, 1940 (1958).
[2] Javan, A., *Phys. Rev. Letters*, **3**, 87 (1959).
[3] Sanders, J. H., *Phys. Rev. Letters*, **3**, 86 (1959).
[4] Maiman, T. H., *Phys. Rev. Letters*, **4**, 564 (1960).

46

Reprinted from *Phys. Rev.* **106:**162–164

MICROSCOPIC THEORY OF SUPERCONDUCTIVITY*

J. Bardeen, L. N. Cooper, and J. R. Schrieffer
Department of Physics, University of Illinois, Urbana, Illinois

SINCE the discovery of the isotope effect, it has been known that superconductivity arises from the interaction between electrons and lattice vibrations, but it has proved difficult to construct an adequate theory based on this concept. As has been shown by Fröhlich,[1] and in a more complete analysis by Bardeen and Pines[2] in which Coulomb effects were included, interactions between electrons and the phonon field lead to an interaction between electrons which may be expressed in the form

$$H_I = \sum_{\mathbf{k},\mathbf{k}',s,s'} \frac{\hbar\omega |M_\kappa|^2}{(E_{\mathbf{k}}-E_{\mathbf{k}'})^2-(\hbar\omega)^2} \times c^*_{\mathbf{k}'-\kappa,s'} c_{\mathbf{k}',s'} c^*_{\mathbf{k}+\kappa,s} c_{\mathbf{k},s} + H_{\mathrm{Coul}}, \quad (1)$$

where $|M_\kappa|^2$ is the matrix element for the electron-phonon interaction for the phonon wave vector $\boldsymbol{\kappa}$, calculated for the zero-point amplitude of the vibrations, the c's are creation and destruction operators for the electrons in the Bloch states specified by the wave vector $\mathbf{k}$ and spin s, and H_{Coul} represents the screened Coulomb interaction.

Early attempts[3] to construct a theory were based essentially on the self-energy of the electrons, although it was recognized that a true interaction between electrons probably played an essential role. These theories gave the isotope effect, but contained various difficulties, one of which was that the calculated energy difference between what was thought to represent normal and superconducting states was far too large. It is now believed that the self-energy occurs in the normal state, and results in a slight shift of the energies of the Bloch states and a renormalization of the matrix elements.

The present theory is based on the fact that the phonon interaction is negative for $|E_{\mathbf{k}}-E_{\mathbf{k}'}|<\hbar\omega$. We believe that the criterion for superconductivity is essentially that this negative interaction dominate over the matrix element of the Coulomb interaction, which for free electrons in a volume Ω is $2\pi e^2/\Omega\boldsymbol{\kappa}^2$. In the Bohm-Pines[4] theory, the minimum value of κ is κ_c, somewhat less than the radius of the Fermi surface. This criterion may be expressed in the form

$$-V = \langle -(|M_\kappa|^2/\hbar\omega) + (4\pi e^2/\Omega\boldsymbol{\kappa}^2)\rangle_{\mathrm{Av}} < 0. \quad (2)$$

Although based on a different principle, this criterion is almost identical with the one given by Fröhlich.[1,3]

If one has a Hamiltonian matrix with predominantly negative off-diagonal matrix elements, the ground state, $\Psi = \sum \alpha_j \varphi_j$, is a linear combination of the original basic states with coefficients predominantly of one sign. A particularly simple example is one for which the original states are degenerate and each state is connected to n other states by the same matrix element $-V$. The ground state, a sum of the original set with equal coefficients, is lowered in energy by $-nV$. One of the authors made use of this principle to construct a wave function for a single pair of electrons excited above the Fermi surface and found that for a negative interaction a bound state is formed no matter how weak the interaction.[5]

Because of the Fermi-Dirac statistics, difficulties are encountered if one tries to apply this principle directly to (1). Matrix elements of H_I between states specified by occupation numbers (Slater determinants) in general may be of either sign. We want to pick out a subset of these between which matrix elements are always of the same sign. This may be done by occupying the individual particle states in pairs, such that if one of the pair is occupied, the other is also. The pairs should be chosen so that transitions between them are possible, i.e., they all have the same total momentum. To form the ground state, the best choice is $\mathbf{k}\uparrow$, $-\mathbf{k}\downarrow$, since exchange terms reduce the matrix elements between states of parallel spin. To form a state with a net current flow, one might take a pairing $\mathbf{k}\uparrow$, $-\mathbf{k}+\mathbf{q}\downarrow$, where $\mathbf{q}$ is a small wave vector, the same for all $\mathbf{k}$ and such that both states are within the range of energy $\hbar\omega$. The occupation of the pairs may be specified by a single spin-independent occupation number, $m_{\mathbf{k}}=0$ or 1. Nonvanishing matrix elements connect configurations which differ in only one of the occupied pairs.[6] It is often convenient to specify occupation in terms of electron pairs above the Fermi surface and hole pairs below.

The best wave function of this form will be a linear combination

$$\Psi=\sum_{\mathbf{k}_1\cdots\mathbf{k}_n} b(\mathbf{k}_1\cdots\mathbf{k}_n)f(\cdots m_{\mathbf{k}_1}\cdots m_{\mathbf{k}_n}\cdots), \quad (3)$$

where the sum is over all possible configurations. In our calculations, we have made a Hartree-like approximation and replaced b by $b(k_1)b(k_2)\cdots b(k_n)$. We have also assumed an isotropic Fermi surface [so that $b(k)$ depends only on the energy ϵ of the Bloch state involved], and that V is the same for all transitions within a constant energy $\hbar\omega$ of the Fermi surface, $\epsilon=0$. A direct calculation gives for the interaction energy

$$W_I=-4[N(0)]^2V\int_0^{\hbar\omega}\int_0^{\hbar\omega}\Gamma(\epsilon)\Gamma(\epsilon')d\epsilon d\epsilon', \quad (4)$$

where $N(0)$ is the density of states at the Fermi surface. The kinetic energy measured from the Fermi sea is

$$W_K=4N(0)\int_0^{\hbar\omega}g(\epsilon)\epsilon d\epsilon, \quad (5)$$

where $g(\epsilon)$ is the probability that a given state of energy ϵ is occupied by a pair, and

$$\Gamma(\epsilon)=\{g(\epsilon)[1-g(\epsilon)]\}^{\frac{1}{2}}. \quad (6)$$

One may interpret the factor $\Gamma(\epsilon)\Gamma(\epsilon')$ as representing the effect of the exclusion principle on restricting the number of configurations which are connected to a given typical configuration. Matrix elements corresponding to $\mathbf{k}\rightarrow\mathbf{k}'$ are possible only if the state $\mathbf{k}$ is occupied and $\mathbf{k}'$ unoccupied in the initial configuration and $\mathbf{k}'$ occupied and $\mathbf{k}$ unoccupied on the final configuration. The probability that this occurs is

$$g(\epsilon)[1-g(\epsilon')]g(\epsilon')[1-g(\epsilon)]=[\Gamma(\epsilon)]^2[\Gamma(\epsilon')]^2. \quad (7)$$

Since matrix elements have probability amplitudes rather than probabilities, the square root of (7) occurs in (4).

A variational calculation to determine the best $g(\epsilon)$ gives

$$W=W_I+W_K=-\frac{2N(0)(\hbar\omega)^2}{\exp[2/N(0)V]-1}. \quad (8)$$

Thus if there is a net negative interaction, no matter how weak, there is a condensed state in which pairs are virtually excited above the Fermi surface. The product $N(0)V$ is independent of isotopic mass and of volume. The energy W varies as $(\hbar\omega)^2$, in agreement with the isotope effect. It should be noted that (8) cannot be obtained in any finite order of perturbation theory. The energy gain comes from a coherence of the electron wave functions with lattice vibrations of short wavelength, and does not represent a condensation in real space.

Empirically, energies are of the order of magnitude of $N(0)(kT_c)^2$, and of course kT_c is much less than an average phonon energy $\hbar\omega$. According to our theory, this will occur if $N(0)V<1$, a not unreasonable assumption. In this weak-coupling limit, the energy may be expressed simply in terms of the number of electrons, n_c, virtually excited in coherent pairs above the Fermi surface at $T=0°K$

$$W=-\tfrac{1}{2}n_c^2/N(0), \quad (9)$$

where

$$n_c=2N(0)\hbar\omega\exp[-1/N(0)V]. \quad (10)$$

It is a great advantage energy-wise to include in the ground state wave function only pairs with the same total momentum. Suppose that instead one had chosen a random pairing, $\mathbf{k}_1\uparrow$, $\mathbf{k}_2\downarrow$, with $\mathbf{k}_1+\mathbf{k}_2=\mathbf{q}$ and consider a typical matrix element $(\mathbf{k}_1,\mathbf{k}_2|H_I|\mathbf{k}_1'\mathbf{k}_2')$ which vanishes unless $\mathbf{k}_1'+\mathbf{k}_2'=\mathbf{q}'=\mathbf{q}$. We shall assume that the $\mathbf{q}$'s of all pairs are small so that if $\mathbf{k}_1$ and $\mathbf{k}_2$ are both within $\hbar\omega$ of the Fermi surface, so are $\mathbf{k}_1'$ and $\mathbf{k}_2'$. If we construct a wave function made up of a linear combination of states with such virtual excited pairs and determine the interaction energy, we would find an expression similar to (4) but with (7) replaced by the much smaller quantity:

$$g(\epsilon_1)g(\epsilon_2)[1-g(\epsilon_1')][1-g(\epsilon_2')]g(\epsilon_1') \times g(\epsilon_2')[1-g(\epsilon_1)][1-g(\epsilon_2)]. \quad (11)$$

The pairing $\mathbf{k}_2=-\mathbf{k}_1$ corresponds to $\mathbf{q}=0$ for all pairs and insures that if $\mathbf{k}_1'$ is unoccupied, so is $\mathbf{k}_2'$. This is also true if all pairs have the same $\mathbf{q}$.

Wave functions corresponding to individual particle excitations may be made of linear combinations of states in which certain occupation numbers, corresponding to real excited electrons or holes, are specified and the rest are used to make all possible combinations of virtual excitations of $\mathbf{k}\uparrow$, $-\mathbf{k}\downarrow$ pairs. Because of the reduction in phase space available to the pairs, the interaction energy is reduced in magnitude. For small excitations the consequent increase in total energy is proportional to the number of excited electrons. This means that a finite energy is required to excite an electron from the ground state. The same applies to real excited **k**, −**k** pairs. If $f(\epsilon)$ is the probability that a Bloch state of energy ϵ is occupied by an excited electron above the Fermi sea, and $1-f(-\epsilon)$ the probability that there is a hole below, one finds for the interaction energy an expression similar to (4) but with $[\Gamma(\epsilon)]^2$ replaced by $g(\epsilon)\{1-[f(\epsilon)]^2-g(\epsilon)\}$. For small excitations above $T=0$°K, the total pair energy may be expressed in the weak-coupling limit as

$$W=-\frac{n_c^2}{2N(0)}\left(1-\frac{4n_e}{n_c}\right), \quad n_e \ll n_c, \tag{12}$$

where n_c is the number of electrons in the virtually excited states at $T=0$ and n_e is the number of actually excited electrons. This leads to an energy gap[7] (i.e., the energy required to create an electron-hole pair):

$$E_G=\partial W/\partial n_e=2n_c/N(0) \quad \text{at} \quad T=0°\text{K}. \tag{13}$$

Taking the empirical $W=-H_c^2/8\pi$ and estimating $N(0)$ from the electronic specific heat, we find $E_G=k\times 13.8$°K for tin. This is to be compared with the experimental value of about $k\times 11.2$°K. Calculations are under way to determine the thermal properties at higher temperatures.

Advantages of the theory are (1) It leads to an energy-gap model of the sort that may be expected to account for the electromagnetic properties.[8] (2) It gives the isotope effect. (3) An order parameter, which might be taken as the fraction of electrons above the Fermi surface in virtual pair states, comes in a natural way. (4) An exponential factor in the energy may account for the fact that kT_c is very much smaller than $\hbar\omega$. (5) The theory is simple enough so that it should be possible to make calculations of thermal, transport, and electromagnetic properties of the superconducting state.

* This work was supported in part by the Office of Ordnance Research, U. S. Army. One of us (J.R.S.) wishes to thank the Corning Glass Works Foundation for a grant which aided in the support of this work.

1 H. Fröhlich, Phys. Rev. **79**, 845 (1950); Proc. Roy. Soc. (London) **A215**, 291 (1952).

2 J. Bardeen and D. Pines, Phys. Rev. **99**, 1140 (1955).

3 For reviews of this work, mainly by Fröhlich and by Bardeen, see J. Bardeen, Revs. Modern Phys. **23**, 261 (1951); *Handbuch der Physik* (Springer-Verlag, Berlin, 1956), Vol. 15, p. 274.

4 See D. Pines, *Solid State Physics* (Academic Press, Inc., New York, 1955), Vol. 1, p. 367.

5 L. N. Cooper, Phys. Rev. **104**, 1189 (1956).

6 This looks deceptively like a one-particle problem, but is not, because of the peculiar statistics involved. The pairs cannot be treated as bosons.

7 Some references to experimental evidence for an energy gap are R. E. Glover and M. Tinkham, Phys. Rev. **104**, 844 (1956); M. Tinkham, Phys. Rev. **101**, 815 (1956); Blevins, Gordy, and Fairbank, Phys. Rev. **100**, 1215 (1955); Corak, Goodman, Satterthwaite, and Wexler, Phys. Rev. **102**, 656 (1956); W. S. Corak and C. B. Satterthwaite, Phys. Rev. **102**, 662 (1956).

8 J. Bardeen, Phys. Rev. **97**, 1724 (1955).

SELECTED BIOGRAPHICAL NOTES

Hannes O. G. Alfvén (1908-), Swedish physicist received the Ph.D. from the University of Uppsala in 1934. He was professor at the Royal Institute of Technology in Stockholm from 1940-1967. More recently he has been a professor at the University of California at San Diego.

Carl D. Anderson (1905-), American physicist, received the Ph.D. from California Institute of Technology in 1927. He has been on the faculty of that institution since 1930.

John Bardeen (1908-), American physicist, received the Ph.D. from Princeton University in 1936. He was professor at the University of Illinois from 1951-1975 and has since been retired.

Hans A. Bethe (1906-) was born in Strasbourg, Alsace-Lorraine, and received the doctorate from the University of Munich in 1928. He came to the United States in 1935 and became a professor at Cornell University in 1937. He became professor emeritus in 1975.

Felix Bloch (1905-), Swiss physicist, received the doctorate from the University of Leipzig in 1928. He came to the United States in 1934 and has been professor of physics at Stanford University since that time.

Niels Bohr (1885-1962), Danish physicist, received the doctorate from the University of Copenhagen in 1911. He was professor of theoretical physics in Copenhagen from 1916 until his death. The University Institute of Theoretical Physics now bears his name.

Max Born (1882–1970) was a native of Breslau in German Silesia (now a part of Poland). He received the doctorate in physics at the University of Göttingen. His teaching and research career took him to various German universities, but his longest tenure was at the University of Edinburgh, where he occupied the chair in natural philosophy from 1936–1953.

Walther Bothe (1891–1957), German physicist, received the doctorate while working under Max Planck at the University of Berlin in 1914. In the post World War I period he worked with Hans Geiger at the Reichsanstalt in Berlin. In 1934 he became director of the Max Planck Institute in Heidelberg, where he remained until his death.

Walter H. Brattain (1902-), American physicist, received the doctorate at the University of Minnesota in 1929. He was a research physicist at the Bell Laboratories from 1929 until his retirement in 1967.

Ferdinand G. Brickwedde (1903-), American physicist, received the doctorate from Johns Hopkins University in 1925 and for many years was on the technical staff of the United States National Bureau of Standards in Washington, D.C. He later held a professorship at the Pennsylvania State University, but is now retired.

L. de Broglie (1892-) was professor of theoretical physics in the Faculty of Sciences, at the Sorbonne in Paris from 1932–1962. He now lives in retirement near Paris.

James Chadwick (1891–1974), English physicist, was educated at the University of Manchester and Cambridge, where he worked with Lord Rutherford. He was professor at the University of Liverpool from 1935–1943 and from 1948–1958 was Master of Gonville and Caius College in Cambridge.

John D. Cockcroft (1897–1967) was a native of Yorkshire. Much of his early professional career was spent in the Cavendish Laboratory in Cambridge working under Rutherford. During the Second World War he was much involved in the work on radar. From 1946–1959 he was director of the United Kingdom Atomic Energy Research Establishment at Harwell. He was Master of Churchill College in Cambridge from 1959 until his death.

Leon N. Cooper (1930-), American physicist, received the Ph.D. at Columbia University in 1954. He has been a professor at Brown University since 1958.

C. J. Davisson (1881-1958) retired from the Bell Laboratories in 1946 and taught thereafter at the University of Virginia until 1954.

P. A. M. Dirac (1902-) received the doctorate from the University of Cambridge in 1926. He served as Lucasian Professor of Mathematics at Cambridge from 1932-1969 and since 1971 has been a professor at the Florida State University in Tallahassee.

Saul Dushman (1883-1954), American physical chemist, was born in Russia, but moved to the United States at the age of 8. He received the Ph.D. at the University of Toronto in 1912. He was on the staff of the General Electric Research Laboratory from 1912 until his death.

Eugene Feenberg (1906-1977), American physicist, received the Ph.D. from Harvard University in 1933. He served as professor at Washington University, St. Louis, from 1946 until his death.

Enrico Fermi (1901-1954) was both a theoretical and experimental physicist. He received the doctorate from the University of Pisa in 1922. When the first chair in theoretical physics in Italy was established in Rome in 1927, Fermi became the first occupant and attracted students from Italy and abroad to his lectures and research projects. He came to the United States in 1938 and worked at Columbia University, the University of Chicago, and Los Alamos. After World War II, he returned to Chicago and remained there for the rest of his life.

Otto R. Frisch, (1904-1979) Austrian physicist, received the doctorate at the University of Vienna in 1927. He was a professor at Cambridge University from 1947 until his death.

George Gamow (1904–1968) was born in Russia and received the Ph.D. at the University of Leningrad in 1928. After carrying out research in Germany, Denmark, and England, he came to the United States in 1934 and was professor of physics at George Washington University in Washington., D.C., from 1934–1956 and at the University of Colorado from 1956 until his death.

Hans Geiger (1882–1945), German physicist, received the doctorate at the University of Erlangen in 1906. From then until 1912 he worked at the University of Manchester in England. Thereafter he held professorships at the universities of Kiel, Tübingen, and Berlin.

James P. Gordon (1928–   ), American physicist, received the Ph.D. at Columbia University in 1955. He has been on the staff of the Bell Laboratories since that time.

Otto Hahn (1879–1968), German physical chemist, received the doctorate from the University of Munich in 1901. He was professor of chemistry at the University of Berlin from 1910–1933 and president of the Max Planck Society for the Advancement of Science from 1946–1960.

Werner Heisenberg (1901–1976) was educated at the universities of Munich and Göttingen. He held professorships at the universities of Leipzig and Berlin. The latter part of his professional career was spent as director of the Max Planck Institute for Physics and Astrophysics at the University of Munich.

Victor F. Hess (1883–1964), Austrian physicist, received the doctorate from the University of Vienna in 1906. He held professorships at the universities of Vienna, Graz, and Innsbruck from 1909 to 1938, when he moved to the United States and became a professor at Fordham University.

Frederic Joliot-Curie (1900–1958), French physicist and husband of Irene Joliot-Curie, received the doctorate from the University of Paris in 1930. He held numerous educational and research positions in France, including the directorship of the Laboratoire Curie de l'Institut de Radium.

Irene Joliot-Curie (1897-1956), French nuclear scientist, was a daughter of Marie Curie. She received the doctoral degree from the University of Paris in 1925. She held numerous research, educational, and governmental positions in France, including a professorship in the Faculty of Sciences of the University of Paris.

Ernst Pascual Jordan (1902-) is a native of Hamburg. He received the doctorate from the University of Göttingen in 1924 and has held professorships at Rostock and Berlin. Since 1957 he has been professor at the University of Hamburg.

Donald W. Kerst (1911-), American physicist, received the Ph.D. from the University of Wisconsin in 1937. He was on the faculty of the University of Illinois from 1938–1957 and has been professor at the University of Wisconsin since 1962.

R. de L. Kronig (1904-) was born in Dresden, Germany. He has held professorships in London as well as in Groningen and Delft in the Netherlands. He now lives in retirement in Switzerland.

Irving Langmuir (1881-1957), American physicist and chemist, received the doctorate from the University of Göttingen in 1906. He was a staff member of the General Electric Research Laboratory from 1907 until his death.

Ernest O. Lawrence (1901-1958), American physicist, received the Ph.D. from Yale University in 1925. He served on the faculty of the University of California at Berkeley from 1928 until his death. The Radiation Laboratory there is named for him.

M. Stanley Livingston (1905-), American physicist, was for many years professor at the Massachusetts Institute of Technology. He now lives in retirement in Santa Fe, New Mexico.

Edwin M. McMillan, (1907-), American physicist, received the doctorate from Princeton University in 1932. He served on the faculty of the University of California at Berkeley from 1931 until his retirement in 1973. He was also director of the Lawrence Radiation Laboratory from 1958–1973.

Theodore H. Maiman (1927–), American physicist, received the Ph.D. from Stanford University in 1955. He has been connected with numerous industrial concerns, but is at present with TRW, Inc. in Los Angeles, California.

Robert J. Maurer (1913–), American physicist, received the doctorate from the University of Rochester in 1939. He has been a professor at the University of Illinois since 1949.

Lise Meitner (1878–1968), Austrian physicist (and aunt of Otto Frisch), received the doctoral degree from the University of Vienna in 1906. She was head of the Physics Department of the Kaiser Wilhelm Institute of Chemistry in Berlin from 1917–1938. From 1938 until her death she served on the staff of the Nobel Institute in Stockholm.

Rudolf L. Mössbauer (1929–), German physicist, received the doctorate from the Technische Hochschule in Munich in 1958. After serving as professor of physics at the California Institute of Technology from 1961–1972, he returned to Germany as professor in the Technische Hochschule in Munich.

George M. Murphy (1903–), American physical chemist, received the Ph.D. from Yale in 1930. He has taught at both Yale and New York University.

Seth H. Neddermeyer (1907–), American physicist, received the doctorate at California Institute of Technology in 1935. He has been on the physics faculty of the University of Washington in Seattle since 1946.

Wolfgang Pauli (1900–1958). After doing his doctoral dissertation in physics under Sommerfeld in Munich, Pauli visited various universities in Europe, including Bohr's Institute, in Copenhagen. He served as Professor of Theoretical Physics at the Swiss Institute of Technology in Zurich, from 1928 to the end of his life.

William G. Penney (1909-), English physicist, received the Ph.D. from Imperial College, London in 1931. He has held numerous governmental positions including the directorship of the United Kingdom Atomic Energy Authority since 1964.

Robert V. Pound (1919-), American physicist who was born in Canada, received his higher education at Harvard University and has taught there since 1948.

Edward M. Purcell (1912-), American physicist, received the Ph.D. from Harvard University in 1938. He has been a professor at Harvard since 1946.

Owen W. Richardson (1879-1959), English physicist, received his higher education at Cambridge University and the University of London. He was professor of physics at Princeton University from 1906-1913 and director of research at King's College London from 1924-1944. He lived in retirement for the rest of his life.

Ernest Rutherford (1871-1937) was a native of New Zealand and received his early education in that country, graduating with the A.B. from Canterbury College in Christchurch in 1892, followed by the M.A. in 1893. He then studied with J. J. Thomson at Cambridge University in England. From 1893 to 1907 he was professor of physics at McGill University in Montreal. It was there that his early work on radioactivity was carried out. From 1907-1919 he was professor at the University of Manchester in England, moving to Cambridge to succeed J. J. Thomson as Cavendish Professor on the latter date. He remained there until his death. He was raised to the peerage in 1931.

J. Robert Schrieffer (1931-), American physicist, received the Ph.D. at the University of Illinois in 1957. He became professor at the University of Pennsylvania in 1964, but more recently has worked at the University of California at Santa Barbara.

Erwin Schrödinger (1887-1961) was a native of Vienna, where he received the doctorate. He held professorships at Stuttgart, Zurich, and Berlin. Schrödinger served as Director of the School of Theoretical Physics in the Institute of Advanced Study in Dublin, Ireland, from 1939 to 1955. He died in Vienna in 1961.

Frederick Seitz (1911-), American physicist, received the Ph.D. at Princeton University in 1934. He was professor at the University of Illinois from 1949-1957, president of the National Academy of Sciences from 1962-1968 and president of Rockefeller University from 1968-1978.

Arnold Sommerfeld (1868-1951) was a native of Königsberg, originally in East Prussia, and received the doctorate from the university in that city in 1891. He was professor of theoretical physics in the University of Munich from 1906-1940. Many outstanding physicists were his pupils there. Though he lectured in the United States in the 1920s and early 1930s, most of his professional life was spent in Munich.

Fritz Strassmann (1902-), German physical chemist, received the doctorate from the Technische Hochschule in Hannover in 1929. For many years he has been a professor of chemistry at the University of Mainz. He also served for a time as director of chemical research in the Max Planck Institute of Chemistry.

G. P. Thomson (1892-1975) was for many years (1930-1952) professor of physics at Imperial College of Science and Technology in London, and later Master of Corpus Christi College in Cambridge University.

Henry C. Torrey (1911-), American physicist, received the doctorate from Columbia University in 1937. He has been a professor at Rutgers University since 1946.

Charles H. Townes (1915-), American physicist, received the Ph.D. from California Institute of Technology in 1939. He was a professor at Columbia University from 1948-1961 and at Massachusetts Institute of Technology from 1961-1967. More recently he has been at the University of California.

Harold C. Urey (1893-1981), American chemist, received the Ph.D. from the University of Chicago in 1923. He held professorships in Columbia University and the University of Chicago from 1929-1958. Thereafter he moved to the University of California, where he remained for the rest of his life.

Robert J. Van De Graaff (1901-1967), American physicist, received the doctorate from Oxford University, England, in 1928. He was on the faculty of the Massachusetts Institute of Technology from 1934-1960. He also served as a director of the High Voltage Engineering Corporation from 1946 until his death.

Ernest T. S. Walton (1903-) is an Irish physicist, educated at the University of Dublin and Cambridge University. He served as professor of natural and experimental philosophy at Trinity College, Dublin, from 1947-1974 and has since lived in retirement.

John A. Wheeler (1911-), American physicist, received the Ph.D. at Johns Hopkins University in 1933. He served as professor of physics at Princeton University from 1947-1976. He is presently professor of physics at the University of Texas in Austin.

Eugene P. Wigner (1907-), mathematical physicist, was born in Hungary. After receiving his advanced scientific education at Berlin he came to the United States in 1930 and has been professor of physics at Princeton University since 1938. He became professor emeritus in 1971.

Charles Thomson Rees Wilson (1869-1959), Scottish physicist, received his principal education at Cambridge University, where he also taught and did research until his death.

Hideki Yukawa (1907-1981), Japanese physicist, received the doctorate from Osaka University in 1938. From 1939 to 1950 he was professor at Kyoto University.

Herbert J. Zeiger (1925-), American physicist, received the Ph.D. from Columbia University in 1951. He has been at the Lincoln Laboratory of Massachusetts Institute of Technology since 1952.

AUTHOR CITATION INDEX

SUBJECT INDEX